T0227889

Making Music
with Computers

Creative Programming in Python

CHAPMAN & HALL/CRC
TEXTBOOKS IN COMPUTING

Series Editors

John Impagliazzo
Professor Emeritus, Hofstra University

Andrew McGettrick
Department of Computer
and Information Sciences
University of Strathclyde

Aims and Scope

This series covers traditional areas of computing, as well as related technical areas, such as software engineering, artificial intelligence, computer engineering, information systems, and information technology. The series will accommodate textbooks for undergraduate and graduate students, generally adhering to worldwide curriculum standards from professional societies. The editors wish to encourage new and imaginative ideas and proposals, and are keen to help and encourage new authors. The editors welcome proposals that: provide groundbreaking and imaginative perspectives on aspects of computing; present topics in a new and exciting context; open up opportunities for emerging areas, such as multi-media, security, and mobile systems; capture new developments and applications in emerging fields of computing; and address topics that provide support for computing, such as mathematics, statistics, life and physical sciences, and business.

Published Titles

Paul Anderson, Web 2.0 and Beyond: Principles and Technologies

Henrik Bærbak Christensen, Flexible, Reliable Software: Using Patterns and Agile Development

John S. Conery, Explorations in Computing: An Introduction to Computer Science

Ted Herman, A Functional Start to Computing with Python

Pascal Hitzler, Markus Krötzsch, and Sebastian Rudolph, Foundations of Semantic Web Technologies

Mark J. Johnson, A Concise Introduction to Data Structures using Java Uvais Qidwai and C.H. Chen, Digital Image Processing: An Algorithmic Approach with MATLAB®

Mark J. Johnson, A Concise Introduction to Programming in Python

Lisa C. Kaczmarczyk, Computers and Society: Computing for Good

Mark C. Lewis, Introduction to the Art of Programming Using Scala

Bill Manaris and Andrew R. Brown, Making Music with Computers: Creative Programming in Python

Henry M. Walker, The Tao of Computing, Second Edition

CHAPMAN & HALL/CRC
TEXTBOOKS IN COMPUTING

Making Music
with Computers

Creative Programming in Python

Bill Manaris

College of Charleston
South Carolina, USA

Andrew R. Brown

Queensland University of Technology
Keperra, Australia

CRC Press
Taylor & Francis Group
Boca Raton London New York

CRC Press is an imprint of the
Taylor & Francis Group, an **informa** business
A CHAPMAN & HALL BOOK

CRC Press
Taylor & Francis Group
6000 Broken Sound Parkway NW, Suite 300
Boca Raton, FL 33487-2742

First issued in hardback 2017

Version Date: 20140402

ISBN-13: 978-1-4398-6791-4 (pbk)
ISBN-13: 978-1-138-46084-3 (hbk)

Library of Congress Cataloging-in-Publication Data

Manaris, Bill.
    Making music with computers : creative programming in Python / authors, Bill Manaris, Andrew R. Brown.
        pages cm. -- (Chapman & Hall/CRC textbooks in computing)
    Includes bibliographical references and index.
    ISBN 978-1-4398-6791-4 (paperback)
    1. Computer composition. 2. Python (Computer program language) I. Brown, Andrew R. II. Title.

MT56.M25 2014
781.3'45133--dc23   2013045978

Visit the Taylor & Francis Web site at
http://www.taylorandfrancis.com

and the CRC Press Web site at
http://www.crcpress.com

Contents

Foreword

THE HUMAN DESIRE TO EXPRESS and communicate has influenced computing almost as long as there have been computers. ENIAC was first turned on in 1947. The first computer music was generated in 1957.

The desire to *say* more with a computer has driven many advances in computer science. Ivan Sutherland invented interactive computer graphics in 1963, and his creation inspired the idea of classes in object-oriented programming. Alan Kay and Adele Goldberg described the computer as human's first *meta-medium*, the first creative medium that could encompass all previous media. Their research group at Xerox's Palo Alto Research Center (PARC) worked in the 1970s to answer the question, "What would a computer used for creative expression look like?" That's what led them to invent the desktop user interface as we know it today. In a real sense, the menus and windows that we use today to access Facebook were invented in order to make the most powerful tool ever for human expression.

Making music on a computer is a natural way to learn more about mathematics, computer science, and music. Bill Manaris and Andrew Brown have created this marvelous book that will engage and inspire you to learn more about the science and art of creating music through computation. They lead us through exploration of fascinating questions. How does music draw on both mathematical patterns and randomness? How did Bach use algorithms to generate canons? How can we turn data about proteins and planets into music? What kinds of new interfaces can you create to make it easier for you and others to make music?

Bill and Andrew offer an accessible path into a wonderful world that is both as modern as your new laptop and as ancient as Plato. In that world of music and mathematics, they constructed a sandbox of computational tools. They encourage you to create, to compose music, and to play with patterns and data. They invite you to continue in the traditions of Ivan Sutherland and Alan Kay to use computing to explore powerful and creative ideas.

Mark Guzdial
Georgia Institute of Technology
July 2013

Preface

THE BOOK IN YOUR HANDS is the result of more than a decade of independent and collaborative effort by the two authors and their computer music associates. Combining computers and music has a long and fruitful heritage. The ideas which underpin the connection between calculating and composing date back centuries. In the 21st century, computers and music are more closely aligned than ever before. In particular, computers have become indispensable in music making, distribution, performance, and consumption.

This book introduces important concepts and skills necessary to make music with computers. It interweaves computing pedagogy with musical concepts and creative activities. It does this while maintaining a natural, steady increase in computational skills that are motivated by creative musical contexts.

This book is intended primarily for introductory computer science courses and for courses in the intersection of computing and the arts. However, it is naturally suited for self-study. It assumes little musical and programming experience; it introduces topics and concepts as they arise through motivating, and hopefully inspiring examples.

CREATIVE PROGRAMMING

"Making Music with Computers" is an introduction to creative software development in the Python programming language. It uses music-making as a vehicle to introduce computer programming and computational thinking to non-traditional audiences. This book helps computer science educators teach students how to synthesize the creativity and design of the arts with the mathematical rigor and formality of computer science.

Initially inspired by Randy Pausch's "head-fake" approach*, we utilize exciting and innovative music-creation activities to ultimately teach

---

* See Randy Pausch's "Last Lecture" (readily available online).

introductory computer science concepts. Our goal is to keep this "game" going throughout the book, just long enough so that the students learn to express themselves algorithmically.

The book covers all concepts found in a traditional "Intro to Computer Programming" (CS1) course. These concepts include data types, variables, assignment, arithmetic operators, input/output, algorithms, selection (if statements), relational operators, logical operators, iteration (loops), lists (arrays), functions, modularization (functions), classes (object-oriented programming). Additionally, the book covers graphical user interfaces (GUIs), event-driven programming, big data, MIDI programming, client-server programming (via OSC messages), recursion, fractals, and complex system dynamics (boids).

TARGET AUDIENCE

This book addresses two trends in computing education: (1) the growing use of the Python language for teaching introductory programming, and (2) the increasing infusion of computational thinking into liberal arts courses, especially interdisciplinary offerings in computing and the arts. It does so by presenting computer music topics in an accessible way for our two main target audiences:

- First- and second-year university students, as well as advanced high school students, who are interested in computer music and wish to learn computer programming in a creative context; and

- Musicians of all levels and backgrounds who wish to expand their creative horizons by modeling musical processes through computer programming, and by applying these processes to create novel and intriguing musical material for composition and live performance.

NAVIGATING THE BOOK

The book may be navigated using one of three narratives, *objects first, procedures first,* or *à la carte:*

- **Objects first** (chapters 1–3, followed by chapters 8–11, with just-in-time introduction of for loops, functions, and if statements). This approach works well with inexperienced students, as it is creatively rich. It includes building graphical user interfaces (GUIs) and interactive musical instruments, and thus motivates hard-to-grasp programming concepts (such as loops, functions, and if statements). To quote one of our students, "students want to do the work, because it is fun." In particular, chapters 1-3 introduce object-based programming (using Notes, Phrases, Parts, and Scores). Chapter 8 introduces GUI objects, event-driven programming, and important human-computer interaction (HCI) ideas, such as how to develop usable interfaces through paper prototyping, usability testing, and iterative refinement. Chapter 9 introduces MIDI and OSC (open sound control) input/output objects, thus enabling programs to connect to traditional musical instruments (e.g., guitars, pianos, etc.)* and physical controllers (e.g., MIDI control surfaces, smartphones, touch-sensitive tablets, etc.). Finally, chapters 10 and 11 introduce Python classes, music from math equations, the harmonograph (a way to visualize and sonify complex, yet beautiful repetitive patterns found in nature, such as planetary orbits), animation, fractals, the golden ratio, recursion, Zipf's law, and chaotic systems (boids). This material is full of musical and other creative possibilities.

- **Procedures first** (chapters 1–11). This is a traditional narrative, which interweaves computational and musical concepts incrementally, from beginning to intermediate level of expertise. In addition to the above topics, it includes randomness and creativity, data sonification, image processing, musical canons (musical puzzles, of which JS Bach was a master), minimalism, and stochastic music, to name a few. It also provides a thorough introduction to programming in Python, including data types, variables, assignment, arithmetic operators, input/output, algorithms, selection (if statements), iteration (loops), lists (arrays), file input/output, modularization (functions), event-driven programming (callback functions), and object-oriented programming (classes).

- **À la carte** (explore topics as desired/needed). This approach is best suited for self-learners, and for musicians (and programmers) who already know some Python (and music), and who wish to explore techniques to enhance their potential for creative expression. If you belong in this group, then the table of contents is your best friend. Study it carefully, looking for items that seem attractive, and go from

---

* Any MIDI-enabled instrument will work, opening the door for some powerful creative possibilities, for building hybrid computer music instruments.

there. If you are lacking some Python background (to fully appreciate the provided examples, without having to read the whole book up to that point), either you can look up Python topics in the book index, or search the Internet (the latter being full of great Python reference material). Please enjoy weaving your own path through this material - we certainly did!

All three narratives are supported by the website provided at http://jythonMusic.org. There you will find additional resources (including more code examples) to enhance your creative exploration and learning. Enjoy!

PEDAGOGY

From the point of view of pedagogy, our primary audience is educators interested in teaching computing and computational thinking in a media-rich context, in conjunction with guidelines such as the CS Principles and Big Ideas.* In particular, this book supports teaching of the 7 CS Big Ideas:

- **Big Idea I**—Creativity. Computing is a creative human activity that engenders innovation and promotes exploration (whole book, and in particular chapters 1, 6, 7, 8, 9, 10, and 11).

- **Big Idea II**—Abstraction. Abstraction reduces information and detail to focus on concepts relevant to understanding and solving problems (whole book, and in particular chapters 2, 3, 7, and 11).

- **Big Idea III**—Data. Data and information facilitate the creation of knowledge (chapters 3, 4, 7, 10, and 11).

- **Big Idea IV**—Algorithms. Algorithms are tools for developing and expressing solutions to computational problems (chapters 4, 5, 6, 7, 10, and 11).

- **Big Idea V**—Programming. Programming is a creative process that produces computational artifacts (chapters 3, 4, 5, 6, 7, 8, 9, 10, and 11).

- **Big Idea VI**—Internet. Digital devices, systems, and the networks that interconnect them enable and foster computational approaches to solving problems (chapter 9).

---

* The College Board, "Computer Science: Principles, Big Ideas and Key Concepts", 2012.

- **Big Idea VII**—Impact. Computing enables innovation in other fields including mathematics, science, humanities, and arts, among others (chapters 1, 8, 9, 10, and 11).

SOFTWARE LIBRARIES

The book comes with a Jython environment and a collection of software examples and libraries, including music, image, graphical user interface, MIDI, audio, and Open Sound Control. The music library is an extension of the jMusic library and incorporates other cross-platform programming tools. This software is available for download on the website associated with this book, http://jythonMusic.org.

We hope that this book will enhance the educational experience of students in entry-level courses in computing and computational thinking. We also hope that it may serve as a reference and text for computer music courses, such as those offered by music technology programs. Finally, we hope this book may serve as a reference and tutorial resource for digital music enthusiasts who wish to expand their creative horizons and learn how to write music software and create algorithmic music compositions.

The Authors

Bill Manaris is a computer science educator, researcher, and musician. He holds a Ph.D. in computer science and is professor of computer science and director of computing in the arts program at the College of Charleston, South Carolina. He has studied music theory, and classical and jazz guitar, and performs live occasionally. He has been active in curriculum development in human–computer interaction, artificial intelligence, and computing in the arts. His teaching experience spans 30 years. His research interests include statistical, connectionist, and evolutionary techniques for modeling human aesthetics and creativity in music and art. He has developed several systems for computer-aided music analysis, composition, and performance, including NEvMuse, Armonique, and Monterey Mirror. For more information visit http://www.cs.cofc.edu/~manaris.

Andrew R. Brown is an educator, musician, digital artist, and computer programmer. He holds a Ph.D. in music and is professor of digital arts at Griffith University, in Brisbane, Australia, where his work explores the aesthetics of process and regularly involves programming software as part of the creative process. In addition to a history of computer-assisted composition and audio-visual installations, Andrew has in recent years focused on real-time artworks using generative processes and musical live-coding. The latter is a practice where the software to generate a work is written as part of the performance. He has been invited to perform live coding in many international venues. His digital media artworks have been shown in galleries across Australia and in China. For more information visit http://andrewrbrown.net.au.

Acknowledgments

W E WOULD LIKE TO THANK John Impagliazzo for recommending and encouraging the writing of this book; Randi Cohen for patiently working through the various stages and ever-extending timeframes that this book required; and our various collaborators and students who assisted in various invaluable ways to complete this project (via code development, API design and review, and testing) including Dana Hughes (GUI library), David Johnson (OSC and MIDI libraries), Kenneth Hanson, J.R. Armstrong, Thomas Zalonis, Patrick Roos, Luca Pellicore, Timothy Hirzel, Brian Muller, William Daugherty, Dallas Vaughan, Christopher Wagner, Semmy Purewal, and Valerie Sessions (Zipf library and metrics). We are particularly indebted to Mark Guzdial for opening the door to introducing computer science concepts via Media Computation. Also of particular importance to the first author was attending the 15-day Workshop in Algorithmic Music Composition (WACM), in 2010, and his interactions with David Cope, Peter Elsea, Paul Nauert, and Daniel Brown, as well as the various workshop participants. The second author is particularly indebted to those who contributed to the development of the jMusic library and tutorials upon which this book builds. These include Andrew Sorensen, Rene Wooller, Tim Opie, Andrew Troedson and Adam Kirby.

We owe a debt of gratitude to the reviewers of this book, especially William Greene and Maximos Kaliakatsos for reading every chapter word-for-word and every code example, and for offering numerous suggestions and improvements. Additional comments and support were provided by David Cope, Daniel Brown, Yiorgos Vassilandonakis, Walter Pharr, and Blake Stevens.

Finally, we want to thank our families for their patience and support as we worked the long nights, weekends, and holidays this book required.

This book has been partially supported through funding by the US National Science Foundation (including DUE-1044861, IIS-0736480, IIS-0849499 and IIS-1049554).

## CHAPTER 1

Introduction and History

Topics: Pythagoras (music, nature, and number), the Antikythera mechanism, Kepler's harmony of the world, cymatics, fractals, electronic music, computers and programming, the computer as a musical instrument, running Python programs.

1.1 OVERVIEW

This chapter provides a quick tour of some of the major technological landmarks in Western music history and computer science. When we think of computer music, we usually imagine electronic technologies, particularly the synthesizer, computer, and sound recording devices. These devices are products of the information age in which we live. This age focuses on computational thinking, that is, using computers in creative ways to manipulate data and perform various tasks, usually involving some form of programming. The introduction of computers and, in particular, computer programming has also expanded the sonic and structural boundaries of music composition and performance.

In the 20th century, the fundamental education of an individual consisted of the three R's—reading, writing, and arithmetic. In the 21st century, with the proliferation of computing devices, this list now consists of four R's, that is, reading, writing, arithmetic, and programming.

Once computer programming is mastered, new vistas of creative expression become available. This new expressive capability is not confined only to computer music—it is available in every area of the arts as well as the sciences. Accordingly, the programming skills you will acquire in this book are not specific only to making music. They may be applied to creative endeavors in all areas of human knowledge and expression.

1.2 CONNECTING MUSIC, NATURE, AND NUMBER

The development of music and mathematics is connected to humanity's early observations of nature, and attempts to explain and formulate aspects of the human experience. The ancient Babylonians, Egyptians, and Greeks investigated the origin of sound and resonance, and developed the notion of musical scale in terms of integer ratios. To them nature was a harmonious artifact, in which humans found themselves exploring and creating. Mathematics was created around that time, and in its early phases, it was intricately connected to nature and music. Even more recent concepts of the golden ratio, Fibonacci numbers, Zipf's law, cymatics, and fractals are all based on this ancient theme. In this book, we let this ancient theme guide us, as we interweave music, number, and computer programming.

1.2.1 Pythagoras—Harmonic Series

The ancient Babylonians, Egyptians, and Greeks were fascinated with the technological nature of music—perhaps even more than we are today. For instance, Pythagoras (c. 570–c. 495 BCE) discovered that musical pitch intervals could be described by numbers. He and his students are credited with the discovery of mathematics, a term which they coined. Pythagoras left Greece at a young age to be educated in Egypt. There he associated himself with Egyptian priests, who at the time studied astronomy, geometry, and religion (all at once, without the divisions we have today). Pythagoras spent close to 20 years in Egypt, but then was captured during a war and was transferred as a slave to Babylon (an area now part of Iraq). There, through his knowledge and intellect, he gained access again to the educated elite and continued his studies in astronomy, religion, geometry, and music.

Pythagoras's contributions helped shape the ideas of subsequent philosophers, mathematicians, and scientists, including Plato, Aristotle, and many more. Aristotle tells us the Pythagoreans discovered that musical harmony can be explained by numbers; they took up mathematics, and "thought its principles were the principles of all things. Since, of these principles, numbers are by nature the first, and in numbers they seemed to see many resemblances to the things that exist and come into being" (Aristotle 1992, pp. 70–71). This observation suggests that everything we experience through our senses can be described (e.g., measured and represented) by numbers, in some way or another, and then it can be turned into music. For instance, consider the music stored on your digital music player (inside the machine, this music is represented by numbers).*

---

* This applies to all information (e.g., text, images) stored on a computer, or the Internet—the term *digital* refers to representing information using numbers (i.e., converting information to data).

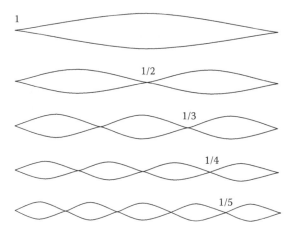

FIGURE 1.1   String resonating at integer ratios.

Also, consider the concept of sonification, that is, the conversion of arbitrary data to sounds, so that they may be perceived more easily.* In other words, music and numbers are interchangeable.

One of the major discoveries contributed by the Pythagoreans, which helped shape the nature of music theory many centuries later, is the observation that strings resonate in simple ratios. In particular, they observed that strings exhibit harmonic proportions—they vibrate at integer ratios of their length, that is, 1/1, 1/2, 1/3, 1/4, 1/5, etc. (see Figure 1.1). The instruments of the era, the *lyra* and the *kithara* (the latter etymologically related to the modern guitar), were most probably used in their experimentations.

This was a major discovery, since it demonstrated that integers emerge from the natural properties (or geometry) of a string. Accordingly, the 19th century mathematician Leopold Kronecker said, "God made the integers; all else is the work of man" (Bell 1986: 477). He argued that arithmetic and mathematical analysis must be founded on integers. The Pythagorean discovery is even more profound when considering the implications of string theory in physics, which states that the universe consists of subatomic particles resembling one-dimensional resonating strings. These ideas are related to the fields of *cymatics* and *fractal geometry* (see the following sections).

Finally, Pythagoras and his students worked on a theory of numbers and explored the *harmony of the spheres*. The harmony of the spheres (or *musica universalis*—music of the spheres) is the philosophical belief that

---

* See Chapter 7 for a more in-depth discussion of sonification.

the planets and stars moved according to mathematical equations. Since numbers are connected to musical notes, the orderly movement of planets was said to create an astronomical symphony. According to different religious/philosophical traditions, this music could be heard only by the most enlightened individuals. However, with the advent of the modern computer (and the knowledge you will accumulate in this book), this music is now accessible to everyone.

One of the major discoveries of this era (first described by Plato, in Timaeus, and then by Euclid, in his Elements) was the golden ratio (or golden) mean. This special proportion, which humans find aesthetically very pleasing, is found in natural or human-made artifacts (Beer 2008; Calter 2008, pp. 46–57; Hemenway 2005, pp. 91–132; Livio 2002). It is also found in the human body (e.g., the bones of our hands, the cochlea in our ears, etc.). The golden ratio reflects a place of balance in the structural interplay of opposites.

1.2.2 The Antikythera Mechanism—The First Known Computer

Ancient astronomical models were well established. They were used to construct the first computing machines approximately 2,100 years ago (Vallianatos 2012). Of these early computational machines, only one survives, in the National Archeological Museum of Greece, in Athens. Interestingly, these machines would have been unknown to us had it not been for the early 20th century discovery of fragments of a working model on a 2,000-old shipwreck near the island of Antikythera (see Figure 1.2).

The Antikythera mechanism uses the same design principles (i.e., employing gear ratios to implement mathematical relations) as the much later (19th century) Difference and Analytical Engines designed by Charles Babbage and Lady Ada Lovelace (see Figure 1.3). The connection between these machines and modern computers is indisputable.

1.2.3 Johannes Kepler—Harmony of the World

The Pythagorean ideas and theories inspired many in the centuries that followed, including Johannes Kepler. In 1619 Kepler wrote his seminal work *Harmonices Mundi* (*Harmony of the World*). In this book, Kepler describes physical harmonies in planetary motion. His work contributed significantly to the scientific revolution that brought us out of the dark ages.

In this book Kepler presents his third law of planetary motion, that the distance of a planet from the sun is inversely proportional to its speed. Based on this result, he also discusses the harmony found in the motions

FIGURE 1.2   Fragment from the Antikythera mechanism.

FIGURE 1.3   Part of Babbage's difference engine.

of the planets. In particular, he discovered that the speeds of consecutive planets approximate musical harmonies. The only exceptions are Mars and Jupiter. However, we now know that this is the result of a missing planet, whose mass is found in the asteroid belt between Mars and Jupiter. This belt was discovered 150 years after Kepler's death.

Saturn   Jupiter

Mars (approx.)   Earth   Venus

Mercury

Here the moon also has a place

FIGURE 1.4 Kepler's study of musical notes representing the motion of the known planets (capturing changes in speed as planets traverse their elliptical orbits around the sun).

Kepler argued that planets can be thought of as "singing" together in near harmony. This harmony fluctuates as planets slow down and speed up (i.e., each has a minimum and maximum angular speed). Only rarely do planets "sing" in perfect concord.

This kind of *sonification* (i.e., turning data into music) has been applied to many natural and human-made phenomena to generate sounds that are not too foreign to our ears, as might initially be imagined (see Figure 1.4). Later in the book, we explore this idea of sonification, so you too can create your own experiments related to the Pythagorean ideas. Recently, geologist John Rodgers and jazz musician Willie Ruff helped materialize Kepler's *Harmonices Mundi* by sonifying actual orbital data of planets in our solar system. This recording can be easily found on the Internet and is very inspiring to listen to.

1.2.4 Cymatics

Cymatics (from the Greek κύμα, "wave") is the study of visible (visualized) sound and vibration in 1-, 2-, and 3-dimensional artifacts. It was influenced significantly by the work of the 18th century physicist and musician Ernst Chladni, who developed a technique to visualize modes of vibration on mechanical surfaces, known as *Chladni plates* (see Figure 1.5).

When drums or gongs are struck, they vibrate in similar ways. That similar modes of vibration relate to musical pitch, rhythmic subdivisions, and sound timbre is interesting and suggests that many aspects of music and sound can be described computationally and controlled through software. Cymatics is an inspiring young field of exploration—for more

FIGURE 1.5 Chladni plates, vintage engraving. Old engraved illustration of Chladni plates isolated on a white background. (From Charton, É. and Cazeaux, E., eds. (1874), Magasin Pittoresque.)

information, see Evan Grant's TED Talk, which demonstrates the science and art of cymatics, through beautiful visualizations of soundwaves (Grant 2009).

1.2.5 Fractals

In the spirit of Pythagoras, mathematical descriptions for musical organization continue to be pursued. The hierarchical nature of music has led many to consider fractal geometry as an interesting candidate for such descriptions. Fractals are *self-similar* objects (or phenomena), that is, objects consisting of multiple parts, with the property that the smaller parts are the same shape as the larger parts, but of a smaller size. Fractals were developed by Benoit Mandelbrot to study harmonic proportions in nature (Mandelbrot 1982). Figure 1.6 displays a fractal tree (also known as a Golden Tree, since it incorporates golden ratio proportions). This fractal is constructed by dividing a line into two branches, each rotated by 60° (clockwise and counter-clockwise), with a length reduction factor equal to the golden ratio (0.61803399…). These smaller lines, again, are each subdivided into two lines following the same procedure. This repetition/subdivision continues on and on (theoretically) to infinity. Interestingly, similar patterns appear extensively in nature (as they maximize the amount of matter that can fit in a limited space, that is, touching but not overlapping). Such artifacts are very easy to construct using a computer.

The Harvard linguist George Kingsley Zipf (1902–1950) was a great influence on the development of fractals. In his seminal book, *Human Behavior and the Principle of Least Effort*, Zipf reports the amazing observation that word proportions in textbooks, as well as notes in musical pieces (among other phenomena), follow the same harmonic proportions (i.e., 1/1, 1/2, 1/3, 1/4, 1/5, etc.) first discovered by Pythagoreans on strings. Zipf proportions have been discovered in a wide range of natural and

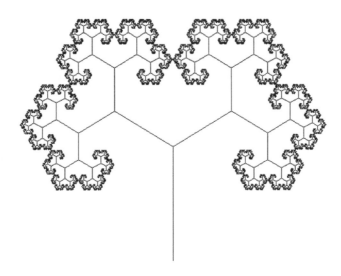

FIGURE 1.6   Fractal tree.

human-made phenomena, including music, city sizes, salaries, subrou-
tine calls, earthquake magnitudes, thicknesses of sediment depositions,
extinctions of species, traffic jams, and visits to websites, among many
others.

Zipf proportions are also known as *pink-noise, harmonic,* and *1/f* proportions and can be considered to be measures of variety or interest. At
one extreme is a random probability of occurrence (i.e., chaos or white
noise, such as radio static) where events are unpredictable and seemingly
unorganized. In the mid range lie fractal and brown-noise that have some
discernable organization. At the other extreme are very monotonous phenomenon (aka black-noise proportions), such as a musical piece consisting mostly of one note. In physics, white-noise, pink-noise, brown-noise,
and black-noise proportions are known as *power laws.* Psychologists have
shown that people prefer music, and other experiences, that have a balance
of predictability and surprise, and so having a computable measure of this
likelihood can be useful in computer music making.

Many interesting attempts have been made to generate music from
fractal artifacts. Conceptually, the process is relatively straightforward—
it involves converting aspects of a fractal object to aspects of a musical
artifact. For instance, the placement and size of a line in Figure 1.6 could
be converted to the pitch and duration of a note. As the fractal object
is being visually generated through a computer program, that same program could output the corresponding musical notes to a MIDI file, thus

generating a fractal music piece. The process of mapping visual elements to audio elements is called *sonification*. Sonification is an art in itself, as there are many possible ways of converting between visual elements and audio elements. (For instance, consider how you might sonify Figure 1.6.) The trick is to identify which visual elements to select, and how to map them to audio, so as to generate the most aesthetically pleasing (or scientifically interesting) audio artifacts. Sonification and fractals are explored later in the book.

1.3 COMPUTER MUSIC HISTORY

Throughout human history, technologies have consistently influenced our societal development, with periods of accelerated influence occurring at times such as the Renaissance, the Industrial Revolution, and the Information Age. This is paralleled by a relatively similar pattern of music technology developments. The earliest harps, horns, and drums are clearly technologies and their development and usage relied on new technologies of their day, very similar to the way computers are applied to music production in our age.

Landmarks in the history of music technologies include the use of written notation from around mid 9th century CE, the development of polyphony in the centuries that followed, and organ building improvements and equal temperament in the Middle Ages.

The Renaissance and Baroque periods saw an obsession with music of the spheres resulting from the newly developed field of astronomy (see above), and a peaking of craftsmanship in the violins of Stradivarius and in compositional technique in the fugues of Bach. The study of alchemy led to 19th century chemistry and physics, which provided new metals and efficient methods to improve instrument fabrication. This surge in instrument development went hand in hand with increases in orchestra size. Also, industrialization was a common underscoring theme in music, such as Wagner's "Der Ring des Nibelungen." Early 20th century landmarks include the automation of music via the player piano, the technological abstractions of electronic and recorded sound in the music of Schaeffer and Stockhausen, and parallel abstractions in the musical structures and notations of Debussy, Stravinsky, Schoenberg, Xenakis, Cage, and others.

This history is continuous in its highlighting of human curiosity and creativity. However, the developments in knowledge and technology are not deterministic and did not follow a simple evolutionary path (i.e., a path of increasing complexity). For example, the interests of Pythagoras resurfaced and inspired (more than one thousand years later) Kepler's explorations of musical patterns in astrological movements and Fourier's investigations into sonic spectra in the later 18th century. In between these investigations were centuries of explorations that followed other technological paths. The path of technological development is in no way straight or predictable in advance, even if such developments appear as a logical sequence with hindsight.

1.3.1 Automated Music

One of the characteristics of the computer as a music machine is that it can be automated by programming. Automatic instruments have existed for a long time, probably since antiquity. One possibility is the *hydraulis* which is attributed to Ctesibius of Alexandria (3rd century BCE). The hyrdraulis, about which little is known (due to the loss of ancient knowledge mentioned earlier), used water pressure to drive air through pipes, thus producing sounds (similarly to later ecclesiastical organs). Another possibility is the wind organ developed by Heron also of Alexandria (1st century CE), which was driven by a wind wheel. These and later designs were passed through Byzantine and later Arab scholars to Italy around the Renaissance period.

These designs contributed to later automated instruments, such as the barrel organ of Henry VIII built in 1502. It was manually driven, but the course of the following century led to fully autonomous instruments driven by clockwork mechanisms (similar to the Antikythera mechanism, whose design principles were also passed through Byzantine and later Arab scholars).

In order to increase the repertoire used in automated music machines, an alternative was sought to barrel organs that used replaceable barrels, which were expensive to produce and on which playing time was limited. A solution to both these problems presented itself in the 18th century in the form of the punch card technologies employed by Jacquard weaving looms. Scores were made in the form of holes punched in paper tape or cards. The cards could be strung together to create long sequences. This became a new form of musical notation, which was not efficient for human reading but quite efficient for machine reading. The machine became the interpreter of these machine-specific scores. Such instruments constitute more than an amusement even if their quality of performance was quite low. They enabled musical performances to be captured and transported, to be reproduced on demand, and replayed time and again for closer inspection.

Perhaps the most sophisticated (and certainly the most popular) automated instrument was the *player piano*. Although its development historically paralleled the gramophone, its sonic quality was far superior for quite a while and brought music on demand into many homes in the first half of the 20th century. The availability of automated musical performances in the home changed the role of the audience, affecting (not always detrimentally) concert attendance and the social status of musical performance skills. The player piano, more than electronic recording technologies, was the parent of MIDI sequencing in choosing to capture pitch, duration, and force (*velocity*) for each note. The piano rolls were editable and so "near perfect" performances could be created, and composers were not slow to realize that piano rolls could produce music beyond that humanly performable. In this way the composer first became a nonperforming producer, involved in all the steps from conception to final sounding.

1.3.2 Early Computer Music

The first public performance of computer music was programmed by Geoff Hill and Trevor Pearcey and generated by CSIRAC (Council for Scientific and Industrial Research Automatic Computer) at the Australian Computer Conference in August 1951. At this time, computer music was little more than a computational barrel organ playing popular tunes of the time; however, to do so at that time was no easy task, especially given the fickleness of the valve components, the timing constraints of the memory using mercury delay lines, and awkward punched paper tapes for describing programs. CSIRAC was the first computer in Australia and a machine intended purely for scientific research, so the achievement is a remarkable example of how quickly people turn any technology to musical purposes.

Computer-based music composition had its start in the mid-1950s when Lejaren Hillier and Leonard Isaacson did their first experiments with computer-generated music on the ILLIAC computer at the University of Illinois. They employed both a rule-based system utilizing strict counterpoint and also a probabilistic method based on Markov chains (also employed by Iannis Xenakis around the same time). These procedures were applied variously to pitch and rhythm, resulting in "The ILLIAC Suite," a series of four pieces for string quartet published in 1957.

The recent history of automated music and computers is densely populated with examples based on various theoretical rules from music theory and mathematics. While The ILLIAC Suite used known examples of these, developments in such theories have added to the repertoire of intellectual technologies applicable to the computer. Among these are the Serial music techniques, the application of music grammars (notably the Generative Theory of Tonal Music by Fred Lerdahl and Ray Jackendoff), sonification of fractals and chaos equations, and connectionist pattern recognition techniques based on work in neuropsychology and artificial intelligence.

Arguably, the most comprehensive of the automated computer music programs is David Cope's Experiments in Music Intelligence (EMI), which performs a feature analysis on a database of a particular composer's works (Cope 2004). Using this analysis, EMI can then compose any number of pieces in that composer's style (e.g., J.S. Bach, Chopin, etc.). The term *style*, here, is a function of many musical aspects, including melody and harmony.

In terms of melody, EMI captures repeated ideas that run through the works in its database, that is, common melodic material that a composer tends to use. Finding such repeated ideas is a complicated task, as the same melodic idea can be presented in a variety of different ways within a single piece: notes can be added to it or removed, it can be sped up or slowed down, and it can be played over different harmonies. This requires sophisticated pattern recognition within a complicated context and is one of the major accomplishments of Cope's research.

In terms of harmony, EMI extracts chords from the works in its database and then constructs its own chord progressions. These progressions do not replicate the ones in the database's works; they are novel. However, they are stylistically similar to (i.e., follow similar construction rules as) the analyzed works. In this way, the composer's "harmonic style" is also replicated in EMI's new compositions.

EMI's database can, actually, be loaded with the works of more than one composer. When this is done, its resulting compositions blend the styles of those composers. The results are sometimes odd, but quite often surprisingly clever and beautiful.

1.3.3 Electronic Music

After Thaddeus Cahill's relatively unsuccessful attempt at creating a massive organ-like device using early American telephone technologies called the *Telharmonium*, one of the first electronic performance instruments was Leon Thérémin's device invented in the 1920s in Moscow. The *Theremin*, as it was known, was played by positioning each hand at a varying distance from two antennae. The location of the hands changed the electromagnetic fields generated by electricity passing through the antennae, one controlling volume, the other the pitch of a constant and relatively pure tone. The Theremin made quite an impact, with pieces being written for it by Aaron Copeland and Percy Grainger, although the most popularly known example is in the opening of the Beach Boys' hit "Good Vibrations."

The first popular electric keyboard instrument was the Hammond organ, invented in 1935 by Laurens Hammond using electromagnetic components to generate sinusoidal waveforms which could be combined in various combinations using drawbars. The drawbars acted similarly to pipe organ stops, but rather than simply turning on or off oscillators, they controlled their degrees of loudness. The B3 model, first produced in 1936, has become legendary in gospel, jazz, and rock music. It provided a relatively affordable and portable keyboard instrument for music performance, and the timbral variety "synthesized" through drawbar settings gave to keyboard players a taste of customizable timbre that would later be expanded by the synthesizer.

The solid body electric guitar was developed after some initial production of semiacoustic electric models in the 1930s. Following early experiments by Adolf Rickenbacker and Les Paul, the first production models appeared in the early 1950s from the Gibson and Fender companies. The major technical hurdle was the refinement of the pickups to eliminate noise and provide a clear signal, which was solved largely by the development of the twin-coil "humbucking" pickup.

The early development of recording technologies by Thomas Edison was done with mechanical technologies around the turn of the 20th century. It was not until electronic amplifiers became available in the form of vacuum tubes that the minute etchings of the recording process could be played back with any fidelity. Even then the making of recorded cylinders was tedious and specialized. Building on this research, the first commercial magnetic tape recorder was introduced in 1948. The ability to record, not only play back, was the shift necessary to motivate musicians to use this technology creatively.

In Paris in the late 1940s after World War II, Pierre Schaeffer developed a compositional use for the previously reproduction-focused tape recorder. The compositional technique Musique Concrète, as it became known, used recorded sounds of both instrumental and environmental origin, manipulated them through variations in timbre, pitch, duration, and amplitude, then collaged these sounds into a polyphonic musical form.

Tape-based compositional works were produced by Karlheinz Stockhausen in Cologne from the mid-1950s, which he called Elektronische Musik (Electronic Music). As well as treating recorded sounds, Stockhausen and contemporaries such as Edgard Varèse focused on synthesizing new timbres using oscillators, filters, and amplifiers.

The successful commercialization of synthesizers came with the release in 1964 of the Moog synthesizers. The technical breakthrough that made these instruments possible was the use of transistors instead of vacuum tubes, which dramatically reduced the instrument's size and increased the stability of voltage control. One of the more popular early recordings using the Moog synthesizers was Wendy Carols's "Switched-on Bach," which was a notable achievement at the time, but created a legacy of imitative thinking which still haunts synthesizer usage, as more recently reinforced in the *General MIDI* specification. The most popular of Robert Moog's synthesizers was the Mini Moog, one of the first portable all-in-one synthesizers, still highly regarded 50 years after its release (see Figure 1.7).

The use of recording as a compositional and synthesis tool did not change much from the days of musique concrète until the late 1970s, when the development in Australia of the Quasar and M8 digital synthesizers by Tony Furse influenced the commercially successful *Fairlight* CMI developed by Peter Vogel and Kim Ryrie, and at the same time the New England Digital *Synclavier* was developed in New Hampshire by

FIGURE 1.7 The Mini Moog synthesizer.

Sydney Alonso, Jon Appleton, and Cameron Jones. The *Fairlight* and the *Synclavier* introduced sampling technologies (short-duration digital recording) to commercial music making in 1979. Both instruments were also capable of sound synthesis processes and used keyboard controllers for performance, attached to computer systems for storage, display, and editing of waveforms. A version of the *Fairlight* is now available for the Apple iPad, which highlights how much computing power and expense has changed in the last half a century or so.

Digital technologies made their way into synthesizers first as memory banks for presets, most famously in the Sequential Circuits *Prophet V*, and later in the sound synthesis engine itself, notably with the Yamaha DX7 (see Figure 1.8). The release of the DX7 in 1983 coincided with another significant event in electronic music history: the introduction of the Musical Instrument Digital Interface (MIDI) standard.

Developed by Dave Smith of Sequential Circuits, and with input from other major manufactures of the time, notably Roland and Yamaha, the MIDI standard replaced the plethora of interconnecting standards such that equipment from different manufacturers could communicate. MIDI began as a note-based live performance protocol, intellectually indebted to music notation and player-piano technologies. The MIDI standard has expanded over the years to include file formats, sample transfer protocols, the General MIDI standard sound set, a music XML format, and a range of other musical and operational parameters.

The synthesizer, in its keyboard form, has remained quite stable since the 1980s, with some controller extensions modeled on other instruments including guitar, woodwind, and percussion. Research continues into new instrument designs, as it always has, with STEIM in the Netherlands and the HyperInstrument group at MIT's Media Lab contributing significantly during the 1990s, but with developments expanding quite broadly since then. Many of the latest research developments are evident

FIGURE 1.8 The Yamaha DX7 synthesizer.

in the proceedings of the annual New Interfaces for Musical Expression (NIME) conference.

Along with advances in MIDI and the digital synthesizers, the 1980s also saw an accelerating increase in personal computer ownership and with it the expansion of music software. Most significant from a commercial aspect was the rise of the MIDI sequencer software. Building on the techniques of earlier electronic sequencers to repeat short series of notes, software sequencers continue to provide more comprehensive musical transformations.

Alongside sequencing, music notation programs were also appearing at this time, although it took the desktop publishing revolution of the early 1990s for all the appropriate technologies to fall into place, notably the postscript font-description language and laser printing. Computer music publishing is now the norm rather than the exception. The first program to successfully combine both sequencing and notation was C-Lab's *Notator* on the Atari computer, which proved the rule that you only need one "must have" program to sell a computing platform. Over the years, this program has transformed into Apple's Logic Pro software.

As personal computer power increased in the late 1990s, synthesis software (long the domain of expensive systems such as the *Fairlight* or computer workstations) became accessible. This is evident in the current popularity of hard disk recording systems, such as *Pro Tools*, as well as real-time signal processing systems, which are becoming practical on mobile computers for reverb and equalization, and even real-time synthesis as complex as frequency modulation, granular, and physical modeling.

The integration of many of these technical threads in computer-based composing, recording, publishing, and multimedia occurred around the late 1990s, and now digital music systems provide rich and expressive tools for the musician. Around the turn of the 21st century the increases in computing power reached a threshold where personal computers were powerful enough to manage most audio and some video processes in real time. This saw the concentration of computer music systems in software or "virtual" versions of what had been previously separate hardware components. The laptop computer had become the one-stop digital music workspace and an instrument for live performance. This process of concentrated computing power continues, with mobile devices such as smartphones and tablet computers increasingly becoming the site for computer music practices.

1.3.3.1 Reflection Questions

1. What were the dominant technological drivers of the past few centuries?

2. Where do you think the current borders of technical innovation are that will affect music making?

3. Given that new materials such as iron and aluminum have shaped the development of acoustic instruments, what changes have driven electronic/computer instrument development?

4. What have been the major developments in automated music described in this chapter?

5. The use of electronics has shaped music making over the past 100 years. Who were some of the musicians to first pioneer the use of electronic devices for music?

6. What has been the impact of audio recording on music making?

7. What was the basis of the compositional technique known as musique concrète?

8. What changes occurring during the 1990s are described in this chapter?

1.4 ALGORITHMS AND PROGRAMMING

Computers have been traditionally programmed to calculate solutions to numerical problems (the name "computer" itself reflects this—modern computers were viewed as a replacement for human computers in the military). This view, of course, is very restrictive, as any normal computer user can attest. Computers are wonderful for playing games, searching the Internet, and for making music. In this book, we introduce computer programming in the context of connecting number, music, and nature.

One of the more significant advantages of the computer for music making is its ability to be programmed: its ability to automatically do a series of tasks and to do them quickly. This is, of course, the basis for all software development but can also be the basis for a music making practice. Algorithmic music using a computer takes advantage of this ability to automate a series of instructions (an algorithm) to musical ends.

Definition: An algorithm is a series of steps (or instructions) for performing a task.

Examples of algorithms include instructions for assembling a bookshelf (assembly instructions sheet), steps for making spaghetti sauce (a recipe), and instructions for performing a musical piece (a musical score).

Computers can be programmed to follow such series of instructions using a programming language. When programming computer music, the challenge is to write instructions that lead to interesting and expressive music. Musical algorithms can describe how each of the musical elements is specified and varied as the piece proceeds. This can include control over the pitch, duration and loudness of notes, the timbre of sounds, the use of structural features such as repetition and variation, as well as tempo, volume, balance and so on.

The ability of computers to run algorithmic processes (programs, or sequences of steps) gives the impression that computers have autonomy and are possibly "smart." At its most advanced levels this autonomy is referred to as Artificial Intelligence (AI), most well known through systems such as IBM's Deep Blue for playing chess, and popularized through science fiction systems such as *Hal* in the science fiction film *2001—A Space Odyssey* and androids such as *R2D2* in *Star Wars* or robots such as *Walle* in the film of the same name.

In algorithmic music systems the intention and possibilities are generally far more modest, even though some comprehensive systems, such as Experiments in Musical Intelligence by David Cope, can construct complex and complete pieces. Generally, algorithmic composition systems are used for more mundane purposes, such as generating a tonal melody of a few bars, creating valid variations in a 12 tone row, suggesting possible chorale harmonization, or sonifying mathematical structures such as fractals or artificial life worlds by converting the numbers generated by formulae into pitches, rhythms, and form.

Many algorithmic systems deal with music at the note level, specifying or manipulating attributes such as pitch, duration, and dynamic. This is historically the most prevalent way of thinking about music and is the basis for common practice notation, so it is not surprising that note-based generative systems are common. Algorithmic processes can be applied in many ways to notes. Small pitch changes at the frequency level can be used for microtonal music, or loudness may be controlled by a function introducing a kind of jitter or instability to the note which, if subtle, may add some life to an otherwise static electronic performance. Similarly, subtle changes can be applied to the dynamic levels of a repeated phrase in order to provide variety which masks the machine-like repetition to some degree; we will explore this example later. Algorithmic systems can be used to generate note-level scores for either acoustic or electronic realization.

Algorithmic processes can be applied to music at a structural level to manipulate measures, phrases, or sections of music.

1.5 THE COMPUTER AS A MUSICAL INSTRUMENT

There are many ways to make music using computers. Some musicians prefer ready-made production software, such as GarageBand, Audacity, and Ableton Live (to name a few). Other musicians prefer more versatility and power—they utilize music programming environments, such as CSound, SuperCollider, Extempore, PureData, and Max/MSP. This book prepares for the second approach, by introducing a simple, yet powerful programming language (Python) and several programming libraries for generating sounds, processing images, and building graphical user interfaces.

The ability of computers to follow arbitrary musical (or other) processes makes it possible to design and implement new musical instruments, running on regular computing platforms, such as a laptop or a smartphone. For instance, through Python and the libraries provided with this textbook, we can develop many different types of computer music instruments. These instruments may have graphical user interfaces, which increases usability. Such instruments may be used to compose and perform classical, popular, or avant garde musical pieces.

Part of the rationale for thinking about computer musical instruments is that it takes years to master playing a guitar or violin; computer musical instruments, on the other hand, can be much easier to learn for beginners. The goal is not to replace traditional instruments—there will always be a need for them. Instead, we wish to allow more people to engage in musical performance. This is similar to the ease with which someone can play a game of football on an XBox, as opposed to a real, physical game of football. The computer, through its constraints and affordances, makes it easier to play (or compose) music (Magnusson 2010). Computer-based musical instruments may be used by a single performer or by many performers in ensembles, such as in laptop or iPad orchestras. Additionally, one could mix traditional instruments and computer instruments. Finally, computer musical instruments can be designed to do things beyond the capabilities of traditional instruments or human performers. This creates exciting new musical possibilities.

As an instrument, the computer becomes a vehicle through which you express musical ideas. Any music instrument (or music technology) provides new capabilities, but also comes with constraints. For instance, acoustic instruments are easier to play in certain pitch ranges, and many orchestral instruments can play only single melodies (not harmony or chords).

Similarly, computers as musical instruments have unique characteristics and limitations. For instance, using MIDI they can generate 128 different sounds, 16 of which can be simultaneous. Each of these 16 instruments can play many simultaneous notes. So, conceivably, a single computer can play as many notes as a medium size orchestra and perform musical pieces of arbitrary complexity (e.g., pieces that a human orchestra could not perform with traditional instruments). They can also play back arbitrary sound recordings and loop them or mix them to construct an arbitrary soundscape. In terms of limitations, computers are "dumb" if not programmed, perhaps even more musically dumb than a drum, which can at least make an interesting noise when struck. To make computers come to life as musical instruments, someone needs to "play" them, by writing and using software, in order to generate music.

So, like other musicians, developers of musical software utilize their knowledge of an instrument to achieve musical tasks with certain aesthetic objectives or goals in mind.

When viewed as a musical instrument, the computer becomes a partner in the music making process. Thus, music making moves beyond simple human-directed activity to become a collaboration with the computer; by assisting, the computer has an influence. However, the more responsibility a computer has over the music production, the greater the expectations on the programmer and composer using the system. This might seem counterintuitive but it is true. A greater emphasis on the unintelligent, generic computer requires more of the intelligence and ingenuity of the human programmer and composer.

Finally, as with traditional musical instruments such as the guitar, intimate engagement with the computer as an instrument requires that you increase your skills; for the computer this means your programming skills. As with any musical instrument, the greatest satisfaction will result when you learn to *play* the computer well. This involves studying the principals involved, regular practice, and an immersion in the computer music culture through listening, reading, and discussion.

This book will assist you to compose musical processes for offline playing (e.g., generating a musical piece and saving it as a MIDI file) and using the computer for real-time performance. The musical possibilities are endless.

Let's begin.

1.6 SOFTWARE USED IN THIS BOOK

This book teaches the programming language Python. Also, it comes with a software package which contains several libraries for music-making in particular and creative computing in general, such as a library for music (MIDI and audio), one for images, and one for graphical user interface (GUI) development.

Python is a general purpose programming language designed to be easy to read and to use. It includes a large and comprehensive library of functions for common computing tasks. Python is widely used by leading computing companies such as Google, and thus many resources are available (including this book) to assist in learning it. The version of Python used in this book is called *Jython*, which is implemented on top of the Java Virtual Machine (JVM). Since the JVM is a truly portable programming environment (i.e., it runs on all popular computing platforms), any code you develop using this book will run identically on different computing platforms. This portability is very desirable, since it allows you to share your algorithmic music compositions with collaborators around the world, regardless of what computer system they are using (e.g., Windows, Mac, or Linux).

The music library provided supports arbitrary music programming tasks through Python. It provides a music data structure based upon note/ sound events, and methods for creating, organizing, manipulating, and analyzing that musical data. Generated music scores can be rendered as MIDI or audio files for storage and later processing or playback. Also, the music library allows the playback of arbitrary notes and sounds in response to user-initiated events (see GUI library below). The music library is used and incrementally presented through the remaining chapters, in conjunction with traditional computer science concepts and Python programming skills. After all, the purpose of this book is to teach you programming in Python through music making. The audio library is used later in the book when we build interactive musical instruments in software.

The image library allows the reading and writing of digital images. These digital images may originate from your digital camera or be downloaded from the Internet. Once an image is read into a program, the image library allows accessing and manipulating image elements (i.e., pixels). For instance, one could read in a image and use its varying colors (or luminosity) to drive a musical process. The image library is covered in Chapter 7, the chapter on sonification.

The GUI library allows development of graphical user interfaces to drive (or be driven by) arbitrary musical processes with an emphasis on musical performance. The idea is that, through this library, you may develop arbitrary musical instruments to be used in performance (as well as in composition). The GUI library is covered in Chapter 8, the chapter on interactive musical instruments.

This software is available for download on the website associated with this book. Installation instructions are provided alongside the software.

1.6.1 Case Study: Running a Python Program

There are two ways to write Python code: directly into the interpreter, and using an editor. The first way is easier for small programs that you intend to run only once. For example, if you want to perform a quick calculation, or if you want to create a melody consisting of a few notes, you may use the interpreter directly.

For a more substantial program (or a program that you intend to run many times) you should use an editor to type your code. See some suggestions on the textbook website. Actually, any text editor will do—some of them are better because they color-code different parts of a Python program (e.g., comments are green, strings are red, reserved words are purple, and so on). This makes programs easier to read.

Using your favorite editor, type the following program:

```
# playNote.py
# Demonstrates how to play a single note.

from music import *            # import music library

note = Note(C4, HN)            # create a middle C half note
Play.midi(note)                # and play it!
```

Save this program in your music programming folder under the filename "playNote.py".

Observe the following points:

- Python is case sensitive. For example, there is a big difference between "note" and "Note" (see line 6 above).

- Be very careful when typing. Most of your errors will likely be caused by a typo.

- Empty lines (vertical space) are NOT important to Python. For example, lines 3 and 5 are there to improve readability.

- Comments allow humans to see the algorithmic process involved, so they are very important. They are ignored by Python. For example, see the comments above following a "#" up to the end of the line.

- Line up comments whenever possible—it is considered good style (e.g., lines 4, 6, and 7). If a comment is too long to fit on one line, you may put it above the statement(s) it explains, as in lines 1 and 2 above.

When finished typing, run it.*

Running this program will ensure that you understand the basics of the software development process using Python and that everything has been installed properly. The program should generate a single note. The note has pitch C4 (i.e., middle C), and duration HN (i.e., half note). Always make sure your volume is adjusted properly before running programs that generate sound.

If you have reached this point, congratulations! You have written and run your first musical program in Python. The rest is incremental and straightforward. Enjoy!

1.7 SUMMARY

This chapter introduced the fundamental ideas and concepts that we will explore throughout the rest of the book. It talked about the early origins of music, mathematics, and computing. It revealed some inspiring areas of exploration, including the harmonic series, the golden ratio, Kepler's *Harmony of the World*, Zipf's law, cymatics, and fractals. It presented some of the pioneers in computer and electronic music. Also, it introduced useful terms, such as algorithm, that we will see again and again. Finally, it introduced the idea that nature, music, and number are all somehow intertwined, i.e., that one can be transformed into another. These inspiring ideas and concepts will guide our creative explorations throughout this book, and hopefully beyond it.

---

* How you run your program depends on the editor used. To run your programs, see setup instructions or follow your instructor's directions.

## CHAPTER 2

Elements of Music and Code

Topics: Fundamentals of music, the Python music library, notes, rests, variables, integers and floats, arithmetic operations, input and output, coding a program.

2.1 OVERVIEW

This chapter provides an overview of music representations, and corresponding ways to represent data and information in Python. It is mainly for people with little or no background in music or computer programming. What we call sound results from vibrations of air molecules. The air molecules are moving forwards backwards a short distance, then forwards and backwards again, repeatedly. There are other instances of the phenomenon of repeated forwards and backwards motions in nature. An electrical current is the movement of electrons through a wire or other conductor. For the alternating current (AC) we use in our households, electrons are moving forwards and backwards.* Such forwards and backwards motions are termed cyclic (or periodic). A cycle means the completion of one forward movement followed by its complementary backward movement. The frequency of cyclic motion means the number of cycles per unit of time. The common measurement of frequency is the Hertz (Hz), which is one cycle per second.

2.2 MUSIC IS SOUND AND ...

Music consists of sound and silence. Its common elements include pitch, duration, dynamics, timbre, texture, and form. Additional concepts

---

* For the direct current (DC) produced by a common battery, the electrons move in one direction only.

FIGURE 2.1   Sound represented as a waveform (one of the authors saying "sound").

include melody, harmony, tempo, meter, and articulation. Below is a basic definition:

Definition: Music is sound organized in time.

There are various ways to describe musical sounds. They may be described using high-level terms, such as *tuned, untuned, mellow, harsh,* etc. They may also be described as belonging to instrument families such as strings, woodwind, brass, electronic, percussion, and so on.

Musical sounds, when recorded, can be represented as an audio wave-form (see Figure 2.1). A waveform depicts the vibrations, that is, changes in air pressure over time. Waveforms are at the core of how musical (and other) sounds get generated, captured, and transported through different natural and electronic media.

While it is possible for computer musicians to create sound through synthesis and sampling, in this book we focus mostly on how sound events (e.g., notes) are structured and played back. We generally defer sound creation to the computer's internal sound synthesizer. However, in later chapters we will explore some sampling possibilities.

Finally, although musical expression varies from culture to culture, in this book we focus mainly on the Western traditions, including classical, popular, jazz, and experimental. Also, we rely on common practice nota-tional conventions as a foundation for music representation in Python.

2.3 NOTES

Before the invention of audio recording (and long before digitizing, which allowed sound to be seen as waveforms), people represented musical events as notes, often drawn as "dots" on paper.

Definition: A note in Python consists of pitch, duration, dynamic (volume), and panning position (within a stereophonic field).*

Typically each note corresponds to an event, such as the pressing of a piano key, the singing of a syllable, or the plucking of a guitar string.

2.3.1 Musical Notation

Musical notation traditionally consists of musical notes written in standard (Western) music notation on a *musical staff* or *stave*. This is also known as *common practice notation.*

A staff consists of five horizontal lines and four spaces, each corresponding to a musical pitch (e.g., see Figure 2.2).

In this notation, time goes from left to right. Vertical lines (or bar lines) mark out a *measure*. Measures divide music into groups of a particular number of beats—typically 2, 3, or 4 beats per measure. In simple common, or 4/4, time, a beat is one quarter note long.

In this chapter we will learn enough of this notation to be able to transcribe pieces of music into Python programs. Knowing how to read this notation allows you to translate it into the (much simpler) notation used by the Python music library. You will not be performing this music—you will instruct your computer to do that.

Frè - re   Jac - ques.   Frè - re   Jac - ques,   dor - mez vous?   Dor - mez vous?

Son-nez les   ma - ti - nes!   Son-nez les   ma - ti - nes!   Din, dan, don.   Din, dan, don.

FIGURE 2.2   Music example in standard music notation.

---

* Traditionally, notes consist only of pitch, duration, and dynamic. The addition of panning (i.e., placement of notes in the stereo field—left, right, or somewhere in between) has become available, since modern computers come with stereo audio cards (i.e., two audio channels—left and right).

2.3.2 Pitch

Pitch is an important characteristic of musical notes. Pitch relates to the frequency (speed) of vibration of the sounding object. Faster vibrations equate to a higher frequency and pitch.

Definition: The pitch of a note specifies how high or low the note sounds.

In standard notation, pitch is represented by the vertical placement of a note on the staff. Notes may be placed on a line or in a space (see Figure 2.3). In Python we can label a pitch by its letter name and octave number (e.g. C4).

2.3.2.1 Pitches Are Integers

In computing pitches are often represented as integers. They range from 0 to 127.

For example, middle C (on a piano) is 60. This follows the MIDI standard, which represents pitches from 0 (lowest pitch) to 127 (highest pitch).* MIDI supports a total of 10 octaves. For comparison, a standard, 88-key piano has range of about 7 octaves.

Definition: An octave is the pitch range equal to a doubling in frequency.

An octave ranges from one pitch (e.g., C4) to the next of the same letter name (e.g., C3 down or C5 up). An octave is typically divided into 12 different pitches, 7 with standard letter names (A, B, C, etc.) and 5 with sharps/flats.

FIGURE 2.3   Standard music notation and Python pitch constants.

---

* MIDI stands for Musical Instrument Digital Interface. It is an industry-wide standard used in digital musical instruments and computing devices to exchange musical data.

C4	CS4 DF4	D4	DS4 EF4	E4 FF4	ES4 F4	FS4 GF4	G4	GS4 AF4	A4	AS4 BF4	B4	BS4 C5
60	61	62	63	64	65	66	67	68	69	70	71	72

FIGURE 2.4 Python pitch constants and MIDI numbers for middle C octave (boldfaced constants correspond to black piano keys).

For convenience, Python also defines meaningful names (constants) for pitches. Pitch constants consist of a letter (C, D, E, F, G, A, or B) followed by the octave (or register) of the pitch. For example, middle C is C4. Appendix A lists all available pitch constants.

Pitches can be made flat (*b*) or sharp (#).

To make a pitch sharp, add 1 to the MIDI number. To make a pitch flat, subtract 1 from the MIDI number.

Figure 2.4 shows pitch constants for the octave above middle C. Notice the natural, flat (F), and sharp (S) pitches. Also notice how some pitches have two names—for example, 61 is both a C sharp and D flat.[*]

All of the relationships between numbers and names for pitch come together in Figure 2.5, which presents different representations of pitch. You may use it as a Rosetta Stone of sorts.[†]

2.3.3 Duration

The duration (or note length) is the second characteristic of musical notes.

Definition: The duration of a note in Python specifies its length over time.

In standard notation, the duration of a note is represented by the type of note *head* used (filled in or hollow), and by the attached vertical lines, called *stems*, which may have *tails* or *beams*. Figure 2.6 shows common durations, including whole (1/1), half (1/2), quarter (1/4), and eighth (1/8) notes.

---

[*] Such pitches are called enharmonic.

[†] The Rosetta Stone is an ancient artifact (circa 196 BCE). Its discovery helped us understand Egyptian hieroglyphs. It has a message in three different languages: Egyptian hieroglyphs, Demotic Greek, and Ancient Greek. Since linguists already knew the latter two, they were able to decipher Egyptian hieroglyphs.

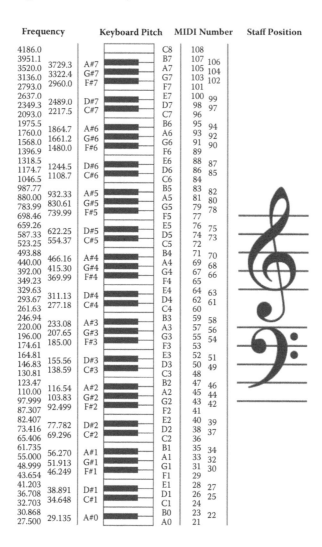

FIGURE 2.5   Correspondence among notes, MIDI numbers, and frequencies.

WN = 4.0, DHN = 3.0, HN = 2.0, DQN = 1.5, QN = 1.0, DEN = 0.75, EN = 0.5, DSN = 0.375, SN = 0.25

FIGURE 2.6   Python duration constants in standard music notation.

2.3.3.1 Durations Are Real Numbers

In Python, durations are real numbers (or floats).* A quarter note is 1.0. Everything else is calculated accordingly.

For example, a half note is 2.0, and a whole note is 4.0; also an eighth note is 0.5, a dotted eighth note is 0.75, and a sixteenth note is 0.25.†

Musicians add personal expression or interpretation when they play notes in a musical score. Sometimes they make note durations slightly shorter or longer. Using real numbers (as opposed to integers) allows you to specify more precise note durations, such as 4.3 or 2.667. This way our code can generate music that sounds less mechanical and more "natural," human-like.

For convenience, Python also defines constants for common durations. Figure 2.6 shows some of these constants together with their standard music notation paired above them, such as WN (whole note), HN (half note), and QN (quarter note). See Appendix A for a complete listing.

2.3.4 Dynamic

Dynamic (or volume) is the third characteristic of musical notes.

Definition: The dynamic of a note specifies how loudly the note sounds.

In Python, dynamic values are integers. They range from 0 to 127. This follows the MIDI standard, which represents dynamics (or velocity) from 0 (silence) to 127 (loudest).

For convenience, Python also defines constants for common dynamic values using traditional (Italian) musical terms, such as *forte, fortissimo, mezzo_forte, mezzo_piano, pianissimo,* and *silent* (see Appendix A).

2.3.5 Panning

Panning (or panoramic) position is the fourth characteristic of musical notes and relates to the location of the sound in space.

---

* *Float* is the computer science term for real numbers. It comes from the term "floating point," which has to do with how real numbers are represented in computer memory. We will talk more about floats soon.

† The duration of a dotted note is one and a half the duration of the original (undotted) note.

Definition: The panning position of a note specifies its placement in the stereo field.[*]

In Python, panning values are real numbers. They range from 0.0 to 1.0. For example, 0.0 is extreme left, 0.25 is center-left, 0.5 is center, and 1.0 is extreme right.

With two ears, humans can discern sound direction with great precision. Using real numbers (as opposed to integers) allows you to specify the exact note location in the stereophonic field (on speakers or headphones).

For convenience, Python also defines the following constants for common panning values: PAN_LEFT, PAN_CENTER, and PAN_RIGHT.

2.3.6 Creating Notes

The most basic musical structure in the Python music library is a *note*. Python notes have the following attributes:

- pitch—an integer from 0 (low) to 127 (high)

- duration—a positive real number (quarter note is 1.0)

- dynamic—(or volume) an integer from 0 (silent) to 127 (loudest)

- panning—a real number from 0.0 (left) to 1.0 (right)

To create a note, we specify its pitch, duration, dynamic, and panning position, as follows:

```
Note(pitch, duration, dynamic, panning)
```

where dynamic and panning are optional. If omitted, dynamic is set to 85 and panning to 0.5 (center).

For example, this Python statement creates a middle C quarter note and stores it in variable n.[†]

```
n = Note(C4, QN)
```

---

[*] Practically, MIDI prohibits different pannings for simultaneous notes. Keep this in mind as your music becomes more complex.

[†] Python can remember values for later use, by letting us give them names (e.g., n above). This name is called a variable. A variable is really an alias for a memory location (similar to a calculator's memory button). Variables are an important building block for programs. They are introduced more formally later in the chapter.

Here we do the same using numerical values (since C4 is 60 and QN is 1.0):

```
n = Note(60, 1.0)
```

Here we create the same note, but as loud as possible (127):

```
n = Note(C4, QN, 127)
```

Here we do the same, but placed to the very left side (in the stereo field):

```
n = Note(C4, QN, 127, 0.0)
```

Good style: Whenever possible use named constants (e.g., C4, QN) when creating notes because they are easier to identify and read. Use numbers only for nonstandard values.

2.4 RESTS

Music consists of sounds and silence. Here we see how to generate silence.

Definition: A rest is a silent note, used to create empty space in a melody.

Figure 2.7 shows common practice notation for rests and their affiliated Python constant values, including whole (1/1), half (1/2), quarter (1/4), and eighth (1/8) rests.

Rests are special notes with only one attribute:

- Duration—a positive real number (quarter note is 1.0)

For example, a half note is 2.0, and a whole note is 4.0; also an eighth note is 0.5, a dotted eighth note is 0.75, and a sixteenth note is 0.25.[*]

WN = 4.0　　　HN = 2.0　　　QN = 1.0　　　EN = 0.5　　　SN = 0.25

FIGURE 2.7　Python rest constants in standard music notation.

---

[*] The duration of a dotted note is one and a half the duration of the original (undotted) note.

For convenience, we may use Python constants for common durations. This includes WN (whole note), HN (half note), and QN (quarter note).

2.4.1 Creating Rests

Musical rests in Python only have a duration value. For example, the following creates a quarter note rest:

```
n = Rest(QN)
```

We can also create rests as notes with pitch REST.

```
n = Note(REST, QN)
```

This is useful in programs where notes and rests are grouped together. We will see this in future case studies.

2.4.2 Case Study: Playing a Note

Now that you have learned more about notes, let's revisit our first program from Chapter 1:

```
# playNote.py
# Demonstrates how to play a single note.

from music import *      # import music library

note = Note(C4, HN)      # create a middle C half note
Play.midi(note)          # and play it!
```

Let's explore this program in detail. The first part of the program consists of comments. They explain what the program does. Good style dictates that all programs begin with such comments.

```
# playNote.py
# Demonstrates how to play a single note.
```

Next, we import the music library. *

```
from music import *      # import music library
```

_____

* This library needs to be imported in every program that makes music.

The Python music library defines various names (e.g., constants for pitches and common durations). Also, it defines various functions and classes to be used by our programs.

Fact: Functions and classes are containers provided by Python to group related functionality.

Functions and classes allow us to access useful functionality by using a single line of code, as follows. They make programming convenient and efficient.*

Next, we create a single note and store it in a variable. The note has pitch C4 (i.e., middle C) and HN duration (i.e., half note). The variable is called "note".

```
note = Note(C4, HN)    # create a middle C half note
```

Remember that Python is case sensitive. So "note" and "Note" are two different names. The first "note" is the name of the variable. The second "Note" is a Python music function, which creates a new note.

The assignment statement (" = ") instructs the computer to take the value from the right-hand side and store it in the variable on the left-hand side. This allows us to use the note later.

The last statement instructs the computer to play the note:

```
Play.midi(note)    # and play it!
```

"Play" is a another class defined in the music library. The second part, "midi()", is a function in the "Play" class.

The statement "Play.midi(note)" calls this function and gives it the variable containing the note to be played.

2.4.2.1 Comments

Notice how the comments describe the code. They provide a narrative for other programmers to read.

---

* For now we will learn how to use various functions and classes provided in the music library. Later, we will learn more about them and how to define new functions and classes to package our own code.

Good style: Write comments first and then translate them to code.

Good programming style dictates that we write comments first. It may seem faster or more productive to write code first. But writing comments helps you concentrate on design and consider alternatives before committing them to code. This saves time in the long term. Remember: "Two hours of design can save you 20 hours of coding."

2.4.3 Exercise

Change the above program to play different notes. Create more elaborate notes, specifying pitch, duration, dynamic, and panning position. In particular, have your program play

- a whole A0 note, at half volume, placed at center panning field;

- a 32nd C9 note, at full volume, placed at the extreme left;

- a half G5 note, at full volume, placed at the extreme right;

- a whole note rest; and

- a double whole BF3 note, at full volume, center (*Hint:* durations are numbers, so they can be added when creating notes).

2.5 VARIABLES AND ASSIGNMENT

Similarly to a calculator's memory button, Python can remember values for later use. The values have to be stored in memory locations. To differentiate among different memory locations, Python allows us to give them names.

Unlike mathematics, where a variable is a symbol for a value we do not know yet, a variable in programming is a symbol for a memory location used to store known values.

Definition: A *variable* is a named memory location used to store a known value.

Fact: To define a variable, we use an *assignment statement*.

Assignment statements in Python have the following format:

```
<var> = <expr>
```

where <var> is the name of the memory location, and <expr> is the expression that generates the value we want to save.

For example,

```
x = 3
```

stores the integer 3 in variable x. If this is the first time we use x, Python finds a new memory location and names it x, then stores 3 in it. If x already exists, its contents are overwritten by 3.

From now on, when we refer to x, it stands for 3 (until we change its value again).

Here is another example:

```
y = 5 + 2
```

stores the integer 7 in variable y. Python acts like a calculator, in that it first evaluates the expression on the right-hand side of the "=" (assignment) sign. Then it stores the resultant value, that is, 7, in the variable y (on the left-hand side).*

Fact: Variable names may consist of letters, numbers, and underscores, but they need to start with a letter.† Variable names can be as long as you wish.

Good Style: Variable names should describe what is stored in them.

This makes programs easier to read.

2.5.1 Examples

The following examples of assignment statements are typed interactively into the Python interpreter (hence the ">>>" prompt).

```
>>> number1 = 3
```

---

* A common misconception among beginner programmers is that Python will store the actual expression (e.g., 5 + 2) in the variable. It does not. It evaluates the expression and stores the result (e.g., 7) in the variable. Actually, it is very easy to use the Python interpreter (as shown above) to do quick calculations.

† There are exceptions to this, but that's outside the scope of this book.

This assignment statement stores 3 into variable number1. If this is the first time we use number1, Python creates a new variable called number1 and stores 3 in it. If number1 already existed, it now contains 3, that is, the previous content is overwritten.

From now on, when we refer to number1, it stands for 3 (until we change its value again):

```
>>> number1
3
>>> number1 = 4
4
>>> number2 = number1 + 1
>>> number2
5
```

Here we see how we changed the value of number1 to 4, and then we used it in an expression to calculate a value for number2.

Next, let's see examples with musical notes:

```
>>> note1 = Note(C4, QN)
>>> note1
<NOTE (Pitch = 60)(Duration = 1.0)(Dynamic = 85)(Length = 0.9)(Pan =
0.5)>
```

First we create a note and store it in variable note1. Then we see that note1 indeed contains a note with pitch 60 (i.e., C4), duration 1.0 (i.e., QN), and some other default values.

When trying to get the value of a variable that does not exist, the interpreter gives us an error:

```
>>> note2
Traceback (most recent call last):
  File "<stdin>", line 1, in <module>
NameError: name 'note2' is not defined
```

You should read error messages from the bottom up. Usually, the last two lines are most helpful. Here, the message tells us that note2 is not defined. This is a common error. It is usually caused by typos in our programs.

2.5.2 Reserved Words

The following words are reserved by Python. You cannot use them as variable names.

and	del	from	not	while
as	elif	global	or	with
assert	else	if	pass	yield
break	except	import	print	
class	exec	in	raise	
continue	finally	is	return	
def	for	lambda	try	

Also, you should not redefine names used by the music library (see Appendix A).

2.6 NUMBERS

In addition to notes and rests, Python can represent numbers. This is necessary, since Python is a *general-purpose programming language*. This means that Python programs can perform general computing tasks, not only musical ones.

2.6.1 Integers

Python can handle arbitrarily long integers. In Python, integers are called int.

The following examples are typed interactively into the Python interpreter:

```
>>> x = 3
>>> x
3

>>> y = -45
>>> y
-45

>>> z = 0
>>> z
0
```

Python can handle positive, negative, and zero integer values.

```
>>> w = 12345678901234567890
>>> w
12345678901234567890L
```

The "L" at the end of the number above indicates it is a long integer. Python can handle arbitrarily large integers. If the number is too large to fit in a memory location, Python will handle it automatically. The "L" just informs you that this happened. Many programming languages cannot handle large integers—they just give you an error. This is one of the many special features that makes Python such an easy programming language to learn and use.

2.6.2 Floats

Python can also handle real numbers. The difference between integers and real numbers is that real numbers have a decimal part (for more accuracy).

Definition: In Python, a real number is called a float.

The following examples are typed interactively into the Python interpreter:

```
>>> x = 1.53
>>> x
1.53

>>> y = -2345678.901234
>>> y
-2345678.901234

>>> z = 2222222222222222222222222222222222222.3
>>> z
2.222222222222223e+36
```

Python can handle arbitrarily large float numbers. In the last example, it automatically switches to scientific notation, that is, $2.222222222222223 \times 10^{36}$.

2.6.3 Arithmetic Expressions

Python supports arithmetic, such as addition, multiplication, and so on. This is not surprising, since programming languages were initially developed for performing calculations and other mathematical tasks (hence the term "computer").

Arithmetic expressions contain values and arithmetic operators. The values can be either literal (e.g., 1) or variables (e.g., a). For example:

```
>>> a = 2
>>> b = 3

>>> (a * a) + b - 1
6
```

The following arithmetic operations are supported (in the examples, assume a is 2 and b is 3):

Operator	Description	Example
+	Addition—adds two values together	a + 1 evaluates to 3
–	Subtraction—subtracts one value from another	b – 2 evaluates to 1
*	Multiplication—multiplies two values	a * b evaluates to 6
/	Division—divides one value by another	10/a evaluates to 5
		10/b evaluates to 3
%	Modulo—calculates remainder of division	10%a evaluates to 0
		10%b evaluates to 1
**	Exponent—raises a value to a power	10 ** 2 evaluates to 100
		a ** b evaluates to 8
()	Parentheses—grouping for higher precedence	(2 + 3) * 4 evaluates to 20
		2 + 3 * 4 evaluates to 14
abs(x)	The absolute value of x	abs(– 3) evaluates to 3
		abs(3) evaluates to 3
round(x)	The rounded value of x	round(0.6) evaluates to 1.0
		round(0.4) evaluates to 0.0
		round(0.5) evaluates to 1.0

Notice above how integer division returns an integer result.[*] The modulo (%) operator returns the integer remainder. If you use integers, Python assumes you want an integer result. For example,

```
>>> 5/2
2
>>> 5%2
1
```

To get more accuracy, use floats. For instance, float division returns a float result. For example,

```
>>> 5.0/2.0
2.5
>>> 5.0%2.0
1.0
```

If you mix floats and integers, Python assumes you want accuracy and returns float results (like normal calculators).

---

[*] Unlike normal calculators, which routinely switch between integers and floats, as needed.

Division by zero is a problem (as it results to infinity). Python raises an error, as follows:

```
>>> 5.0/0.0
Traceback (most recent call last):
File "<stdin>", line 1, in <module>
ZeroDivisionError: float division
```

Finally, parentheses can be used to change the normal precedence of operators (as in algebra). Normally, multiplication and division are done before addition and subtraction.

2.7 INPUT AND OUTPUT

Python programs can communicate with the outside world (i.e., outside of the running program) through various means. These may include:

- a MIDI device (say a MIDI keyboard, a MIDI control surface, or a MIDI synthesizer);

- an external file (e.g., a file containing weather data patterns from the 20th century across the country, a file containing biosignals such as heart rate, brain-wave patterns, or skin conductance, recorded from a person while, say, listening to jazz music, or a file containing the text of "The Gold-Bug" by Edgar Allan Poe);

- a smartphone or tablet (e.g., iPhone, iPad, or Android phone) through a computer network protocol, such as Open Sound Control (OSC); and

- the physical keyboard and mouse connected to your computer.

Below we introduce input/output (I/O) through the computer keyboard and screen. This is the most basic interaction available to our Python programs.

2.7.1 Input from the Keyboard

Python programs, while running, can receive information interactively from the user, through keyboard input. This allows end-users to provide information to a running program. This is done through this special assignment statement

```
<var> = input(<prompt>)
```

where <var> is a variable to receive the input from the user, and <prompt> is the prompt to show to the user to let them know what to enter. For example,

```
x = input("Enter an integer:")
```

This is a special assignment statement. It assigns a value to variable x. This value comes from the keyboard. Let's see this interaction with the Python interpreter:

```
>>> x = input("Enter an integer:")
Enter an integer: 3
>>> x
3
```

When the input statement is executed, the computer prompts the user (i.e., "Enter an integer: "). Here the user typed 3, followed by <RETURN>. Hitting <RETURN> tells the computer to take this value and store it in variable x.

The expression "Enter an integer:" is called a string.

A string is another Python data type, like int and float. A string contains a sequence of characters. Strings are used to represent text data. They are enclosed in quotes. There will be more about strings in Chapter 7.

2.7.2 Output to the Screen

Python programs can display information on the screen using the print statement.

Print has the following format[*]:

```
print <expr>
print <expr>,..., <expr>
print <expr>,..., <expr>,
```

where <expr> may be a value, a variable, or an arbitrary expression.

For example, if we run a program with these print statements:

```
x = 3
print 2 + 3
print 2, 3, 2 + 3
print 2,
print 3
print x, x + 1
print
print "The value of x is", x
```

---

[*] The book uses Python 2.x syntax. There is a newer, slightly different syntax used in Python 3.x.

it will generate this output (try it!):

```
5
2 3 5
2 3
3 4
```

The value of x is 3.

How does this work? Well, the first print statement outputs the value of the expression 2 + 3. Python will first evaluate any arithmetic or other expression and then output its result. The second print statement outputs a list of expressions. The third print statement outputs 2, but does not advance to the next line, so the fourth print statement outputs 3 on the same line. The fifth print statement outputs the value of x and x + 1. The sixth print statement outputs an empty line (useful for improving the readability of output).

The last print statement outputs a string, followed by the value of variable x. Strings can be used to create readable output.

2.8 DATA TYPES

So far we have seen various Python data types, namely, integer, float, and string. Additionally, in terms of music, we have seen notes and rests.

Definition: A *data type* describes a distinct type of value that can be handled by the computer (e.g., integer, float, etc.) and the particular operations that are associated with it (e.g., addition, subtraction, etc.).

Python has several additional data types, which will see in later chapters as needed by our creative endeavors.

2.8.1 The type() Function

The type() function tells us the data type of a particular value or variable. For example,

```
>>> type(3)
<type 'int'>

>>> x = 3
>>> type(x)
<type 'int'>

>>> type(3.1)
<type 'float'>
```

```
>>> type("hi")
<type 'str'>

>>> n = Note(C4, QN)
>>> type(n)
<type 'jm.music.data.Note'>
```

Knowing the data type of a variable or value can be useful at times. For now, let's use the type() function to make sure that a variable is what we think it is, when necessary. The last example shows that notes are special objects that come from the music library.

The type() function is one of many useful functions provided by Python to help you with programming. We will see more such functions soon.

2.8.2 Case Study: Finding the Octave of a Pitch

Arithmetic expressions can be very useful in music. Actually, for computers music is numbers, and numbers are music (as we see throughout this book).

The following program automates the process of calculating the octave of a MIDI pitch.

```
# findPitchOctave.py
# Given a MIDI pitch integer, find its octave.

from music import *     # import music library

# get input from user
pitch = input("Please enter a MIDI pitch (0 - 127): ")

# an octave has 12 pitches, and octave numbering starts -1, so
# division by 12 and subtracting 1 gives us the octave (e.g. 4th)
octave = (pitch/12) - 1

# output result
print "MIDI pitch", pitch, "is in octave", octave
```

Let's explore the program in detail. The first part of the program consists of the usual comments. After importing the music library, we prompt the user to enter a MIDI pitch.

```
pitch = input("Please enter a MIDI pitch (0 - 127): ")
```

Notice the prompt, "Please enter a MIDI pitch (0 - 127):". It tells the user what the program expects as input; also, it gives an idea of the valid range of input values. Contrast this with a prompt like "Enter a number:", or worse, no prompt. Which one would you like to see, as a user?

Good Style: Provide meaningful user prompts. They make programs more useable.

The next statement,

```
octave = (pitch/12) - 1
```

takes into account that an octave contains 12 pitches so dividing by 12 gives us the octave of the particular pitch. Actually, MIDI octaves start at –1, so we need to subtract 1 to produce the correct result. Try this calculation by hand. Does it work, say, for C4 (i.e., 60)?

Since we do not know ahead of time (when we write the program) what the user will enter (at runtime), we use variable pitch to store the particular pitch chosen by the user.

The last statement,

```
print "MIDI pitch", pitch, "is in octave", octave
```

outputs the result to the user. Notice how we combine strings and numbers to build a meaningful message.

If you followed this example, you understand the essence of programming.

Fact: Most programs follow this outline: input, process, and output.

2.8.3 Testing Programs

To test the program, pick three values—the two extremes (0 and 127) and something in the middle (say, C4 or 60). Determine the result by hand (i.e., 0, 9, and 4, respectively). Then run the program and test its answers.

2.8.4 Exercise

Change the above program to also output the step within the octave, for a particular pitch. Again, an octave has 12 steps (numbered 0 to 11). *Hint:* Division by 12 gives us the octave. Remainder (modulo) gives us the step within the octave. Test the program with different values to make sure it works correctly.

2.9 SUMMARY

This chapter introduced the building blocks of music making and Python programming. In particular, we learned how notes are represented using Western notation. We also learned how to create notes in Python programs, namely, by using values for pitch, duration, and possibly for dynamic (volume), and panning (placement in the stereo field). We also learned about rests. Although rests denote the absence of sound, a decision was made when designing the music library to explicitly represent rests as a special (silent) type of note. This design decision simplifies creating long sequences of notes, in that we do not have to worry about explicitly specifying note start times (relative to the beginning of the piece). As we will see in Chapter 2, we simply place notes in the order we want them (including any rests), specifying only their duration. This way we can process long strands (phrases) of notes easily (e.g., move them around) without worrying about having to change their start times. This allows for the efficient creation of musical material.

Additionally, we learned about different useful data types in Python, namely, integers, floats, and strings, and some of their operations. We will learn more, as needed, in later chapters. We also learned about variables and how to store data values for use in our programs. Finally, we learned how to do basic input and output.

## CHAPTER 3

Organization and Data

Topics: Notes, phrases, parts, scores, Python lists, Ludwig van Beethoven, scales, MIDI instruments, Harold Faltermeyer, chords, Bruce Hornsby, 2Pac, Joseph Kosma, drums and other percussion, drum machines, Deep Purple, top-down design, reading and writing MIDI files.

3.1 OVERVIEW

In this chapter, we introduce Python data structures for music making. In particular, we look at some of the Python music data objects and how they represent musical information. It is assumed that you already know how to create notes and rests, and that you can write basic Python programs that play a note, or that output a value, as introduced in Chapter 2.

3.2 MUSICAL ORGANIZATION

Music is made up of structured sounds and silence. But music, as sound or ideas, is not a concrete thing that we can see, grasp, or capture. In order to do so, we create musical representations. Typically, these representations stand for particular musical events and describe the characteristics of these events. But what are music representations for? They serve a number of functions. First, they make music concrete and stable so that it can be stored and manipulated; we can capture music as scores, data structures, and files that we can process and modify. Second, they allow music to be communicated in ways other than aurally; we can transfer files to other software packages, send the musical representations over the Internet, and print them out using music-publishing software. Third, a representation is a conception of music, a way of understanding it, a way to think about it. The way we represent music says a lot about what we conceive music to be, and it influences the types of musical transformations that will be evident and available to us. Importantly, the music representation is *not* the music, and so the way a representation is interpreted by software, hardware, or human beings can vary the outcome significantly by making assumptions about aspects not represented and amplifying any distortions present in the representation system.

Over time there have been many schemes for representing music, and in Western cultures the most prevalent has been the use of the notated score. The data structure in the Python music library takes this tradition as a starting point and builds upon it. A computer representation of music, or anything else, needs to be quite detailed because the computer's interpretation is very literal, even moronic. As we proceed through this book the level of detail in the music representations will become more fine-grained, but for now many assumptions are made to keep the task manageable.

3.2.1 Music Data Structure

Musical information in Python (through the music library) is stored in a hierarchical fashion, mimicking a conventional score on paper.

Score (contains any number of Parts)

|

+— — Part (contains any number of Phrases)

   |

   +— — Phrase (contains any number of sequential Notes—e.g., a melody)

      |

      +— — Note (contains a single musical event)

This is a highly orthodox structure where a piece of music is represented as a score, that score has several parts (e.g., a flute part and a percussion part), each part contains phrases (e.g., melodies, riffs, grooves, sequences, patterns), and each phrase is made up of a series of one or more notes (individual sound events). The structure and naming conventions are designed to be familiar to those with some musical knowledge however, this hierarchy is only a starting point and, as will become clear later, you can represent any style of music within this structure.

Notes (and rests) are the most important and most basic music event. Actually, notes (or note objects) are created from the Note class (as seen in the previous chapter):

```
n = Note(C4, QN)
```

Below we see how we can organize notes into larger structures, using more classes from the music library, to organize and perform realistic musical pieces.

3.3 PHRASES

In music theory, a phrase refers to a grouping of consecutive notes. A phrase typically contains a melody—a linear list of notes that acts as a musical unit.

Definition: A phrase consists of a sequence of notes, the start time (in the piece), timbre (instrument used), and tempo (speed).

The Python music library provides the Phrase class, which allows storing and manipulating a series of Note objects. Phrase objects have other attributes, including:

- start time, that is, when it is supposed to be played in a piece (a real number—0.0 or greater)

- instrument—the instrument number used to play the phrase (0 to 127)

- tempo—the speed of playback (a real number—0.0 or greater, in quarter note beats per minute)

Notes in a phrase are played sequentially (i.e., one after the other). If a gap is desired between two notes, then a Rest note should be introduced. Phrases may also contain chords (i.e., sets of concurrent notes).

3.3.1 Creating Phrases

We can create a new phrase like this:

```
phr = Phrase()    # create an empty phrase
```

Alternatively, a new phrase can be created using an argument, the start time.

```
phr = Phrase(0.0)
```

A phrase created like this is set to start at time 0.0, that is, the beginning of the piece.

Fact: Musical time in Python is measured in quarter notes (i.e., time 1.0 equals the length of a quarter note).

There is a subtle, but big difference between Phrase() and Phrase(0.0). If no start time is specified, then when added to a part (more on this later) the phrase starts at the end of the previous phrase (or at the beginning of the piece, if this is the first phrase). On the other hand, phrases with start times are placed within the part at exactly the time specified. While it is convenient to create phrases without start times, when creating larger songs it is necessary to create phrases with start times.

As mentioned earlier, the music library provides constants for common time values, such as QN for 1.0 (one quarter note), or WN for 4.0 (four quarter notes). These constants are listed in Appendix A.

3.3.2 Adding Notes

Once an empty phrase has been created (as above), you can add notes to it using the addNote() function.

```
n = Note(C4, HN)
phr.addNote(n)
```

Since adding notes to phrases is very common, you can combine these two statements into one:

```
phr.addNote(C4, HN)
```

For example, here is the first theme of Beethoven's fifth symphony:

```
phr = Phrase()            # create an empty phrase

# the first note is an eighth note rest, so
r = Rest(EN)              # create it
phr.addNote(r)            # and add it to the phrase

g = Note(G4, EN)          # create a G4 note eighth note
phr.addNote(g)            # and add it to the phrase three times
phr.addNote(g)
phr.addNote(g)

ef = Note(EF4, QN)        # create an EF4 note
phr.addNote(ef)           # add it to the phrase

phr.addRest(r)            # add the eighth note rest again
```

```
f = Note(F4, EN)           # create an F4 note
phr.addNote(f)             # and add it to the phrase three times
phr.addNote(f)
phr.addNote(f)

d = Note(D4, QN)           # create a D4 note
phr.addNote(d)             # add it to the phrase

Play.midi(phr)             # play phrase
```

Notice how we create a note once and may add it several times to the phrase. This results in several notes of the same pitch and duration played back to back.

Also, notice how time consuming it becomes to add notes like this. Given that many more notes are needed to construct a piece, the music library provides a much more economical way for adding notes by using lists.

3.4 PYTHON LISTS

In addition to integers, floats, and strings, Python has a very useful data type, called a list. As its name suggests, the list data type is a collection of values in a specific order. To create a list, simply put the values in square brackets and separate them by commas. For example, here is a list of integers assigned to the variable x:

```
>>> x = [1, 2, 3]
>>> x
[1, 2, 3]
```

Here is a list of strings:

```
>>> y = ["Python", "is", "fun!"]
>>> y
['Python', 'is', 'fun!']
```

A list may contain values of different data types, for example:

```
>>> z = [2, "be or not", 2.0, "be?"]
>>> z
[2, 'be or not', 2.0, 'be?']
```

In this chapter, we use lists to store pitch and durations. For example,

```
>>> pitches = [C4, E4, G4]
>>> pitches
```

```
[60, 64, 67]
>>> durations = [QN, HN, EN]
>>> durations
[1.0, 2.0, 0.5]
```

By the way, notice how the interpreter replaces music library constants (e.g., C4) with the corresponding numerical values (e.g., 60).

3.4.1 List Concatenation

Python uses the "+" operator with lists for concatenation (i.e., joining of two lists). If you think about it, this makes sense—concatenation is equivalent to addition; that is, we are adding lists together. For example,

```
>>> [1, 2] + [3, 4]
[1, 2, 3, 4]

>>> [2, "be or not"] + [2.0, "be?"]
[2, 'be or not', 2.0, 'be?']

>>> [2] + ["be or not"] + [2.0, "be?"]
[2, 'be or not', 2.0, 'be?']
```

3.4.2 List Repetition

Python uses the "*" operator with lists for repetition (i.e., duplicating a list a number of times). In a way, this makes sense, as "multiplying" a list results in many copies of the list concatenated together. This is consistent with the concept of multiplication with numbers, that is, multiplication is a shorthand for multiple additions. For example,

```
>>> [0] * 10
[0, 0, 0, 0, 0, 0, 0, 0, 0, 0]

>>> [1, 2] * 4
[1, 2, 1, 2, 1, 2, 1, 2]

>>> ["hello"] *6
['hello', 'hello', 'hello', 'hello', 'hello', 'hello']

>>> [2, 'be or not', 2.0, 'be?'] * 2
[2, 'be or not', 2.0, 'be?', 2, 'be or not', 2.0, 'be?']
```

Finally, we can mix list concatenation ("+") and repetition ("*") operators. Repetition operations are performed first, followed by concatenation (similarly to multiplication and addition with numbers). For example,

```
>>> ["Python"] + ["is"] + ["fun"]*3
['Python', 'is', 'fun', 'fun', 'fun']

>>> [0]*2 + [1]*3 +[2]*5
[0, 0, 1, 1, 1, 2, 2, 2, 2, 2]

>>> [WN] + [HN]*2 + [QN]*4
[4.0, 2.0, 2.0, 1.0, 1.0, 1.0, 1.0]
```

Notice how the duration constants WN, HN, and QN are converted to the corresponding numbers before list operations are performed.

3.5 ADDING NOTES WITH LISTS

A faster way to enter notes is to use Python lists. Instead of adding one note at a time to a Phrase, you can store the pitch and durations for each note in parallel lists:

```
pitches = [E4, E4, E4, C4, REST, D4, D4, D4, B3]
durations = [ENT, ENT, ENT, HN, QN, ENT, ENT, ENT, HN]
```

Then, in one step, you can add all these pitches and durations into a Phrase, as follows:

```
phr.addNoteList(pitches, durations)
```

The first list contains the pitches of the notes, while the second list contains the corresponding durations. In other words, the first note has a pitch of E4 (the first element of the pitches list) and a duration of ENT (the first element of the durations list), and so on. The last note has a pitch of B3 and HN. The number of items in both lists must be the same, otherwise you get an error.

Definition: Two lists are called parallel if every item of the first list corresponds to the item, at the same position, in the second list.

To demonstrate how much faster this way is, here is the opening of Beethoven's fifth again, this time using lists of pitches and durations:

```
phr = Phrase()          # create an empty phrase

# create parallel lists of pitches and durations
pitches = [E4, E4, E4, C4, REST, D4, D4, D4, B3]
durations = [ENT, ENT, ENT, HN, QN, ENT, ENT, ENT, HN]
```

```
# add notes to phrase
phr.addNoteList(pitches, durations)

Play.midi(phr)          # play phrase
```

Compare the number of lines (and typing effort) between this and the earlier example.

Fact: The addNoteList() function supports up to four parallel lists, namely, pitches, durations (these two are required), dynamics, and pannings (these two are optional, but in this order).

Good Style: In parallel lists, align the corresponding pitches and durations.

As seen above, aligning the corresponding pitches and durations by adding additional white space improves the readability of your musical data but does not affect the list.

Appendix B contains additional Phrase functions. These will be very useful when creating algorithmic music in later chapters (or in your own endeavors).

3.6 CASE STUDY: LUDWIG VAN BEETHOVEN—"FÜR ELISE"

Now that we have learned about creating phrases and adding notes to them, let's create a theme from one of the most popular classical pieces, Ludwig van Beethoven's Bagatelle No. 25 in A minor for solo piano, commonly known as "Für Elise" ("For Elise").

```
# furElise.py
# Generates the theme from Beethoven's Fur Elise.

from music import *

# theme has some repetition, so break it up to maximize economy
# (also notice how we line up corresponding pitches and durations)
pitches1 = [E5, DS5, E5, DS5, E5, B4, D5, C5, A4, REST, C4, E4, A4,
            B4, REST, E4]
durations1 = [SN, SN, SN, SN, SN, SN, SN, SN, EN, SN, SN, SN, SN, EN,
              SN, SN]
pitches2 = [GS4, B4, C5, REST, E4]
durations2 = [SN, SN, EN, SN, SN]
pitches3 = [C5, B4, A4]
durations 3 = [SN, SN, EN]

# create an empty phrase, and construct the theme from the above motifs
theme = Phrase()
```

```
theme.addNoteList(pitches1, durations1)
theme.addNoteList(pitches2, durations2)
theme.addNoteList(pitches1, durations1) # again
theme.addNoteList(pitches3, durations3)

# play it
Play.midi(theme)
```

When looking at the musical score for this piece (available online), you may notice that the theme has some repetition in it. As we have done in this example, try to take advantage of such repetitions in musical material to reduce the effort in representing it in code. You can create a list of notes one time and add it many times. In the above, pitches1 and durations1 were constructed to capture such material, and thus are added twice.

3.6.1 Exercise

Transcribe one of your favorite melodies into Python. It should be at least as long as "Für Elise."

3.7 MUSICAL SCALES

The collection of notes from one pitch to the next one of the same name, say C4 to C5, is an octave. All the notes within an octave (including sharps or flats) make up the notes of the "chromatic" scale (12 notes).

Definition: A *scale* is a set of pitches within an octave.

Definition: The first pitch of a scale is called the *root*, or the *tonic*.

Any set of pitches within an octave can be a scale. Some are more consonant (pleasant) or dissonant than others. This is partially due to cultural conditioning (i.e., the types of music we were born into); however, there are some aspects that are controlled by the structure of our inner ear.*

Scales can be described by specifying the process of playing them on a piano (or guitar). All you need to do is specify the distances (or intervals) between consecutive pitches in the scale.

---

* The cochlea in the inner ear (the place where sound waves are broken up into individual frequencies) is constructed in terms of the golden ratio. This clearly incorporates certain "biases" as to what sounds go together well. Ear anatomy and its aesthetic implications are beyond the scope of this book. But we invite you to explore more. Musical beauty is partially in the ear of the beholder (and the fact that we all share the same general ear design).

Definition: An *interval* is the distance between two pitches.

Intervals are specified in terms of the number of chromatic steps (piano keys or guitar frets) you have to move through to get from one note to the other.

Fact: The music library uses integers to describe intervals.

For example, the interval between C and D is 2 steps (i.e., two piano keys, or two guitar frets). The interval between E and F is 1. Can you say why?

Patterns of interval steps can form scales. In Western music we have two very common scales, called the "major" and "minor" scales.

3.7.1 The Major Scale

The major scale is very important in Western music. For example, the *C major scale* consists of the seven white keys on a piano starting on C (i.e., C, D, E, F, G, A, B). In other words, to step through the C major scale, you have to start at C, move up two keys to D (interval of 2), move up two more keys to E (again, interval of 2), move up one key to F (interval of 1), and so on. Put together, this pattern of intervals is 2, 2, 1, 2, 2, 2, and 1 (the last interval to bring us to the root pitch, one octave higher, for closure/completeness). This pattern (2, 2, 1, 2, 2, 2, and 1) gives this scale its "major" sound characteristic. If the pattern is *transposed* (i.e., shifted) to start on D (i.e., D, E, F#, G, A, B, C#), it gives us the *D major* scale, and so on, for each note.

Definition: The *relative intervals* in a major scale are 2, 2, 1, 2, 2, 2, 1.

Another useful way to describe a scale is in terms of absolute intervals, that is, intervals from the root. To do so, for every pitch in the scale, we state the distance from the root. For example, using absolute intervals, the C major scale becomes 0 (i.e., C—the root), 2 (i.e., D), 4 (i.e., E), 5 (i.e., F), and so on.

Definition: The *absolute intervals* in a major scale are 0, 2, 4, 5, 7, 9, 11.

These intervals hold for any major scale, regardless of the root note. Accordingly, the Python music library defines a major scale as follows:

```
MAJOR_SCALE = [0, 2, 4, 5, 7, 9, 11]
```

This representation is convenient, because it allows us to easily create a major scale from any given root.

3.7.2 The Minor Scale

The minor scale is another common scale in Western music. The *minor* scale follows a slightly different pattern. All the white keys starting from A (A, B, C, D, E, F, G) make up a minor scale—the *A natural minor* scale. Transposing this pattern to start at C generates the *C natural minor* scale (C, D, Eb, F, G, Ab, Bb), and so on.

Definition: The *relative intervals* in a natural minor scale are 2, 1, 2, 2, 1, 2, 2.

Definition: The *absolute intervals* in a natural minor scale are 0, 2, 3, 5, 7, 8, 10.

These intervals hold for any natural minor scale, regardless of the root note. Accordingly, the Python music library defines a natural minor (or Aeolian) scale as follows:

```
AEOLIAN_SCALE = [0, 2, 3, 5, 7, 8, 10]
```

3.7.3 Other Scales

For convenience, the Python music library defines several scales as constants (see Appendix A, *Scale and Mode Constants*). These include BLUES_SCALE, CHROMATIC_SCALE, DORIAN_SCALE, MIXOLYDIAN_SCALE, and PENTATONIC_SCALE.

These scales do not contain specific notes (e.g., C5, D5, etc.). Instead, they are defined as absolute intervals. The first interval is always 0, denoting the root. Subsequent intervals are calculated by subtracting MIDI pitches of scale notes from the root. There is wisdom in this representation; it is very efficient. It allows you to use any MIDI note (pitch) as the starting note and, through arithmetic, find the other notes in the scale. For example, the following piece of code outputs the pitches for a major scale starting at C4*:

---

* This example uses a Python construct called a for-loop. We will explore for-loops in Chapter 5.

```
# printScale.py
# Outputs the pitches of the major scale pitches,
# starting at a given root.

from music import *

root = C4                  # starting pitch
scale = MAJOR_SCALE        # scale

# output the pitches in this scale
print "The major scale starting at pitch", root, ":"
for interval in scale:
   pitch = root + interval  # add the interval to the root
   print pitch


# after the loop
print "Done."
```

The scales in Appendix A are not exhaustive by any means. There is a wide variety of scales used in contemporary as well as non-Western styles of music. By representing a scale using a list of intervals, the music library makes it possible to define any scale you wish in your programs.

3.7.4 Exercise

In the above example, change the C4 to another pitch, or change MAJOR_SCALE to another scale (see Appendix A), to get the pitches of that scale.

Can you tell the difference between major and minor scales when you hear them? Some people can describe this difference, in terms of the feelings each one generates—can you? Give it a try.

3.8 MUSICAL INSTRUMENTS

Musical instruments are as old as human culture. The musical instruments used to perform a composition dictate its sound. Each instrument has a specific timbre, or tone quality. On a computer we use electronic (virtual) instruments to play back our music. These can have sounds sampled (recorded) from acoustic instruments or produced by synthesizers.

Definition: *Timbre* is the sound color or tone quality that distinguishes different instruments.

Computers support playback of music with a wide variety of instruments, as discussed below.

3.8.1 MIDI Instruments

In the 1980s the electronic musical instrument industry developed the MIDI (Musical Instrument Digital Interface) standard. MIDI specifies how computers and other digital musical instruments can generate and exchange musical data in a standard, unified way. A numbered list of instruments and sounds is part of the General MIDI specification. Your computer is equipped with a music synthesizer, which plays the notes generated by your Python programs with sounds corresponding to this instrument list.

Fact: The Play.midi() function plays music through your computer's MIDI synthesizer.

A couple of decades ago, such a synthesizer was a very expensive, avant garde instrument. Today, it is routinely packaged with standard computers. This powerful synthesizer remains hidden from the casual user. It is through your Python programs (and the music library) that you can easily access this synthesizer to generate algorithmic music.

The Python music library provides 128 different instruments as specified by the General MIDI standard. These instruments are divided into 15 different families (Table 3.1).

The General MIDI specification also supports a set of drum and percussion instruments, specifically for drum tracks and drum-machine programs. We'll address these in more detail in the next chapter.

3.9 SETTING THE INSTRUMENT

Phrases normally are played on a MIDI piano instrument. You can change the instrument for some phrase, phr, by calling its setInstrument() function:

```
phr.setInstrument(CLARINET)
```

The music library provides constants for all MIDI instruments (see Table 3.1.). These include piano, guitar, bass, and violin. They also include vibraphone, marimba, and tubular_bells. Other possibilities include organ, bandoneon, jazz_guitar, contrabass, voice, trombone, alto_sax, clarinet, and ocarina. Finally, examples of more exotic instruments include ice_rain, goblins, shamisen, reverse_cymbal, seashore, bird, helicopter, and applause.

TABLE 3.1   The 128 MIDI Instruments (grouped by family).

Family	Instruments
Piano	Acoustic grand piano, bright acoustic piano, electric grand piano, honky-tonk piano, electric piano 1 (Rhodes), electric piano 2 (Dx7), harpsichord, clavinet.
Chromatic Percussion	Celesta, glockenspiel, music box, vibraphone, marimba, xylophone, tubular bells, dulcimer.
Organ	Drawbar organ, percussive organ, rock organ, church organ, reed organ, accordion, harmonica, tango accordion.
Guitar	Acoustic guitar (nylon), acoustic guitar (steel), electric guitar (jazz), electric guitar (clean), electric guitar (muted), overdriven guitar, distortion guitar, guitar harmonics.
Bass	Acoustic bass, electric bass (finger), electric bass (pick), fretless bass, slap bass 1, slap bass 2, synth bass 1, synth bass 2.
Strings and Timpani	Violin, viola, cello, contrabass, tremolo strings, pizzicato strings, orchestral harp, timpani.
Ensemble	String ensemble 1, string ensemble 2, synth strings 1, synth strings 2, choir aahs, voice oohs, synth voice, orchestra hit.
Reed	Soprano sax, alto sax, tenor sax, baritone sax, oboe, English horn, bassoon, clarinet.
Pipe	Piccolo, flute, recorder, pan flute, blown bottle, shakuhachi, whistle, ocarina.
Synth Lead	Lead 1 (square), lead 2 (sawtooth), lead 3 (calliope), lead 4 (chiff), lead 5 (charang), lead 6 (voice), lead 7 (fifths), lead 8 (bass + lead).
Synth Pad	Pad 1 (new age), pad 2 (warm), pad 3 (polysynth), pad 4 (choir), pad 5 (bowed), pad 6 (metallic), pad 7 (halo), pad 8 (sweep).
Synth Effects	Fx 1 (rain), fx 2 (soundtrack), fx 3 (crystal), fx 4 (atmosphere), fx 5 (brightness), fx 6 (goblins), fx 7 (echoes), fx 8 (sci-fi).
Ethnic	Sitar, banjo, shamisen, koto, kalimba, bag pipe, fiddle, shanai.
Percussive	Tinkle bell, agogo, steel drums, woodblock, taiko drum, melodic tom, synth drum, reverse cymbal.
Sound Effects	Guitar fret noise, breath noise, seashore, bird tweet, telephone ring, helicopter, applause, gunshot.

See Appendix A for the complete list of MIDI instrument constants.

3.9.1 Exercise

Experiment with different sounds (MIDI instruments). Which ones do you like better?

- Identify four favorite instruments.

- Create three unique lists of instruments that go together well, for different styles of music. Be creative, experiment, discover.

- Remember that these computer instruments (similarly to the physical instruments they represent) have a "sweet spot," that is, a range of pitches they were intended to play (usually encompassing a couple of octaves). Try getting out of those ranges. For instance, how does *voice* sound at very low or very high pitches? Do the same for other instruments. Find combinations of unusual timbres and write them down.

This exploration may provide inspiration in your future music-making activities.

3.9.1.1 Setting the Tempo

To play back the phrase at a faster or slower speed, you can set the tempo.

The tempo is specified in beats per minute (BPM). Higher values indicate faster playback. The default tempo is 60 BPM, which is quite slow—but has the advantage that 1 beat equals one quarter note duration and lasts for 1 second. Typical musical tempos range from 80 to 150 BPM.

To set the tempo of a phrase, use the setTempo() function. For example,

```
phr.setTempo(120)
```

sets the phrase's tempo to 120 BPM.

3.10 CASE STUDY: HAROLD FALTERMEYER—"AXEL F"

Now that we have seen how to set the instrument and tempo of a phrase, here is an example where they are useful: Harold Faltermeyer's electronic instrumental theme from the 1984 film Beverly Hills Cop.

```
# axelF.py
# Generates Harold Faltermeyer's electronic instrumental theme
# from the film Beverly Hills Cop (1984).

from music import *

# theme (notice how we line up corresponding pitches and durations)
pitches1 = [F4, REST, AF4, REST, F4, F4, BF4, F4, EF4, F4, REST, C5,
            REST]
durations1= [QN, QN, QN, EN, QN, EN, QN, QN, QN, QN, QN, QN, EN]
pitches2 = [F4, F4, DF5, C5, AF4, F4, C5, F5, F4, EF4, EF4, C4, G4, F4]
durations2= [QN, EN, QN, QN, QN, QN, QN, QN, EN, QN, EN, QN, QN, DQN]

# create an empty phrase, and construct theme using pitch/duration data
theme = Phrase()
theme.addNoteList(pitches1, durations1)
theme.addNoteList(pitches2, durations2)
```

```
# set the instrument and tempo for the theme
theme.setInstrument(SYNTH_BASS_2)
theme.setTempo(220)

# play it
Play.midi(theme)
```

Notice how the theme has been divided into two groups of pitches and durations—nothing special here. Also, notice the setting of the instrument and tempo at the bottom. This new functionality allows you to write code to play any musical theme possible.

3.10.1 Exercises

Select two diverse but well-known musical themes.

- Write code that generates them.

- Try to match the instruments and tempos as closely as possible.

- Using the more exotic MIDI instruments (e.g., see Synth Pad, Synth Effects, and Sound Effects families), create an interesting musical theme of your own. You might explore possibilities using a regular musical instrument (if you know how to play one), or your voice, before you implement your code. Remember, 2 hours of design (in this case, about 5 minutes) can save you 20 hours of coding (i.e., about 50 minutes of work).

3.11 CHORDS

In music, chords occur when two or more notes sound together. Instruments that can play many notes at one time, such as guitars and pianos, often play chords. So far in our introduction to Phrase objects, we have focused on melodies where only one note sounds at one time—monophonic music. Here, we discuss chords and how to enter them—homophonic music.

Definition: Monophonic music refers to a sequence of nonoverlapping notes, that is, melody.

Definition: Homophonic music refers to two or more simultaneous sequences of notes with different pitches, but the same durations, that is, chords.

Definition: Polyphonic music refers to two or more simultaneous sequences of notes having both distinct pitches and durations.

Harmony is produced when two or more notes sound together. Therefore, both homophonic and polyphonic music produce harmony.

Chords are often made up of three or more notes. The chord is often named after the lowest note (also known as the root). For example:

- The C major chord consists of note pitches C, E, and G.

- The G major chord consists of note pitches G, B, and D.

- The A minor chord consists of note pitches A, C, and E.*

Figure 3.1 shows some commonly used chords and the pitch constants to use in your programs.

A common chord progression is C, F, G, C (many folk singers, e.g., Woody Guthrie and Bob Dylan, made their early careers using only these chords). Another common chord progression is C, Am, F, G, C.†

Many other possible chord progressions exist. Some locations within a progression sound more stable than others. Composers use these relative places of rest (stability) and unrest to create interest within a piece. When this is done well, we get musical hits. When this is not done well, we get boring or awkward pieces. Let your ear guide you.

3.11.1 Adding Chords

Earlier we saw that a phrase can contain a grouping of consecutive notes. Typically, this can be a melody or a bass line.

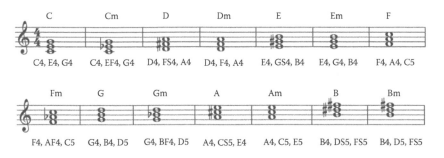

FIGURE 3.1   Common chords in staff notation and music library pitches.

---

* Chord construction and harmony are beyond the scope of this book. If you are unfamiliar with these topics, there are many good references, such as Surmani et al. (1999), and Levine (1995).

† As you may know already, an upper case letter chord (e.g., C) is a major chord. Adding a lower case "m" to the chord name (e.g., Cm) makes it a minor chord. Again, chord construction is beyond the scope of this book.

For convenience, the music library also allows chords to be added to a phrase.* This is done with the addChord() function. For example,

```
phrase = Phrase()                       # create an empty phrase
aMinorChord = [A3, C4, E4]              # create chord pitch list
phrase.addChord(aMinorChord, WN)       # add chord to phrase
Play.midi(phrase)                       # play it back
```

Notice how, when adding a chord to a phrase, we provide a list of pitches and a single duration. This duration applies to all pitches—they will be played together as one block of notes.

We can continue adding chords to a phrase to create a chord progression. Chords added this way will sound sequentially. For example, the following creates a C, F, G, C chord progression:

```
phrase = Phrase()                       # create an empty phrase

C4MajorPitches = [C4, E4, G4]          # C major (4th octave)
F4MajorPitches = [F4, A4, C5]          # F major
G4MajorPitches = [G4, B4, D5]          # G major
C5MajorPitches = [C5, E5, G5]          # C major (5th octave)

# add chord progression to phrase
phrase.addChord(C4MajorPitches, WN)
phrase.addChord(F4MajorPitches, WN)
phrase.addChord(G4MajorPitches, WN)
phrase.addChord(C5MajorPitches, WN)

Play.midi(phrase)                       # play it back
```

The addChord() function allows two additional (yet optional) parameters, namely, dynamic level and panning position. Similarly to the duration, only one value for these is needed per chord.†

Fact: It is possible to mix chords and notes in the same phase. They will be played sequentially in the order added.

Good Style: Keep chords separate from melody, that is, create a separate phrase for each.

Keeping chords and melodic lines separate makes it easier to maintain your work. Music composers do the same thing when writing music on paper.

---

* The alternative would be to create parallel phrases each containing a separate note in a chord. You can still do this, but it is awkward and time consuming.
† Dynamic and panning values are discussed in the previous chapter, under Notes.

One exception is when single notes are part of the chord progression; sometimes, such notes are used as embellishment. This is the case in the following example.

3.11.2 Case Study: Bruce Hornsby—"The Way It Is"

Now that we have seen how to create chords, here is the main chord progression from Bruce Hornsby's "The Way It Is" (1986).

```
# theWayItIs.py
# Plays main chord progression from Bruce Hornsby's
  "The Way It Is" (1986).

from music import *

mPhrase = Phrase()
mPhrase.setTempo(105)

mPhrase.addNote(A4, SN)
mPhrase.addNote(B4, SN)
mPhrase.addChord([A3,E4,G4,C5], DEN)
mPhrase.addChord([A3,E4,G4,C5], DEN)
mPhrase.addChord([E3,D4,G4,B4], HN)
mPhrase.addChord([D4,A4], SN)
mPhrase.addNote(G4, SN)
mPhrase.addChord([D3,D4, FS4, A4], DEN)
mPhrase.addChord([D3, D4, G4, B4], DEN)
mPhrase.addChord([C3, C4, D4, G4], DQN)
mPhrase.addChord([C3, E4], EN)
mPhrase.addNote(D4, SN)
mPhrase.addNote(C4, SN)
mPhrase.addChord([G2, B4, D4], DQN)
mPhrase.addChord([G2, D4, G4, B4], EN)
mPhrase.addChord([D3, E4, A4, C5], DEN)
mPhrase.addChord([D3, D4, G4, B4], EN)
mPhrase.addNote(A4, SN)
mPhrase.addNote(G4, EN)
mPhrase.addChord([C3,C4,D4,G4], DEN)
mPhrase.addChord([C3,C4,D4,G4], DEN)
mPhrase.addChord([G3,B3,D4,G4], HN)

Play.midi(mPhrase)
```

Notice how this particular chord progression includes single notes interspersed into the sequence. Each item (chord or note) will be played in the order added.

Also, notice the repetition of chords in the chord progression above. Adding chords with the addChord() function may be convenient for smaller chord progressions. But it does not allow us to take advantage of chord repetitions to reduce the size of our programs. Below, we address this concern.

3.11.3 Adding Chords with Lists

A faster way to enter chords is to use Python lists. Instead of adding one chord at a time into a Phrase, you can store the pitch and durations for each note in parallel lists. For example, the following code will create a simple melody intertwined with chords:

```
pitches =   [C4, [D4, F4, G4], E4, [F4, A4, C5], G4]
durations = [QN, QN,            QN, QN,          QN]
```

Then, in one step, you can add all these pitches and durations into a Phrase, as follows:

```
phr.addNoteList(pitches, durations)
```

Again, it advisable to keep chords separate from melody, that is, to use two (or more) phrases to store such music. This is accomplished using a Part object, which may contain many Phrase objects. (More on this soon.)

3.11.4 Case Study: 2Pac—"Changes"

Now that we have seen how to use lists to create chord progressions more quickly, here is the main chord progression from 2Pac's (Tupac Shakur's) "Changes" (1998). It so happens that this is the same chord progression as in the last case study. This is convenient, as it allows us to compare the two alternate ways of entering chords.

```
# changesByTupac.py
# Plays main chord progression from 2Pac's "Changes" (1998).

from music import *

mPhrase = Phrase()
mPhrase.setTempo(105)

# section 1 - chords to be repeated
pitches1  = [[E4,G4,C5], [E4,G4,C5],   [D4,G4,B4], A4, G4, [D4, FS4, A4],
            [D4, G4, B4], [C4, E4, G4], E4,         D4, C4, [G3, B4, D4]]
durations1 = [DEN,   DEN,   HN,  SQ,  SQ,  DEN,
            DEN,   DQN,   EN,  SN,  SN,  DQN]

# section 2 - embellishing chords
pitches2   = [A4, B4]
durations2 = [SN, SN]
```

```
mPhrase.addNoteList(pitches1, durations1) # add section 1
mPhrase.addNoteList(pitches2, durations2) # add section 2
mPhrase.addNoteList(pitches1, durations1) # re-add section 1

Play.midi(mPhrase)
```

Clearly, this is a more economical representation compared to adding chords one by one. Also, notice how using lists allows us to further reduce the size of the program, by taking advantage of repetition in the original musical material.

Good Style: When creating long parallel lists, wrap around and indent by one space, to maintain visual connection between pitches and corresponding durations.* Doing this allows us to create very long lists and still keep track of which pitch(es) goes with which duration.

It is possible to add additional parallel lists for dynamics and pannings. These would be constructed similarly to the durations list in the example above. In other words, provide one dynamic and one panning value per chord.

3.12 PARTS

So far we have used notes and phrases for musical organization. Notes are the elemental music objects. Many notes can be stored within a Phrase. Phrases have a start time in a piece. Many phrases can be added to a Part (polyphonic music). Parts can have an instrument associated with them.

In musical arrangements, it is common to refer to a musical part, such as the trumpet part or the guitar part. For this reason, the music library provides the Part class. A Part object contains a set of Phrase objects to be played by a particular instrument. These Phrase objects are played in series (i.e., one after another) or in parallel (i.e., simultaneously) and thus may overlap (according to their start times and durations).

For example, Figure 3.2 shows a Part which consists of three Phrase objects. Two of the phrases are sequential (one follows the other) and contain monophonic material (i.e., a sequence of nonoverlapping

---

* Python is very particular with indentation, except with items in lists. We can use as much vertical or horizontal space as we wish, as long as we keep the brackets (open and close) balanced.

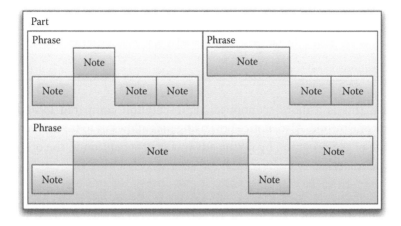

FIGURE 3.2   A Part object contains Phrases, which in turn contain Notes.

notes—melody). The third phrase contains monophonic material. Together, the three phrases make up polyphonic material (i.e., simultaneous sequences of notes having both distinct pitches and durations).

This example demonstrates that, even if the particular instrument does not allow for polyphony (e.g., a flute), a Part using this instrument can have different simultaneous melodies. In other words, a Part can be thought of as a group of several instruments of the same type (e.g., flute section), each playing a different melody (a Phrase).

Part objects have several attributes. These include:

- Title—a descriptive string (e.g., "Violin 1")

- Instrument—an integer from 0 to 127 (or MIDI instrument constant)

- Channel—an integer (0–15)

Parts are also the best place to set the panning for the notes that it contains, because all the notes in a part are assigned for playback on the same MIDI channel, and panning is a channel attribute in the MIDI specification. To change the stereo pan location for the part use Part.setPan(position) where position is a value between 0 and 127, with 64 (center) being the default.

3.12.1 Creating Parts

Creating a part is done using the Part() function, and passing the desired values for a title, instrument, and channel. For example,

```
part1 = Part("a flute part on channel 0", FLUTE, 0)
```

creates a part with the title "a flute part on channel 0", using a FLUTE instrument, and assigns that instrument to MIDI channel 0.

You may also create a part with only instrument and channel. For example,

```
part2 = Part(DISTORTION_GUITAR, 1)
```

creates a part using a DISTORTION_GUITAR instrument, and assigns that instrument to MIDI channel 1.

3.12.2 MIDI Channels

The MIDI standard provides 16 channels (0–15). Each MIDI channel is capable of playing any of the 128 different MIDI instruments, but only one at a time.

Parts may be assigned to a particular MIDI channel.

Good Style: Keep parts with different instruments on different MIDI channels.

It is important to keep parts using different instruments in different MIDI channels. If you assign two parts with different instruments to the same MIDI channel, only the second instrument will be used.

If two parts are using the same instrument, it is OK to assign them to the same MIDI channel.

There is a special case among MIDI channels: MIDI channel 9 is reserved for drum kit and percussion sounds. Regardless of a part's selected instrument, if that part is assigned to MIDI channel 9, its notes will generate percussion sounds, based on the notes' pitches. We will explore MIDI channel 9 and percussion later in this chapter.

3.12.3 Adding Phrases

Once an empty part has been created (as above), you can add phrases to it using the addPhrase() function.

```
n = Note(C4, HN) # create a note

phrase1 = Phrase() # create an empty phrase
phrase1.addNote(n) # add note to phrase - nothing new, so far...

part1 = Part(DISTORTION_GUITAR, 1) # create a part
part1.addPhrase(phr)              # add phrase to part
```

Now, `part1` contains `phrase1`, which in turn contains `note1`. When this part is played, it will generate a C4, a half note using a distorted guitar sound.

Good Style: If you are using parts, set instruments only at the part level.[*]

Although it is possible to add phrases which have been assigned different instruments (e.g., a violin and a piano) to the same part, this is really bad style and may not work as you might expect. Even worse, the part could have been assigned yet another instrument (e.g., a DISTORTION_GUITAR).

What happens in that case? Which instrument takes precedence? The part instrument? Each phrase's own instrument? The fact that we even have to ask this question is proof that such code style would be unnecessarily complicated.

The main reason for having parts is to provide a container for multiple phrases that need to be played by a single instrument. So, if you are using parts, simply create phrases without an instrument and add them to a part. The part takes care of the instrument for these phrases.

3.12.4 Creating Ensembles

Now that we know how to create parts that contain phrases, we can create bands and ensembles with several instruments.

Here is a simple program that demonstrates how to create a string quartet. It plays four overlapping notes (each in a separate phrase), using a MIDI strings sound.

```
# stringQuartet.py
# Demonstrates how to create concurrent musical parts.
# Hayden, Opus 64 no 5

from music import *

stringsPart = Part(STRINGS, 0) # create empty strings part
stringsPart.setTempo(104)

pitches1 =  [A5, REST, A5, REST, A5, REST, A5, A6, E6, D6, D6, CS6,
             D6, D6, CS6, D6, E6]
durations1 = [EN, EN, EN, EN, EN, EN, WN, DHN, EN, EN, HN, DEN, SN,
              DEN, TN, TN, QN]
```

---

[*] In other words, it is stylistically okay to set a phrase's instrument only when you are not planning to add it in a part.

```
violin1 = Phrase(0.0)                       # create a phrase
violin1.addNoteList(pitches1, durations1) # addnotes to the phrase
stringsPart.addPhrase(violin1)              # now, add the phrase to the
                                              part


pitches2 =   [FS4, G4, FS4, E4, D4, REST, G4, A4, G4, FS4, E4]
durations2 = [QN, QN, QN, QN, QN, DHN, QN, QN, QN, QN, QN]
violin2 = Phrase(3.0)                       # create a phrase
violin2.addNoteList(pitches2, durations2) # addnotes to the phrase
stringsPart.addPhrase(violin2)              # now, add the phrase to the
                                              part


pitches3 =   [D4, E4, D4, A3, FS3, REST, E4, FS4, E4, D4, CS4]
durations3 = [QN, QN, QN, QN, QN, DHN, QN, QN, QN, QN, QN]
violin3 = Phrase(3.0)   # create a phrase
violin3.addNoteList(pitches3, durations3) # addnotes to the phrase
stringsPart.addPhrase(violin3)              # now, add the phrase to the
                                              part


pitches4 =   [D2, FS2, A2, D3, A2, REST, A2]
durations4 = [QN, QN, QN, QN, QN, DHN, QN]
violin4 = Phrase(7.0)                         # create a phrase
violin4.addNoteList(pitches4, durations4) # addnotes to the phrase
stringsPart.addPhrase(violin4)              # now, add the phrase to the
                                              part

Play.midi(stringsPart)
```

Notice how the notes begin sounding at different times (i.e., they are not chords). Creating separate phrases with different start times and durations is possible when we create ensembles.

Also, notice that, if we explicitly specify the start time of phrases, the order in which we add them to a part does not dictate their playback order.*

Fact: When phrases have explicit start times (i.e., their start time has been set), the order that you add them to a part is independent of their start time. Such phrases play back according to their start time (and they may overlap).

Fact: When phrases do not have explicit start times, that is, phrase = Phrase(), the order that you add them to a part is critical. Such phrases play back sequentially (one after another), in the order added.

_____

* As opposed to adding notes to a phrase, which always dictates their playback order.

3.13 SCORES

If we consider the music of a jazz trio, with its multiple parts (i.e., instruments), these parts need to be collected together into a score. Scores are essential in multipart music composition and performance.

The Score object is the highest-level musical structure provided by the Python music library. Score objects have several attributes. These include:

- Title—a descriptive string (e.g., "My New Composition")

- tempo—the speed of playback (a positive real number — in quarter note beats per minute)

3.13.1 Creating Scores

Creating a score is done using the Score() function and passing the desired attribute values. There are several Score() functions (see Appendix B). The most common one takes arguments for title and tempo. If the tempo is omitted, it is set to 60 bpm (beats-per-minute) by default. This is quite a slow tempo but has the advantage that each beat (quarter note duration) equals one second of time, which is useful for music that is time-based rather than based on note durations.

This Python statement creates a score with title "Opus 1" and a default tempo of 60 bpm. The score is assigned to the variable s:

```
s = Score("Opus 1")
```

whereas this statement creates a similar score but with a tempo of 135 bpm:

```
s = Score("Opus 1", 135.0)
```

Once a score has been created, to insert parts into it use the addPart() function, as follows:

```
s.addPart(part1)
```

You can also set the tempo of the score with the setTempo() function, as follows:

```
s.setTempo(120.0) # set score's tempo to 120 bpm
```

There are various other functions available for Scores, as well as for Parts, Phrases, and Notes. For more information on this and other music library functions, see Appendix B.

3.13.2 Putting It All Together

The music library data structure (Score, Part, Phrase, Note) has been designed to accommodate a wide variety of musical forms and structures. Note objects describe musical events; they are arranged in a list within phrases. Phrase objects have a start time, which positions them within a part. Part objects can contain many phrases and also have an instrument assigned to them. That instrument plays all notes within the part. Parts are added to a Score object, which also sets the tempo of music within it.

The Python music data structure can be thought of as a set of concentric rings, one within another, as shown in Figure 3.3.

The "target" view of the score is a simplification, because in reality a score can contain many parts, a part many phrases, and a phrase many notes. More realistically, the Python music data structure can also be represented visually as in Figure 3.4.

Creating music of significant complexity involves coordinating several musical lines at one time. Typical examples include multipart vocal chorales or string quartets. However, such multipart coordination is required by music of almost any style. For example, consider a jazz arrangement, where you have to coordinate a melody, a bass part, and a percussive accompaniment. The music library Part and Score classes help with this structuring.

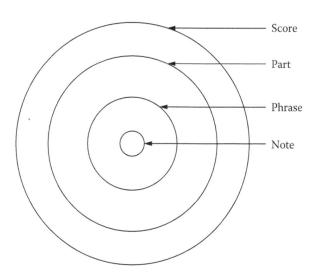

FIGURE 3.3   The Python music data structure.

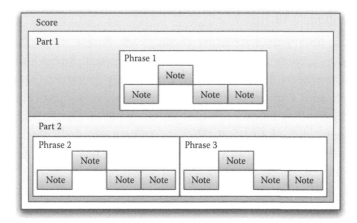

FIGURE 3.4 A view of a simple musical segment represented in the Python music data structure.

Good Style: Reflecting compositional practices in acoustic music, there are two ways to write polyphonic music with the music library data structure:

1. If you are dealing with polyphonic material to be played by a single instrument (e.g., a harpsichord fugue) or by a group of monophonic instruments of the same kind (e.g., a flute ensemble), you can create multiple Phrases (voices) and assign them to a single Part (with that instrument).
2. If you are dealing with polyphonic material to be played by a group of monophonic instruments of different kinds (e.g., a woodwind quintet), you can create a single Phrase and add it to different Parts (one for each instrument) — as demonstrated by the following case study.

Obviously, nothing prevents you from combining both techniques in a computer music piece. In any case, follow the stylistic guidelines and examples of this book.

3.14 A COMPLETE EXAMPLE

3.14.1 Case Study: Joseph Kosma—"Autumn Leaves" (Jazz Trio)

Now that we have seen the all the data structures available for creating music in Python, let's see a complete example. The following program plays the theme from "Autumn Leaves." This song was composed by Joseph Kosma in 1945 and has become a jazz standard. It has been performed by many famous musicians, such as Jo Stafford, Edith Piaf, Nat King Cole, Frank Sinatra, Bill Evans, John Coltrane, Manfred Mann, and Eric Clapton, among others.

```
# autumnLeaves.py
#
# It plays the theme from "Autumn Leaves", in a Jazz trio arrangement
# (using trumpet, vibraphone, and acoustic bass instruments).

from music import *

##### define the data structure (score, parts, and phrases)
autumnLeavesScore = Score("Autumn Leaves (Jazz Trio)", 140) # 140 bpm

trumpetPart = Part(TRUMPET, 0)          # trumpet to MIDI channel 0
vibesPart   = Part(VIBES, 1)            # vibraphone to MIDI channel 1
bassPart    = Part(ACOUSTIC_BASS, 2)    # bass to MIDI channel 2

melodyPhrase = Phrase()   # holds the melody
chordPhrase  = Phrase()   # holds the chords
bassPhrase   = Phrase()   # holds the bass line


##### create musical data
# melody
melodyPitch1 = [REST, E4, FS4, G4, C5, REST, D4, E4, FS4, B4,  B4]
melodyDur1   = [QN,   QN, QN,  QN, WN, EN,   DQN,QN, QN,  DQN, HN+EN]
melodyPitch2 = [REST, C4, D4, E4, A4, REST, B3, A4, G4, E4]
melodyDur2   = [QN,   QN, QN, QN, WN, EN,   DQN,QN, QN, 6.0]

melodyPhrase.addNoteList(melodyPitch1, melodyDur1) # add to phrase
melodyPhrase.addNoteList(melodyPitch2, melodyDur2)


# chords
chordPitches1   = [REST, [E3, G3, A3, C4], [E3, G3, A3, C4], REST,
                   [FS3, A3, C4]]
chordDurations1 = [WN,    HN,               QN,              QN,
                   QN]
chordPitches2   = [REST, [D3, FS3, G3, B3], [D3, FS3, G3, B3]]
chordDurations2 = [DHN,   HN,                QN]
chordPitches3   = [REST, [C3, E3, G3, B3], REST, [E3, FS3, A3, C4],
                   [E3, FS3, A3, C4]]
chordDurations3 = [QN,    QN,              DHN,  HN,
                   QN]
chordPitches4   = [REST, [DS3, FS3, A3, B3], REST, [E3, G3, B3],
                   [DS3, FS3, A3, B3]]
chordDurations4 = [QN,    QN,                DHN,  HN,
                   QN]
chordPitches5   = [REST, [E3, G3, B3], REST]
chordDurations5 = [QN,    HN,          HN]


chordPhrase.addNoteList(chordPitches1, chordDurations1) # add them
chordPhrase.addNoteList(chordPitches2, chordDurations2)
chordPhrase.addNoteList(chordPitches3, chordDurations3)
```

```
chordPhrase.addNoteList(chordPitches4, chordDurations4)
chordPhrase.addNoteList(chordPitches5, chordDurations5)


# bass line
bassPitches1   = [REST, A2, REST, A2, E2, D2, REST, D2, A2, G2, REST,
                  G2, D2, C2]
bassDurations1 = [WN,   QN, EN,    EN, HN, QN, EN,    EN, HN, QN, EN,
                  EN, HN, QN]
bassDurations2 = [EN,    EN, HN, QN,  EN,   EN, HN, QN, EN,    EN, HN,
                  QN]
bassPitches2   = [REST, C2, G2, FS2, REST, FS2, C2, B1, REST, B1, FS2,
                  E2]
bassPitches3   = [REST, E2, E2, B1, E2, REST]
bassDurations3 = [EN,    EN, QN, QN, HN, HN]


bassPhrase.addNoteList(bassPitches1, bassDurations1)  # add them
bassPhrase.addNoteList(bassPitches2, bassDurations2)
bassPhrase.addNoteList(bassPitches3, bassDurations3)


##### combine musical material
trumpetPart.addPhrase(melodyPhrase) # add phrases to parts
vibesPart.addPhrase(chordPhrase)
bassPart.addPhrase(bassPhrase)


autumnLeavesScore.addPart(trumpetPart) # add parts to score
autumnLeavesScore.addPart(vibesPart)
autumnLeavesScore.addPart(bassPart)

Play.midi(autumnLeavesScore) # play music
```

Notice how variables names (e.g., trumpetPart, chordPhrase, etc.) capture two things: the type of information it holds (e.g., instrument or chord) and the data structure used (i.e., score, part, or phrase). By doing this, the name itself tells you everything you need to know.

Good Style: Create meaningful names for variables. Be consistent.

Meaningful variable names minimize confusion and the need to go back and forth in your code to remember the name or purpose of a particular variable. Accordingly, this minimizes programming errors and, in the long run, saves time. The longer the program, the more you will benefit from this stylistic guideline.

Also notice how, in the melody line, we added two durations (HN+EN) to construct an unusual duration value. In the notated score for this piece, this would be called a tie.

Definition: A musical *tie* connects two notes of the same pitch into a single note (with the combined duration).

Fact: In Python, a musical tie is implemented by adding two duration values.

Finally, this piece needs a drum kit track. We will see how to do this in the next section.

3.14.2 Exercise

The form of "Autumn Leaves" is AABC (meaning that theme A repeats once and then it is followed by two other themes, B and C). The case study above includes only theme A. Find a score of the complete piece online and add the remaining code. Follow the above coding style—it will save you time in the long term.

3.15 MIDI DRUMS AND PERCUSSIVE SOUNDS

The MIDI standard allows us to write percussive (e.g., drum) parts. As mentioned above, MIDI has 16 channels (numbered 0 to 15). Of these, channel 9 is reserved for percussion.

When adding notes to a Part object assigned to channel 9, the pitch of the note determines which percussive instrument to play. So, whereas for other channels (0–8 and 10–15) the MIDI pitch corresponds to the note's frequency (or piano key number), for channel 9 the MIDI pitch corresponds to a particular percussive sound (without relationship to pitch). The General MIDI standard suggests a mapping between MIDI pitch values and percussion sounds.

For example, here are MIDI pitches for some percussive instruments found in a drum kit:

36 (C2)	Bass Drum
38 (D2)	Snare Drum
42 (FS2)	Closed Hi Hat
46 (AS2)	Open Hi Hat
49 (DF3)	Crash Cymbal

Appendix A provides the complete list of General MIDI Drum and Percussion constants. For example, this list includes: BASS_DRUM,

SIDE_STICK, SNARE, HAND_CLAP, CLOSED_HI_HAT, OPEN_HI_
HAT, LOW_FLOOR_TOM, CRASH_CYMBAL_1, RIDE_CYMBAL_1,
COWBELL, MARACAS, SHORT_WHISTLE, and many others.

Also, the library provides abbreviated constants, for easier inclusion
in pitch lists. For the percussion instruments above, the corresponding
abbreviations are BDR, STK, SNR, CLP, CHH, OHH, LTM, CC1, RC1,
CBL, MRC, and SWH.

For example, the following program demonstrates how to play a single
drum sound:

```
# drumExample.py
# A quick demonstration of playing a drum sound.

from music import *

# for drums always use a part on channel 9
# when using channel 9, the instrument (2nd argument) is ignored
drumPart = Part("Drums", 0, 9)

note = Note(ACOUSTIC_BASS_DRUM, QN) # a bass drum strike

drumPhrase = Phrase()
drumPhrase.addNote(note)

drumPart.addPhrase(drumPhrase)

Play.midi(drumPart)
```

3.15.1 Exercises

1. Modify the above program to generate different percussive sounds.
   Possibilities include: ACOUSTIC_BASS_DRUM, BASS_DRUM_1,
   ACOUSTIC_SNARE, HAND_CLAP, CLOSED_HI_HAT, OPEN_
   HI_HAT, HIGH_FLOOR_TOM, LOW_TOM, and CRASH_
   CYMBAL_1. See Appendix A for a complete list of constants for
   percussive sounds.

2. Create a program that plays different percussive sounds in sequence
   (*Hint:* Create a phrase where you add a sequence of notes corresponding to the desired instruments.)

3.15.2 Case Study: Drum Machines

Drum machines are electronic musical instruments which can play
rhythmic patterns using drums and other percussion instrument sounds.

Drum machines are popular in many musical genres, such as rock, dance, and electronic music. They are routinely used in practice or jamming sessions, replacing (the sometimes difficult to find) human drummers. Drum machines are also used in live performances, especially in electronic dance music.

Given its power as a musical instrument, a drum machine may form an integral part of a composition or performance, in that it can play complex rhythmic patterns that could not possibly be performed by a single human drummer, or could be hard to perform (without errors) even by a group of drummers. Drum machines can easily be programmed to generate rhythmic patterns, from the simple to the intricate, using a wide variety of percussive sounds. Let's see how.

3.15.2.1 Drum Machine Pattern #1

Here is a program that implements a drum machine pattern consisting of bass (kick), snare, and hi-hat sounds. It uses many notes, three phrases, a part, and a score, with each layer adding additional rhythms.

Notice how the program is structured using comments to follow the general algorithm for creating music. Most importantly, note how the comments capture the calculations of how many notes to have per measure. This style of commenting allows you to do the work once (i.e., calculate how many notes to have per measure) to make sure that everything lines up metrically, and then never have to think about it again. Just read the comments and make appropriate changes to the code below it. Then, if you like the outcome, go back and update the comment accordingly.

```
# drumMachinePattern1.py
#
# Implements a drum-machine pattern consisting of bass (kick),
# snare and hi-hat sounds. It uses notes, three phrases, a part and
# a score, with each layer adding additional rhythms.

from music import *

repetitions = 8             # times to repeat drum pattern

##### define the data structure
score = Score("Drum Machine Pattern #1", 125.0) # tempo is 125 bpm

drumsPart = Part("Drums", 0, 9)   # using MIDI channel 9 (percussion)

bassDrumPhrase = Phrase(0.0)       # create a phrase for each drum sound
```

```
snareDrumPhrase = Phrase(0.0)
hiHatPhrase = Phrase(0.0)


##### create musical data

# bass drum pattern (one bass + one rest 1/4 notes) x 4 = 2 measures
bassPitches = [BDR, REST] * 4
bassDurations = [QN, QN] * 4
bassDrumPhrase.addNoteList(bassPitches, bassDurations)


# snare drum pattern (one rest + one snare 1/4 notes) x 4 = 2 measures
snarePitches = [REST, SNR] * 4
snareDurations = [QN, QN] * 4
snareDrumPhrase.addNoteList(snarePitches, snareDurations)


# hi-hat pattern (15 closed 1/8 notes + 1 open 1/8 note) = 2 measures
hiHatPitches = [CHH] * 15 + [OHH]
hiHatDurations = [EN] * 15 + [EN]
hiHatPhrase.addNoteList(hiHatPitches, hiHatDurations)


##### repeat material as needed
Mod.repeat(bassDrumPhrase, repetitions)
Mod.repeat(snareDrumPhrase, repetitions)
Mod.repeat(hiHatPhrase, repetitions)


##### combine musical material
drumsPart.addPhrase(bassDrumPhrase)
drumsPart.addPhrase(snareDrumPhrase)
drumsPart.addPhrase(hiHatPhrase)
score.addPart(drumsPart)


##### view and play
View.sketch(score)
Play.midi(score)
```

Notice how we use the list repetition operator (*), e.g., [BDR, REST]*4, to repeat list items (instead of typing them out repeatedly). The former is more economical, easier to understand (especially when constructing drum machine patterns), and definitely less error prone.

Notice how we use both list repetition and concatenation operators (* and +) to construct the hi-hat pattern, e.g., [CHH]*15 + [OHH].

Finally, notice the statement

```
Mod.repeat(bassDrumPhrase, repetitions)
```

This has the effect of repeating the notes in bassDrumPhrase a number of times, namely, repetition times (i.e., 8). This and other Mod functions will be covered in the next chapter in more detail.

3.15.2.2 Exercise

Modify the above program to play various other drum machine patterns. Try different dance patterns. Try some exotic rhythms.

3.15.3 Case Study: Deep Purple—"Smoke on the Water"

Now that we know how to create drum parts, we can write programs that create more complete music from various genres. To demonstrate combining drums with other instruments, here is the opening of Deep Purple's "Smoke on the Water," a rock riff from 1972. It combines melody, chords (actually, power-chords constructed from two simultaneous pitches), and drums. Again, the emphasis here is on demonstrating how to combine the various building elements we have seen so far.

```
# DeepPurple.SmokeontheWater.py
#
# Demonstrates how to combine melodic lines, chords, and
# percussion. This is based on the intro of "Smoke on the Water"
# by Deep Purple.

from music import *

##### define the data structure
score = Score("Deep Purple, Smoke on the Water", 110) # 110 bpm

guitarPart = Part(OVERDRIVE_GUITAR, 0)
bassPart = Part(ELECTRIC_BASS, 1)
drumPart = Part(0, 9)   # using MIDI channel 9 (percussion)

guitarPhrase1 = Phrase()    # guitar opening melody
guitarPhrase2 = Phrase()    # guitar opening melody an octave lower
bassPhrase = Phrase()       # bass melody
drumPhrase = Phrase()       # drum pattern

##### create musical data
# guitar opening melody (16QN = 4 measures)
guitarPitches   = [G2, AS2, C3,  G2, AS2, CS3, C3, G2, AS2, C3,  AS2,
                   G2]
guitarDurations = [QN, QN,  DQN, QN, QN,  EN,  HN, QN, QN,  DQN, QN,
                   DHN+EN]
guitarPhrase1.addNoteList(guitarPitches, guitarDurations)

# create a power-chord sound by repeating the melody an octave lower
guitarPhrase2 = guitarPhrase1.copy()
Mod.transpose(guitarPhrase2, -12)

# bass melody (32EN = 4 measures)
bassPitches1    = [G2, G2, G2, G2, G2, G2, G2, G2, G2, G2,
                   G2, G2, G2, G2, G2, G2, G2, G2]
```

```
bassDurations1  = [EN, EN, EN, EN, EN, EN, EN, EN, EN, EN,
                    EN, EN, EN, EN, EN, EN, EN, EN]
bassPitches2    = [AS2, AS2, C3, C3, C3, AS2, AS2, G2, G2,
                    G2, G2, G2, G2, G2]
bassDurations2  = [EN, EN, EN, EN, EN, EN, EN, EN, EN,
                    EN, EN, EN, EN, EN]
bassPhrase.addNoteList(bassPitches1, bassDurations1)
bassPhrase.addNoteList(bassPitches2, bassDurations2)


# snare drum pattern (2QN x 8 = 4 measures)
drumPitches   = [REST, SNR] * 8
drumDurations = [QN,   QN]  * 8
drumPhrase.addNoteList(drumPitches, drumDurations)
##### repeat material as needed
Mod.repeat(guitarPhrase1, 8)
Mod.repeat(guitarPhrase2, 6)
Mod.repeat(bassPhrase, 4)
Mod.repeat(drumPhrase, 2)


##### arrange material in time
guitarPhrase1.setStartTime(0.0)   # start at beginning
guitarPhrase2.setStartTime(32.0)  # start after two repetitions
bassPhrase.setStartTime(64.0)     # start after two more repetitions
drumPhrase.setStartTime(96.0)     # start after two more repetitions


##### combine musical material
guitarPart.addPhrase(guitarPhrase1)
guitarPart.addPhrase(guitarPhrase2)
bassPart.addPhrase(bassPhrase)
drumPart.addPhrase(drumPhrase)
score.addPart(guitarPart)
score.addPart(bassPart)
score.addPart(drumPart)

##### write score to a MIDI
```

3.16 TOP-DOWN DESIGN

Notice how the above code has been written out conceptually from the highest level of detail to the lowest. First, we define the musical data structures, starting from the score, to the parts, to the phrases. Then we create the musical data (notes of the melody, chords, bass line). Finally, we combine the musical material to form the complete song.

Top-down design and implementation is a strategy for constructing programs. It starts from the biggest and goes to the smaller—in our case, from the score, to the parts, to the phrases, and so on. By specifying the highest-level pieces of our program first, and then dividing them into successively smaller pieces, we gain perspective, and the structure of the program is clearly defined. Then it is easy to go back and fix problems or update the program to perform slightly different tasks (e.g., extend a particular song by adding another part, add some more functionality to the program, and so on). Top-down design will make more sense when we introduce functions. Until then, use the above program as a model when constructing larger pieces of music.

Fact: By being systematic in your programming, you save time.

Top-down design and implementation is most beneficial when developing larger programs. But by applying it even to the smallest programs you write, you gain flexibility to make them grow easily, at a later time, without wasting effort on added complexity and the potential logical errors that this added complexity may introduce.

3.17 INPUT AND OUTPUT

In the last line of the previous case study the program saved the composition as a MIDI file. In previous examples we have played back compositions directly. In addition to keyboard input and screen or speaker output (seen earlier), the Python music library allows you to read and write standard MIDI files.

A standard MIDI file is a common format for storing musical compositions. By convention, files saved in this format have a ".mid" suffix appended to their name. Many music software applications can read and/ or write MIDI files. By saving your Python music compositions as MIDI files you will be able to open them for further editing, display, or playback in other software.

3.17.1 Reading MIDI Files

You can read a MIDI file into your program using the Read.midi() function. This function expects an empty score and the name of a MIDI file (saved in the same folder as your program). For example,

```
Read.midi(score, "song.mid")
```

inputs the musical data from the MIDI file "song.mid", and stores them into score. Once the file has been read in, you can manipulate or play back the score. For example,

```
from music import *

score = Score()              # create an empty score
Read.midi(score, "song.mid") # read MIDI file into it

Play.midi(score)             # play it back
```

This program reads the MIDI file called "song.mid" into the variable score. Notice how a score has been already created—it is an empty Score object. The program then plays the score, to demonstrate that the musical data has been successfully read into the program.

The Read.midi() function works only with standard MIDI files.*

Reading musical data into our programs opens the door for some interesting possibilities. We will discuss them below.

3.17.2 Writing MIDI Files

You can create a MIDI file from your program using the Write.midi() function. This function expects a score (part, phrase, or note) and the name of a MIDI file. For example,

```
Write.midi(score, "song.mid")
```

writes the musical data in score into the MIDI file called "song.mid". This file is saved in the same folder as your program. If the MIDI file already exists, it will be overwritten.

For example, if you replace the statement, Play.midi(autumnLeavesScore), in the last case study with the statement, Write.midi(autumnLeavesScore, "autumnLeaves.mid"), your program will save the generated music into a MIDI file.[†]

This is very useful. It is likely that you will want to save your music as a file for storage, playback, or use in another application. Adding Write.midi() to any of the examples so far will give you an external MIDI file that you can share with (e.g., email to) others—without the need for Python.[‡] This way, Python becomes your music production environment—and

---

* If a file does not work, it is probably not a standard MIDI file. Try opening it with a MIDI editing program (there are several freeware programs available), and resave it as a type 0 or type 1 MIDI file.

† Be careful to end the MIDI filename with the extension ".mid". In the above example, if you accidentally use the extension ".py" you would delete the file that contains your program! This brings up another piece of advice. Back up your work often (e.g., in a different folder or backup drive). At some point, you will be happy you did.

‡ To play a MIDI file, simply double click on it. Most computers know how to play MIDI files.

the MIDI file a vehicle with which you share your results. (You could use other external tools, e.g., GarageBand, to convert your MIDI files to MP3 or other formats, if you wish.)

Another possibility is to use writing and reading of MIDI files to modularize/compartmentalize your work. There is no need for one program to do ALL the work. Programs can become short and self-contained, i.e., they do one thing—and they do it well. This makes the idea of design (as discussed in Chapter 1) even more important and relevant, and opens up many creative possibilities for you to explore.

3.17.3 Exercises

1. Modify the "Smoke on the Water" case study above to output a MIDI file. Then play it back. Do you notice any differences in timbre? If so, that's due to the different synthesizers used (Play.midi() uses the Java synthesizer. MIDI files use the operating system's default synthesizer.)

2. Select a simple classical piece consisting of two voices. Transcribe it into a Python music representation using a phrase for each voice.

3. Select a favorite popular song consisting of a chord progression and a melody. Transcribe it, using one phrase for the chords and another for the melody. Ensure that the two phrases line up in terms of note (and chord) durations. If it helps to better organize the music, consider using several phrases, as in the "Smoke on the Water" case study above.

4. Transcribe a favorite piece of music (or an excerpt of it) in Python. You may do so by ear or from a notated musical score. Since most music has repetitions, use these to minimize duplication of effort (data entry). Duplicate material can be entered once and reused (one phrase being a copy of another). Another possibility is to repeat material using Mod.repeat() and, possibly, Mod.transpose()—see the next chapter.

3.18 SUMMARY

This chapter introduced the remaining components of the Python music data structure. This now gives us Notes, Phrases, Parts, and Scores. Together, they allow us to transcribe arbitrary musical material into Python programs. This format is equivalent to (and some may argue easier to learn than) traditional Western musical notation. The end result is the same, that is, a representation of musical material. What makes this Python representation more powerful is that, in addition to being able to listen to the music it describes (via your computer's MIDI synthesizer), it also allows you to write algorithms that implement various musical and other processes that modify and transform this material, as shown in the next chapter. This leads to the development of tools that can contribute to composing and/or performing music (as shown in the rest of the book).

Additionally, we learned about another data type, that is, the Python list, and its operations. We discussed musical scales and explored MIDI instruments. We discussed chords and how to efficiently model them via Python lists. We explored MIDI drums and percussion, and learned how to combine drums with other musical material (through Parts playing on MIDI channel 9). Finally, we discussed how to read and write MIDI files, which allows us to send our musical material to other people (e.g., musical collaborators), so that they may easily play it or use it with standard MIDI editors and sequencers.

## CHAPTER 4

Transformation and Process

Topics: Musical patterns and meaning, minimalism, Steve Reich, Mod functions, musical canon, J.S. Bach, Arvo Pärt, viewing musical material, software development process, computer-aided music composition.

4.1 OVERVIEW

Being able to write down and represent music is an important skill that we have examined in some detail in the previous chapters. In this chapter we focus on another important skill, transforming music to create variations and developing it into longer and more interesting compositions. More broadly, this chapter explores foundational programming skills required to manipulate data. But first, a little background to provide context.

4.2 GESTURES, EMOTION, AND MUSICAL STRUCTURE

The contrast between the composer's and audience's experience of music was discussed by Arnold Schoenberg, a 20th century composer who developed a system of serial (algorithmic) music composition, which avoided tonality by requiring an even distribution of all the 12 chromatic scale tones. On the issue of musical structure existing as space and time, he commented that, "Music is an art which takes place in time. But the way in which a work presents itself to a composer … is independent of this; time is regarded as space. In writing the work down, space is transformed into time. For the hearer this takes place the other way round; it is only after the work has run its course in time that it can be seen as a whole—its idea, its form, and its content" (quoted in Cook 1990: 40).

A common conception of musical fragments is as gestures or sequences of sound with characteristic shape, which appear as movement in pitch and time space. Like animation, a series of notes can provide the illusion of movement even though it is only a series of stationary events. For example, a scale of ascending tones may appear as a rising gesture. These musical impressions emerge from the overall effect of individual scale notes, and even when the pattern is interrupted with a brief descending tone, the overall shape of the line is maintained. This tendency of musical sequences to be perceived as gestural curves was observed by Roger Scruton, who wrote that, "It seems then that in our most basic apprehension of music there lies a complex system of metaphor, which is a true description of no material fact" (Scruton 1983).

This metaphorical aspect of musical structures goes beyond the gestural nature of musical shapes in time and space to include the ways in which these combine to indicate musical styles and emotions. Particular musical patterns can be characteristic of particular musical styles, or even of individual composers. David Cope called such frequently found gestures *stylistic signatures* (Cope 2004). Musical metaphors are even more direct when depicting emotional states, for example, the slow tempo and wide vibrato that imitate the despondent movement and quivering lip of a very sad person, or the violent staccato bowing of Bernard Herman's score for the stabbing featured in the "shower" scene from Alfred Hitchcock's film *Psycho*.

4.2.1 Musical Patterns

Many experiments have been done by psychologists to determine how we hear musical lines. It should be noted that the tendencies in our imaginative perception are not simply choices about how things appear to us, but are the result of how our brain attempts to group events and always seeks to find patterns in what we hear. In musical terms, the patterns that appear in musical material are the basis for the texture and the form of music. These structural elements are designed by the composer and then described in a medium, for example, in written notation or, in our case, Python code. This compositional process requires the imagining of musical structures and then their description. Interestingly, this is the reverse of the listeners' experience, where they hear the music unfold over time, then retrospectively construct the form from their memory of passing events, or more particularly from the lines, shapes, and textures they perceived as the events unfolded.

There is a connection between music's structure and its meaning and enjoyment. Understanding some of the psychology of perception that underpins these sound-gesture metaphors assists the composer to better create a desired musical experience. The roots of psychological concerns about how people perceive patterns in a holistic way can be traced back to the Gestalt psychology movement that originated in Europe in the early decades of the 20th century. The movement's core principle is that the mind tends to impart structure and organization to its experiences, and that a system's functioning cannot be wholly understood solely in terms of its parts. The experiments of Gestalt psychologists were particularly focused on visual perception; however, they also made connections to audible perception of music. The Gestalt psychologists examined a range of phenomenal qualities and derived a series of rules or laws about ways in which humans organize their experiences. These include:

- The **law of proximity**, where objects that are closer (in space, time, pitch, and so on) are grouped together;

- The **law of symmetry**, where things that create a trend or pattern are perceived as a closed region; and

- The **law of good continuation**, where a direction implied by object relations remains expected even in the face of interruption, change, or inconsistency.

The Gestalt psychologists also tended to associate the recognition of complex patterns with intelligence or creativity, seeing the best thinking as being novel rather than reproductive. They tended to believe that the mind operated to impose logical organization upon the world, as it was perceived.

A holistic and schema-based view of pattern recognition was not limited to the Gestalt movement. Other approaches to understanding perception were presented by psychologists such as Jean Piaget and James J. Gibson. In these approaches, mental representations were thought of as interactions with experiences in the world rather than the imposition of organizational schema on the mind of the individual. Piaget's work was in the area of child development, where he conducted experiments and observations to study how the young mind's understanding of the physical world evolved with age. J. J. Gibson's work focused on how understanding arose from interpretation of cues in the environment; these cues he called environmental *affordances*. Gibson argued that many behaviors were based on instinctive or automatic responses, which did not require mental processing at all, but could be refined through learning and experience. More recent work in cognitive science has elaborated on these views (Gardner 1985).

These views of mental models and perceived organization need to be kept in mind by the composer, that is, that the listener will perceive sequences of notes and durations as having higher-order organization and meaning when grouped into phrases, riffs, chords, rhythms, melodies, and so on. This is particularly important for a composer operating at the potentially more abstract level of algorithmic composition. In this chapter we will examine some useful compositional patterns and show how they can be coded in Python.

4.3 MINIMALISM

There are various musical forms and styles. One of the most intriguing ones is minimalism.

Definition: *Minimalism* refers to an artistic movement that uses a limited number of elements to create the strongest possible effect.

Minimalism emerged in the mid to late 20th century as a reaction to the complexity (and some would say, intellectual arrogance) of early 20th century music (e.g., 12-tone music). As a movement, it is a rebooting of sorts, that is, an attempt to return to basics.

Minimalist music usually employs repetition and layering of simple musical patterns to generating intricate musical textures and structures.[*] Minimalist composers include La Monte Young, Terry Riley, Steve Reich, and Philip Glass. Minimalist music is sometimes known as systems music—this refers to the system of rules created by the composer, which is applied to simple musical material to generate intricate musical experiences. Sometimes, this may result in sparse works (such as those of La Monte Young and Philip Glass, which feature drones and slowly overlapping phrases) or very repetitive works (such as Harold Budd's "Lovely Thing," which consists of one chord played softly over and over for about 15 minutes, or Erik Satie's "Vexations," where a short section is repeated hundreds of times). Another famous minimalist piece is Terry Riley's "In C,", which features repeated material with very limited harmonic changes over about an hour of performance.

Music from non-Western cultures, which relies heavily on repetition, has had an influence on minimalist composers. Steve Reich, for example,

---

[*] In a way, this mimics the simple processes of nature and how these (through repetition and transformation) generate the complex world we live in.

was inspired by the repetition of simple phrases and rhythms in the music of Bali and Ghana.

4.3.1 Repetition and Phasing

Repetition is a common musical technique that can be used to extend music or to create slowly evolving changes like those used by minimalist composers. Computers excel at mundane repetitive tasks and so this section will explore how repetition can be put to effective musical use. Computers have already found a significant role in popular music styles, particularly electronic dance music, where the use of repetition is pervasive. While some exact repetition provides music with coherence and structure, too much repetition can often be dull or boring. Music that relies on repetition often introduces variety in one parameter to provide sufficient interest that we continue to pay attention to the evolving change. Electronic dance music, for example, uses small timbral changes to maintain interest—usually achieved by sweeping a highly resonant low-pass filter over the sound source.

Phasing is a technique notably used by Steve Reich that adds interest to repetition. Phasing involves taking a musical phrase (melodic or rhythmic) and overlaying it with copies of itself that are not aligned. Copies can be played back at different speeds, causing changes in alignment over time, which generate various perceptual artifacts.

4.3.2 Case Study: Steve Reich, "Piano Phase" (1967)

This case study demonstrates how to recreate Steve Reich's "Piano Phase," a minimalist piece for two pianos involving tempo differences and repetition.

One way of achieving phasing is to independently vary the playback speed of two identical looped phrases. This is a subtle effect and, if the speed variation is only minor, adds timbral variety through phase cancellation to the more obvious rhythmic evolution. This technique emerged from Reich's early experiments with tempo phasing using tape players. Each tape machine used the same tape loop of pre-recorded material. However, one tape player was set to a slightly faster playback speed than the other. Once Reich understood this musical process well, he transferred it to compositions for acoustic instruments.

In "Piano Phase," two pianists have the same short phrase to play on separate pianos. They begin together, then after some time one pianist speeds up slightly, causing the phase effect. When the faster part is one quarter note ahead (i.e., aligning the first and the second note of the overlaid musical material), the faster pianist reverts to playing at the original tempo. This process is repeated several times, with the phase distance increasing each time. This provides a staircase effect of stable and phasing sections.

In the code example below, we recreate a section of Steve Reich's "Piano Phase," where the two parts play at different speeds. The speed difference is quite small, 0.5 beats-per-minute (quarter notes per minute), so the phase shift is quite gradual, but nevertheless the result is dramatic.

```python
# pianoPhase.py
# Recreates Steve Reich's minimalist piece, Piano Phase.

from music import *

pianoPart = Part(PIANO, 0)        # create piano part
phrase1   = Phrase(0.0)           # create two phrases
phrase2   = Phrase(0.0)

# write music in a convenient way
pitchList    = [E4, FS4, B4, CS5, D5, FS4, E4, CS5, B4, FS4, D5, CS5]
durationList = [SN, SN, SN, SN, SN, SN, SN, SN, SN, SN, SN, SN]

# add the same notes to both phrases
phrase1.addNoteList(pitchList, durationList)
phrase2.addNoteList(pitchList, durationList)

Mod.repeat(phrase1, 41)    # repeat first phrase 41 times
Mod.repeat(phrase2, 41)    # repeat second phrase 41 times

phrase1.setTempo(100.0)    # set tempo to 100 beats-per-minute
phrase2.setTempo(100.5)    # set tempo to 100.5 beats-per-minute

pianoPart.addPhrase(phrase1)    # add phrases to part
pianoPart.addPhrase(phrase2)

Write.midi(pianoPart, "pianoPhase.mid")    # save music to a MIDI file
Play.midi(pianoPart)                       # and play it.
```

Notice how the two parts start in synchrony and then slowly fall out of phase, creating different melodic and rhythmic artifacts exploiting our brain's need to organize unrelated events into meaningful patterns. Steve Reich's "Piano Phase" could be viewed as a statement about our perceptions and experiences in the world we live in (e.g., how much of this is real and how much of it is an artifact of our mind's need to organize—or does it make a difference)? Quite a powerful effect.

The program also demonstrates how to use the function Mod.repeat() to repeat a phrase. This function can also be used to repeat parts and scores. There will be more on the Mod class and its functions later in this chapter.

The above example demonstrates how, through computer programming, we can easily surpass the musical capabilities of a single performer on a traditional instrument.* The rest of the book is full of such examples.

4.4 MODIFYING MUSICAL MATERIAL (MOD FUNCTIONS)

Musical transformations are widely used in Western music composition. The music library contains many common compositional transformations for phrases, parts, and scores. These functions are included in the Mod class.

We have already been using functions from other classes in the music library such as Note, Phrase, Part, and Score. For example, addNote() is a function of the Phrase class, and setInstrument() is a function of the Part class.

The Mod class has functions that allow transformation (hence the name "Mod"—short for "modify") of musical material stored in phrases, parts, and scores. While the arguments for (i.e., the values passed to) each function vary slightly, the general structure of calling Mod functions is:

```
Mod.functionName(data, argument1, argument2,...)
```

Here are two examples:

```
Mod.append(phrase1, phrase2)
Mod.repeat(phrase1, 4)
```

The first example modifies phrase1 by adding to its end all the notes from phrase2. The second example repeats phrase1 so that it plays four times (i.e., this is the same as appending phrase1 to itself three times — why?).

All the functions in the Mod class follow the same syntactic pattern. The first argument is always the musical data to be modified. Subsequent arguments describe the way in which the first argument should be modified; subsequent arguments are never altered. For example, in Mod.append(phrase1, phrase2) the second argument, phrase2, is not altered. It is the first argument, phrase1, that is altered to include the notes stored in phrase2.

---

* It should be noted that there are some exceptionally skilled pianists who can play both piano parts in Steve Reich's "Piano Phase" at the same time —one part per hand.

Many functions in the Mod class come in different flavors, that is, several functions have the same name but work with different arguments. For example, the Mod class contains several append functions.

```
Mod.append(phrase1, phrase2)
Mod.append(part1, part2)
Mod.append(score1, score2)
```

In most cases the function names are self-explanatory. Below we present some useful Mod functions for Phrase objects. Most of these functions also apply to Part and Score objects. We organize them into functions that modify volume, functions that modify length, functions that modify pitch, and functions that use randomness to modify different musical aspects.

4.4.1 Modifying Volume

The following functions modify the volume (or dynamic) of notes in a phrase.

Function	Description
Mod.accent(phrase, meter)	Increase volume (accent) of the first quarter note beat of each measure in phrase, where meter is the number of beats per measure (e.g., 4.0).
Mod.accent(phrase, meter, beats, amount)	Same as above, except we provide a list of beats (e.g., [2, 3]) to accent and amount of accent (e.g., 15).
Mod.compress(phrase, ratio)	Compress (or expand) the dynamic range of phrase, by the provided compression (or expansion) ratio (e.g., 4.0).*
Mod.crescendo(phrase, time1, time2, vol1, vol2)	Increase volume in phrase from vol1 at time time1, to vol2 at time time2.
Mod.fadeIn(phrase, length)	Starting from zero volume, increase volume gradually over length quarter notes to reach the normal phrase volume.
Mod.fadeOut(phrase, length)	Starting from the normal phrase volume, decrease volume gradually over length quarter note to reach zero volume.
Mod.normalize(phrase)	Increase note volumes proportionally in phrase, so the loudest note is at maximum level.

* Use cycle() when you wish to fit a particular length. Use repeat() when you wish to repeat a number of times, regardless of length. When repeating overlapping phrases of different lengths, cycle() will guarantee they all end at approximately the same time.

4.4.2 Modifying Duration

The following functions modify the length of a phrase in various ways.

Function	Description
Mod.append(phrase1, phrase2)	Modifies phrase1 by appending to it the notes of phrase2.
Mod.elongate(phrase, factor)	Change (elongate) the length of notes in phrase by the provided scaling factor. If the factor is less than 1.0 it shrinks the duration (e.g., 0.5 reduces the duration to half—which is the same as doubling the speed or tempo).
Mod.changeLength(phrase, beats)	Change (scale) the duration in phrase, so its total duration equals the provided number of beats (quarter notes).
Mod.cycle(phrase, beats)	Repeats phrase until it spans the specified number of beats (quarter notes). Repetition stops with the note that reaches (or exceeds) the provided length (i.e., the last repetition of phrase may be incomplete).
Mod.palindrome(phrase)	Doubles the length of phrase, by appending a copy of itself in reverse order.
Mod.repeat(phrase, times)	Same as cycle(), expect we specify the number of times to repeat the phrase regardless of length (i.e., the last repetition of phrase is always complete).*
Mod.quantize(phrase, quantum)	Rounds (quantizes) the start time and duration of each note in phrase to fit multiples of quantum (e.g., 1.0). The smaller the quantum, the lesser the amount of change. A large quantum results in music that sounds mechanical.

* Use cycle() when you wish to fit a particular length. Use repeat() when you wish to repeat a number of times, regardless of length. When repeating overlapping phrases of different lengths, cycle() will guarantee they all end at approximately the same time.

4.4.3 Modifying Pitch

The following functions modify the melody (note pitches) of a phrase in various ways.

Function	Description
Mod.invert(phrase, pitch)	Mirror (invert or flip) the pitch of all notes in phrase relative to pitch.
Mod.transpose(phrase, interval)	Shift the pitch of every note in phrase by interval (number of chromatic steps, e.g., −12 lowers by an octave; 12 raises by an octave).
Mod.retrograde(phrase)	Reverse the order of the notes in phrase.
Mod.rotate(phrase, times)	Move (rotate) the notes around the phrase (i.e., last note becoming first, first note becoming second, second note becoming third, and so on), where times indicates how many times (i.e., how many notes) to rotate.

4.4.4 Modifying with Randomness

The following functions transform musical material using randomness in various ways.

Function	Description
Mod.shuffle(phrase)	Randomly reorganize the order of the notes in phrase.
Mod.mutate(phrase)	Mutate phrase by changing one pitch and one duration value. The new pitch is selected randomly between the lowest and the highest note of phrase. The random duration is selected from those in the existing notes.
Mod.randomize(phrase, pitchAmount)	Randomizes pitches in phrase by a value within +/- the specified pitchAmount.
Mod.randomize(phrase, pitchAmount, durationAmount)	Same as above, plus specify random variation of the duration (plus or minus durationAmount).

See Appendix B for additional Mod functions and descriptions of what they do. Soon we will discuss how to define our own functions and classes. This way we can implement additional transformations, as we desire. Learning how to do this is part of the essence of programming music.

4.5 MUSICAL CANON

A canon is a structurally intriguing example of polyphonic music, and one of the oldest musical forms. Composers create canons by taking a musical theme (also known as subject) and making one or more copies of it and then overlaying them offset in time. If done well, the end result can be quite amazing. Although not traditionally thought of as such, Steve Reich's "Piano Phase" is a type of musical canon.

Definition: A *musical canon* (or *round*) is a musical structure where a melody is played against one or more displaced copies of itself.

One well-known and simple example is the nursery rhyme, "Row Your Boat" — we will see it next. As you might imagine, there are many more canons in music history — some quite elaborate. Later in the chapter, we will see two canons by J. S. Bach. Another interesting example is the first movement of Béla Bartók's "Sonata for Two Pianos and Percussion."

4.5.1 Case Study: Traditional "Row Your Boat"

The following program illustrates how to create a musical canon. It also demonstrates how, through programming, your computer is transformed into a musical instrument that exceeds the capabilities of a single human performer on a traditional instrument. This computer music instrument can render a quite complicated piece of music. Also, unlike other playback environments (e.g., your digital music player), these pieces can be easily modified by making changes in the code. This is very enabling; imagine being able, through the press of a button, to change the musical parameters of a piece (i.e., when a certain voice starts, or which sounds to use to play a particular melody or chords).

The program in this section creates a melody as a round (canon) in three parts. It uses the Phrase.copy() function to create a copy of a musical phrase. This is very useful when you have several musical phrases that differ in small ways. Instead of having to create them all from scratch (which can be time-consuming), you can create one phase and then make copies of it. This way you can modify individual copies any way you like, without affecting the original phrase.

In our example below, the phrases differ in two ways:

1. The start time (i.e., when they start in the piece) and

2. The instrument associated with them.

This program also demonstrates how to use functions from the Mod class. As mentioned earlier, classes in Python are one way to organize (modularize) functionality and related code. Below, we use the Mod.transpose() function to move all the pitches in a phrase up or down by a number of chromatic steps. Using a positive number with Mod.transpose() moves the whole phrase up by that number of chromatic steps (semitones), i.e., this number is added to the pitch values. A negative number moves the whole phrase down by that number of chromatic steps, i.e., the number is subtracted from the pitch values.

Notice how comments are used to elucidate or describe the musical process. Again, when you program, it is advisable to first write good comments, which capture your thought process, and then write the corresponding code. This will save you considerable time in the long run.

```
# rowYourBoat.py
# Demonstrates how to build a musical canon.
```

```
from music import *

# Create the necessary musical data
rowYourBoatScore = Score("Row Your Boat", 108.0)  # tempo is 108 bpm

flutePart    = Part(FLUTE, 0)        # flute part on channel 0
trumpetPart  = Part(TRUMPET, 1)      # trumpet part on channel 1
clarinetPart = Part(CLARINET, 2)     # clarinet part on channel 2

themePhrase  = Phrase(0.0)           # theme starts at the beginning

# "Row, row, row your boat gently down the stream"
pitches1    = [C4, C4, C4,  D4, E4, E4,  D4, E4,  F4, G4]
durations1 = [QN, QN, DEN, SN, QN, DEN, SN, DEN, SN, HN]

# "merrily, merrily, merrily, merrily"
pitches2    = [C5, C5, C5, G4, G4, G4, E4, E4, E4, C4, C4, C4]
durations2 = [ENT, ENT, ENT, ENT, ENT, ENT, ENT, ENT, ENT, ENT, ENT, ENT]

# "life is but a dream."
pitches3    = [G4,  F4, E4,  D4, C4]
durations3 = [DEN, SN, DEN, SN, HN]

# add the notes to the theme
themePhrase.addNoteList(pitches1, durations1)
themePhrase.addNoteList(pitches2, durations2)
themePhrase.addNoteList(pitches3, durations3)

# make two new phrases and change start times to make a round
response1Phrase = themePhrase.copy()
response2Phrase = themePhrase.copy()

response1Phrase.setStartTime(4.0)    # start after 4 quarter notes
response2Phrase.setStartTime(8.0)    # start after 8 quarter notes

# play different parts in different registers
Mod.transpose(themePhrase, 12)       # one octave higher
Mod.transpose(response2Phrase, -12)  # one octave lower

# play each phrase twice
Mod.repeat(themePhrase, 2)
Mod.repeat(response1Phrase, 2)
Mod.repeat(response2Phrase, 2)

# add phrases to corresponding parts
flutePart.addPhrase(themePhrase)
trumpetPart.addPhrase(response1Phrase)
clarinetPart.addPhrase(response2Phrase)

# add parts to score
rowYourBoatScore.addPart(flutePart)
rowYourBoatScore.addPart(trumpetPart)
rowYourBoatScore.addPart(clarinetPart)

# play score as MIDI
Play.midi(rowYourBoatScore)
```

```
# write score to MIDI file
Write.midi(rowYourBoatScore, "rowYourBoat.mid")
```

Notice how the comments provide the lyrics — this a helpful way to recognize different melodic lines. Also, notice how comments are used to elucidate or describe the musical process.

4.5.1.1 Exercise

Extend the above code to include a fourth part. This part should start after the third part (with the distance equal to that between the last two parts). Use an instrument of your choice to play this part, and a pitch register of your choice (using Mod.transpose(), as demonstrated in the code above).

4.5.2 Analyzing the Musical Process

The musical process for creating canons is a formalized and well-understood case of transforming a theme and combining it with variations. Studying it will give us insights into, and practice with, capturing a musical process in code. This is the basis for creating arbitrary computer music programs.

As seen in the code above, the steps in creating a canon are as follows:

1. Create a melodic phrase (theme).

2. Make copies of that phrase.

3. Transform the copies of the phrase; in "Row Your Boat" above, we change when phrases start in the piece.[*]

4. Add each phrase to a different part with its own instrument.

Humans develop processes (or algorithms) for accomplishing important tasks , such as creating a musical canon. As we discussed in Chapter 1, any field of knowledge, from cooking to building houses, to healing people, to gardening, to solving math problems, to designing bridges, to repairing automobiles, to playing games, to making music involves algorithms for achieving tasks within that particular domain.

---

[*] All canons involve some form of systematic transformation of copies of the musical theme. Earlier in the chapter, we saw the Mod class, which gives us a wide variety of systematic transformation functions to potentially use with canon construction.

Definition: An algorithm is a sequence of steps for accomplishing a given task—in our case, a musical task.

We will explore musical algorithms in more detail in Chapter 5. For now let's examine the interesting sections of "Row Your Boat" code in more detail.

4.5.2.1 Creating Musical Material

The first thing in programming any musical process is to create the necessary musical data objects. The code below initializes a score, the three different parts/instruments, and the theme phrase.

```
# Create the necessary musical data
rowYourBoatScore = Score("Row Your Boat", 108) # tempo is 108 bpm

flutePart   = Part(FLUTE, 0)      # flute part on channel 0
trumpetPart = Part(TRUMPET, 1)    # trumpet part on channel 1
clarinetPart = Part(CLARINET, 2)  # clarinet part on channel 2

themePhrase = Phrase(0.0)              # theme starts at the beginning
```

The score object is created with a title and tempo. Then three parts are created and set to different instruments and channels. This will allow us to use three instruments to play the round. The phrase `themePhrase` starts at the beginning of the piece.

The next step is to create the actual notes. As seen in the previous chapter, the music library allows us to enter notes quickly using lists, as follows:

```
# "Row, row, row your boat gently down the stream"
pitches1   = [C4, C4, C4, D4, E4, E4, D4, E4, F4, G4]
durations1 = [QN, QN, DEN, SN, QN, DEN, SN, DEN, SN, HN]
```

The first note is a C4, QN (i.e., see the first element in both the `pitches1` and `durations1` lists); the second note is also a C4, QN; the third note is a C4, EN; and so on. Again, both lists need to have the same number of elements, otherwise we get an error.

Once we have specified the two lists, we can add them in a phrase using the `addNoteList()` function of the Phrase object called `themePhrase`.

```
themePhrase.addNoteList(pitches1, durations1)
```

4.5.2.2 Making Copies of Musical Material

To create a canon we need to start the melody again and again with a delay between each entry. In this example, we want a three-part round, so we create three phrases each starting at a different position. The first phrase starts at the beginning of the piece.

```
themePhrase = Phrase(0.0)          # theme starts at the beginning
```

After we add a note list to the above phrase, we make copies using the Phrase.copy() function. This allows us to modify the copies in desired ways without affecting the original.

```
response1Phrase = themePhrase.copy()
response2Phrase = themePhrase.copy()
```

4.5.2.3 Shifting Musical Material in Time

At this point, response1Phrase and response2Phrase are identical copies of themePhrase, including their start time. To adjust their start time as prescribed by this canon, we use the Phrase.setStartTime() function, which accepts a value indicating the new start time. The start time is in quarter notes (and fractions of them). Again, the beginning of the music is at 0.0 time.

```
response1Phrase.setStartTime(4.0)    # start after 4 quarter notes
response2Phrase.setStartTime(8.0)    # start after 8 quarter notes
```

The code above sets response1Phrase to start at the fifth quarter note position, and response2Phrase at the ninth quarter note position.

4.5.2.4 Transposing Musical Material

In order to hear the parts a little more clearly, we set them to different registers (octave ranges) by transposing them. Transposition means shifting a melody up or down in pitch. This can be done easily with the Mod.transpose() function.

```
Mod.transpose(themePhrase, 12)       # one octave higher
Mod.transpose(response2Phrase, -12)  # one octave lower
```

Here themePhrase is transposed up one octave (twelve chromatic steps), and response2Phrase transposed down one octave.

4.5.2.5 Combining Music Material

Up to now we have created the score, three parts and three phrases. However, they exist independently (see Figure 4.1).

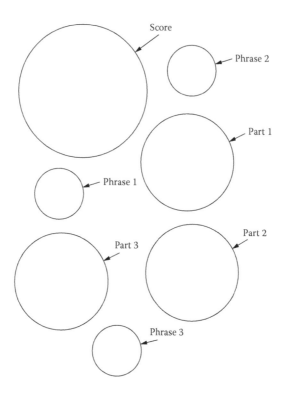

FIGURE 4.1   An unorganized collection of music library data objects.

We need to combine the musical material into a single score (to create the hierarchical structure shown in Figure. 3.2). This way we can listen to all the music simultaneously, when we play the score, that is, Play.
midi(score). We can do this in any order, as long as we put phrases into the
appropriate part(s), and the parts into the score.

In the above code, we first put the phrases into the parts. We use the
addPhrase() function of the flute part to store the theme in it (theme-
Phrase). We use the same functions of the trumpet and clarinet parts to
put in them the first and second responses, respectively.

```
# add phrases to corresponding parts
flutePart.addPhrase(themePhrase)
trumpetPart.addPhrase(response1Phrase)
clarinetPart.addPhrase(response2Phrase)
```

Then we store the three parts in the score, using the score's addPart()
function.

```
# add parts to score
rowYourBoatScore.addPart(flutePart)
```

```
rowYourBoatScore.addPart(trumpetPart)
rowYourBoatScore.addPart(clarinetPart)
```

4.5.2.6 *Saving and Playing Musical Material*

Finally, we play the score in real time and also save it as a MIDI file:

```
# play score as MIDI
Play.midi(rowYourBoatScore)

# write score to MIDI file
Write.midi(rowYourBoatScore, "rowYourBoat.mid")
```

Now it is time to look at a more elaborate example of a musical cannon.

4.5.3 Case Study: J.S. Bach—Goldberg Ground, Canon 1 (BWV 1087)

In 1741, Johann Sebastian Bach (1685–1750), one of the most important composers of all time (and an expert in connecting numbers and music), wrote the Goldberg Variations (BWV 988), for harpsichord. This work consists of an aria and a set of 30 variations, and is an amazing example of the transformation of a theme to create variations.

Then, in 1974 an unknown manuscript was discovered in France, written in Bach's hand, containing 14 canons on the first eight notes of the Goldberg aria. These canon variations were previously unknown (except for canons 11 and 13). These canons are very important. They encapsulate Bach's ideas for how music, number, and the universe are all connected together.

Bach did not provide the complete music for these canons. Instead, he specified the theme and then wrote cryptic notations (riddles) for the reader to solve. Solving the riddle unlocked the music by providing the musical process to generate it. Generally, Bach's canons are constructed by layering copies of the theme that are transformed using one or more of the following rules (canonic devices):

- Shifting in time

- Retrograde (reversing the order of notes—horizontal mirroring)

- Inversion (mirroring the notes around a pivot note—vertical mirroring)

- Augmentation (stretching the duration of notes, e.g., double their duration)

- Diminution (shrinking the duration of notes, e.g., halve their duration).

FIGURE 4.2   J. S. Bach's seal consisting of his initials and a canonic interweaving of them under a crown.

Canon symmetries and rules were so important to Bach that he used them in his personal seal (see Figure 4.2). This seal is characteristic of the ornamental style of the Baroque period and consists of his initials (the letters J, S, and B), their mirror, an interweaving of them (possibly representing his musical work, as a reflection of the symmetries found in nature), and a crown at the top (possibly representing the royal and/or religious context within which this work took place).

These canonic rules are implemented, through the music library, as follows:

- Create a phrase with the melody (theme).

- Create copies (`phrase.copy()`) of the original phrase.

- Use `phrase.setStartTime()` or `Mod.retrograde()`, `Mod.invert()`, or `Mod.elongate()` to modify each copy.

- Put resulting phrases into a Part and Score.

The following program demonstrates how to use the Mod functions to "unlock" (or create) the first of these 14 canons. It is built from a theme consisting of eight notes and reversed version (retrograde) played simultaneously. For more information, see (Smith 1996).

```
# JS_Bach.Canon_1.GoldbergGround.BWV1087.py
#
# This program recreates J.S. Bach's Canon No. 1 of the Fourteen on
# the Goldberg Ground.
#
# This canon is constructed using the Goldberg ground as the subject
# (soggetto) combined with the retrograde of itself.

from music import *
```

```
# how many times to repeat the theme
times = 6

# define the data structure
score   = Score("J.S. Bach, Canon 1, Goldberg Ground (BWV1087)", 100)
part    = Part()
voice1  = Phrase(0.0)

# create musical material (soggetto)
pitches   = [G3, FS3, E3, D3, B2, C3, D3, G2]
durations = [QN, QN, QN, QN, QN, QN, QN, QN]
voice1.addNoteList(pitches, durations)

# create 2nd voice
voice2 = voice1.copy()
Mod.retrograde(voice2)                   # follower is retrograde of leader

# combine musical material
part.addPhrase(voice1)
part.addPhrase(voice2)
score.addPart(part)

# repeat canon as desired
Mod.repeat(score, times)

# write score to a MIDI file
Write.midi(score, "JS_Bach.Canon_1.GoldbergGround.BWV1087.mid")
```

4.5.4 Case Study: Trias Harmonica canon (BWV 1072)

J.S. Bach (1685–1750) created many beautiful canons (see Figure 4.3). We present the implementation of another one, namely, the Trias Harmonica (BWV 1072). Like all canons, the Trias Harmonica involves repetition and layering of melodic phrases; it consists of two parts with four separate voices each—a total of eight overlaid phrases.

This canon was particularly significant to Bach. The name trias ("τριάς", or triad) refers to the triunity of the divine in the Christian tradition, that is, "three in one and one in three" (Bach was a devout Lutheran). The Trias Harmonica canon interweaves a major triad (C4, E4, G4—representing the divine) with the minor triad (D4, F4, A4—representing man). Also, it uses a dotted quarter note (i.e., 1.5) as the duration for the major triad, and an eighth note (i.e., 0.5) for the minor triad, emphasizing the 1/3 relationship between the trinity and man (i.e., the son in Christian theology). It is believed that here Bach attempted to musically represent the harmony he believed existed between the natural and spiritual world (Smith 1996).

FIGURE 4.3   J. S. Bach holding the Canon triplex a 6 vocibus, BWV 1087, one of the goldberg canons (painted by Haussmann in 1746).

```
# JS_Bach.Canon.TriasHarmonica.BWV1072.py
#
# This program creates J.S. Bach's 'Trias Harmonica' BWV 1072 canon.
#
# This canon is constructed using two parts (choirs), each consisting
# of four voices. The first part's voices use the original theme,
# each separated by a half-note delay. The second part's voices use
# the inverted theme delayed by a quarter note, each separated by
# a half-note delay.
#

from music import *

# define the theme (for choir 1)
pitches1   = [C4,  D4, E4,  F4, G4,  F4, E4,  D4]
durations1 = [DQN, EN, DQN, EN, DQN, EN, DQN, EN]

# define the inverted theme (for choir 2)
pitches2   = [G4,  F4, E4,  D4, C4,  D4, E4,  F4]
durations2 = [DQN, EN, DQN, EN, DQN, EN, DQN, EN]

# how many times to repeat the theme
times = 8

# choir 1 - 4 voices separated by half note
choir1 = Part()

voice1 = Phrase(0.0)
voice1.addNoteList(pitches1, durations1)
```

```
Mod.repeat(voice1, times)          # repeat a number of times
voice1.addNote(C4, DQN)            # add final note


voice2 = voice1.copy()             # voice 2 is a copy of voice 1
voice2.setStartTime(HN)            # separated by half note


voice3 = voice1.copy()             # voice 3 is a copy of voice 1
voice3.setStartTime(HN*2)          # separated by two half notes


voice4 = voice1.copy()             # voice 4 is a copy of voice 1
voice3.setStartTime(HN*3)          # separated by three half notes


choir1.addPhrase(voice1)
choir1.addPhrase(voice2)
choir1.addPhrase(voice3)
choir1.addPhrase(voice4)


# choir 2 - 4 voices inverted, delayed by quarter note,
# separated by a half note.
choir2 = Part()


voice5 = Phrase(QN)                # delayed by quarter note
voice5.addNoteList(pitches2, durations2)
Mod.repeat(voice5, times)          # repeat a number of times
voice5.addNote(G4, DQN)            # add final note


voice6 = voice5.copy()             # voice 6 is a copy of voice 5
voice6.setStartTime(QN + HN)       # separated by half note


voice7 = voice5.copy()             # voice 7 is a copy of voice 5
voice7.setStartTime(QN + HN*2)     # separated by two half notes


voice8 = voice5.copy()             # voice 8 is a copy of voice 5
voice8.setStartTime(QN + HN*3)     # separated by three half notes


choir2.addPhrase(voice5)
choir2.addPhrase(voice6)
choir2.addPhrase(voice7)
choir2.addPhrase(voice8)


# score
canon = Score("J.S. Bach, Trias Harmonica (BWV 1072)", 100)
canon.addPart(choir1)
canon.addPart(choir2)


# write it to a MIDI file
Write.midi(canon, "JS_Bach.Canon.TriasHarmonica.BWV1072.mid")
```

4.5.5 Exercises

Write code that reconstructs more J. S. Bach canons:

1. J. S. Bach's "Canon a 2. perpetuus" (BWV 1075).

2. J. S. Bach's "Canon Super Fa Mi a 7 Post Tempus Musicus" (BWV 1078).

3. Find additional canons online (e.g., Smith 1996).

4.5.6 Case Study: Arvo Pärt—"Cantus in Memoriam" (1977)

Arvo Pärt's "Cantus in Memoriam Benjamin Britten" is a minimalistically composed, yet very powerful piece for string orchestra and bell, based on Pärt's tintinnabuli style. It was written to mourn the passing of Benjamin Britten, an English composer, whom Pärt considered a kindred spirit. Pärt, being in Estonia at the time (on the wrong side of the Iron Curtain), never met Britten, who died in England in 1976.

This version of the piece is slightly simplified compared to the original score. It introduces a single voice descending in stepwise motion the A aeolian (natural minor) scale. This voice is repeated at half tempo several times, creating beautiful harmonic combinations stemming from all permutations of aeolian-scale notes.

Although Pärt's score lists all the notes spelled out, below we construct the piece using Bach's canon approach. In other words, we state the theme and apply canonic rules (copying, shifting in time, and elongation) to generate the final piece. This implementation allows us to better appreciate Arvo Pärt's compositional style (as reflected in this piece); also it may provide inspiration for things to try in your own music composition explorations.

```
# ArvoPart.CantusInMemoriam.py
#
# Recreates a variation of Arvo Part's "Cantus in Memoriam Benjamin
# Britten" (1977) for string orchestra and bell, using Mod functions.

from music import *

# musical parameters
repetitions = 12      # length of piece
tempo = 112           # tempo of piece
bar = WN+HN           # length of a measure
```

```
# create musical data structure
cantusScore = Score("Cantus in Memoriam Benjamin Britten", tempo)


bellPart = Part(TUBULAR_BELLS, 0)
violinPart = Part(VIOLIN, 1)


# bell
bellPitches   = [REST, A4,    REST, REST, A4,    REST, REST, A4]
bellDurations = [bar/2, bar/2, bar,  bar/2, bar/2, bar,  bar/2, bar/2]


bellPhrase = Phrase(0.0)
bellPhrase.addNoteList(bellPitches, bellDurations)
bellPart.addPhrase(bellPhrase)


# violin - define descending aeolian scale and rhythms
pitches = [A5, G5, F5,  E5,  D5,  C5,  B4,  A4]
durations = [HN, QN, HN, QN, HN, QN, HN, QN]


# violin 1
violin1Phrase = Phrase(bar * 6.5)  # start after 6 and 1/2 measures
violin1Phrase.addNoteList(pitches, durations)


# violin 2
violin2Phrase = violin1Phrase.copy()
violin2Phrase.setStartTime(bar * 7.0)  # start after 7 measures
Mod.elongate(violin2Phrase, 2.0)        # double durations
Mod.transpose(violin2Phrase, -12)       # an octave lower


# violin 3
violin3Phrase = violin2Phrase.copy()
violin3Phrase.setStartTime(bar * 8.0)  # start after 8 measures
Mod.elongate(violin3Phrase, 2.0)        # double durations
Mod.transpose(violin3Phrase, -12)       # an octave lower


# violin 4
violin4Phrase = violin3Phrase.copy()
violin4Phrase.setStartTime(bar * 10.0) # start after 10 measures
Mod.elongate(violin4Phrase, 2.0)        # double durations
Mod.transpose(violin4Phrase, -12)       # an octave lower


# repeat phrases enough times
Mod.repeat(violin1Phrase, 8 * repetitions)
Mod.repeat(violin2Phrase, 4 * repetitions)
Mod.repeat(violin3Phrase, 2 * repetitions)
Mod.repeat(violin4Phrase, repetitions)


# violin part
violinPart.addPhrase(violin1Phrase)
violinPart.addPhrase(violin2Phrase)
violinPart.addPhrase(violin3Phrase)
violinPart.addPhrase(violin4Phrase)
```

```
# score
cantusScore.addPart (bellPart)
cantusScore.addPart (violinPart)

# fade in
Mod.fadeIn(cantusScore, WN * repetitions)

# view and write
View.sketch(cantusScore)
Play.midi(cantusScore)
Write.midi (cantusScore, "ArvoPart.CantusInMemoriam.mid")
```

Notice how the above is similar to the canons we have seen. Also, notice the introduction of function View.sketch(), which generates a visual representation of the musical material. More about this and other View functions after the next exercise. For now, observe the visual structure of the piece over time. Seasoned composers work at the macro level, that is, they first make a high-level, visual sketch of a piece before they decide the smaller-scale details. Having seen this structure, can you also hear it in the piece?

4.5.7 Exercises

1. Create a variation of the above piece using the C major scale.

2. Modify the code so that the voices move upwards.

3. Instead of starting with the fastest moving voice, start with the slowest moving one. Then add the faster voices, while keeping relative proportions and start times.

4. The musical outcome of (3) may remind you one of the pieces we have already studied. Which one and why? What are the common elements? Explore the history of the two composers and hypothesize on how this creative connection may have come about.

4.6 VIEWING MUSIC

The case study above demonstrates another music library class, namely, View. This class contains several functions which help visualize musical material. The View functions work with Phrase, Part, and Score objects. There are several different views available, namely, View.notation(), View. pianoRoll(), View.internal(), View.sketch(), each with its own benefits and limitations.

FIGURE 4.4 Notation output from the "Theme and Variations" program (seen later in the chapter).

4.6.1 Notation Display

The View.notation() function provides a limited notational facility for displaying musical material in Western notation. For example, to view the notation of a phrase stored in variable phrase, you would use View.notation(phrase). This function will select a treble, bass, or piano staff, as required, to display the music. The time signature will default to 4/4 and the key to C major, unless otherwise specified (e.g., see Figure 4.4). Accidentals (black notes on the piano) will be shown with a sharp (#) or flat (♭) symbol before them.

An interesting feature of this view is that it allows you to edit the music material interactively. You can move notes by dragging the note heads. Also, you can add new notes by clicking on an empty place on the staff. Although it is not a full-blown music notation editor, View.notation() allows you to do some exploration and experimentation.* Once modified, the edited musical material may be saved as a MIDI file through the available menus.

One possibility is to create initial musical material through a Python program. Then use View.notation() to experiment with changes. If you like the outcome, you can save the modified material to a MIDI file and use a full-blown editor to finalize the piece.

It is important to note that View.notation() cannot save these modifications back in your Python code. If you would like your code to generate the modified material, you will have to make these changes manually.

4.6.2 Piano Roll Display

The View.pianoRoll() function displays music in a piano roll view. This notation resembles the piano rolls used in player pianos, that is, long strips of paper with holes where notes are meant to be played. By loading

---

* For anything other than a few notes, you will be better served by regular music notation editors, such as Noteflight and Sibelius.

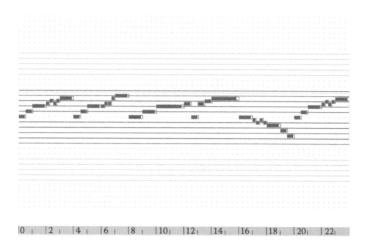

FIGURE 4.5   Piano roll output from the "Theme and Variations" program.

a piano roll into a player piano, the piano would automatically play these notes.*

In the output produced by View.pianoRoll() notes appear as rectangles on a staff. The dark horizontal lines make up the standard treble and bass staffs, while the gray staffs indicate upper treble and lower bass ranges two octaves away. Accidentals (black notes on the piano) are shown with a sharp (#) sign in front, just as on the notation display. Phrase boundaries are drawn as rectangles around note groups, and notes in each part are shown in the same color. The depth of color indicates the dynamic level of the note, darker shades for louder notes. The ruler along the bottom displays the timeline and can be used to adjust the time scale by dragging. Quarter note position numbers are displayed when ruler size permits (Figure 4.5).

4.6.3 Internal Values Display

While a visual display of the score provides a graphical overview of the score, sometimes it is useful to see the precise data values. This is done with the View.internal() function. It outputs a textual list of the important values of the musical material, organized in terms of score, parts, phrases, and notes (as provided). This output can be especially handy for debugging code, if you are getting unexpected musical results (see Figure 4.6).

---

* One could think of player pianos as programmable music boxes. The piano roll provided the instructions on what music to play and when. If you think about it, this is very similar to the function of your Python program, that is, your programs are like elaborate piano rolls for your computer to play. Additionally, these programs allow for functionality that piano rolls cannot possibly support (such as selection, iteration, randomness, interaction—as we will see soon).

```
<- - - PART: 'Untitled Part' contains 5 phrases.- - ->
Channel = 0
Instrument = 0
<- - - PHRASE: 'Untitled Phrase' contains 8 notes. Start time: 0.0- - | - ->
<NOTE (Pitch = 60) (Duration = 0.5) (Dynamic = 0.5) (Length = 127) (Pan = 0.45) (Pan = 0.5) >
<NOTE (Pitch = 64) (Duration = 0.5) (Dynamic = 0.5) (Length = 127) (Pan = 0.45) (Pan = 0.5) >
<NOTE (Pitch = 67) (Duration = 1.0) (Dynamic = 1.0) (Length = 127) (Pan = 0.9) (Pan = 0.5) >
<NOTE (Pitch = 69) (Duration = 0.25) (Dynamic = 0.25) (Length = 127) (Length = 0.225) (Pan = 0.5) >
<NOTE (Pitch = 71) (Duration = 0.25) (Dynamic = 0.25) (Length = 127) (Length = 0.225) (Pan = 0.5) >
<NOTE (Pitch = 69) (Duration = 0.25) (Dynamic = 0.25) (Length = 127) (Length = 0.225) (Pan = 0.5) >
<NOTE (Pitch = 71) (Duration = 0.25) (Dynamic = 0.25) (Length = 127) (Length = 0.225) (Pan = 0.5) >
<NOTE (Pitch = 72) (Duration = 1.0) (Dynamic = 1.0) (Length = 127) (Length = 0.9) (Pan = 0.5) >

<- - - PHRASE: 'Untitled Phrase copy' contains 8 notes. Start time: 4.0- - | - ->
<NOTE (Pitch = 60) (Duration = 0.5) (Dynamic = 0.5) (Length = 127) (Pan = 0.45) (Pan = 0.5) >
<NOTE (Pitch = 64) (Duration = 0.5) (Dynamic = 0.5) (Length = 127) (Pan = 0.45) (Pan = 0.5) >
<NOTE (Pitch = 67) (Duration = 1.0) (Dynamic = 1.0) (Length = 127) (Pan = 0.9) (Pan = 0.5) >
<NOTE (Pitch = 67) (Duration = 0.25) (Dynamic = 0.25) (Length = 127) (Length = 0.225) (Pan = 0.5) >
<NOTE (Pitch = 69) (Duration = 0.25) (Dynamic = 0.25) (Length = 127) (Length = 0.225) (Pan = 0.5) >
<NOTE (Pitch = 69) (Duration = 0.25) (Dynamic = 0.25) (Length = 127) (Length = 0.225) (Pan = 0.5) >
<NOTE (Pitch = 72) (Duration = 0.25) (Dynamic = 0.25) (Length = 127) (Length = 0.225) (Pan = 0.5) >
<NOTE (Pitch = 74) (Duration = 1.0) (Dynamic = 1.0) (Length = 127) (Length = 0.9) (Pan = 0.5) >
. . .
```

FIGURE 4.6 Internal values output (excerpt) from "Theme and Variations" program.

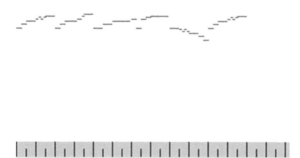

FIGURE 4.7   Sketch output from the "Theme and Variations" program.

4.6.4 Sketch Display

The last display provided by the View class is the View.sketch() function. This function provides a graphical representation that resembles a piano roll, except that individual notes are drawn with thin lines. What makes the display particularly interesting is that you can draw music with the mouse (by clicking on the display and dragging). The type of music generated is very particular given the constraints and affordances of the sketch interface. Similarly to all other graphical notations, the modified musical material can be saved into a MIDI file, for later processing, if desired (see Figure 4.7).

4.6.5 Exercises

Write a program that opens up a blank sketch display. Experiment drawing various lines and hearing the result. Try to produce an interesting music composition this way. Save the musical material into a MIDI file. Write a second program that reads in this MIDI file using the following:

```
score = Score()
Read.midi(score, "nameOfMidiFile.mid")
```

where "nameOfMidiFile.mid" is the name of the MIDI file containing your music material generated through the Sketch display. Explore the application of various Mod functions to this score, such as Mod.randomize() and Mod.quantize(), to transform this musical material in interesting ways.

4.7 THE SOFTWARE DEVELOPMENT PROCESS

Just as musical composition is a creative act, so too is software development. Both also benefit from systematic planning and processes. The lifecycle of a computer program starts from when we come up with the original idea and extends to designing the algorithm, to writing some code, to testing and refining it, to using the program for its intended task, to updating the program to serve new needs, and to reusing some of the code in newer programs. Most programs, once written, are (re)used in some way; they are rarely thrown away. Keep this in mind as you embark on your software development journey.

The software development process consists mainly of design, implementation, testing, and documentation. These steps are not necessarily sequential. As in most creative processes, they are interwoven into the lifecycle of a program.

4.7.1 Design

An early phase in developing software is to identify what it is we are building. This can be a simple statement or a quite elaborate description. It is OK to try out some ideas in code first, but to make significant progress a software developer needs to use a piece of paper (or a whiteboard) to draw out ideas and steps required to achieve these ideas. The following maxim captures this:

Advice: Two hours of design can save you 20 hours of coding.

Computers need to be told what to do; it is our responsibility to first identify these steps and then translate them to a language (such as Python) that the computer understands. Although at times we might feel otherwise (and it is this mystification that makes computers so exciting), at the end of the day, computers are just following instructions; they have no creative intent. Computer programs reflect the creativity of the software developer and user.

Fact: A software developer is like a music composer.

When programming and composing are combined (i.e., to write software for algorithmic music), the challenge for the musician is to write instructions that lead to interesting and expressive music. Musical algorithms can describe how each of the musical elements is specified and varied as the piece proceeds. This can include control over the pitch, duration, and loudness of notes, the timbre of sounds, the use of structural features such as repetition and variation, as well as tempo, volume, balance, and so on.

4.7.2 Implementation

Once we have come up with the steps (or algorithm) to create a musical artifact, we can then translate it to a programming language. In this book we use Python (a textual programming language which is simple and thus great for beginners). We have extended this language with several libraries for generating sounds and music, as well as for processing images, and for creating graphical user interfaces. These activities are necessary for modern music making using computers. These libraries will be presented in later chapters.

4.7.3 Testing

Once some code has been written, it can be tested to see if it works as intended. This is very important. A program should not only run (i.e., without crashing) it also needs to perform the intended task precisely. If it does not, we modify the design of the program until we get it to work appropriately.

Normally, there are two types of programming errors that testing reveals:

- Syntax errors

- Logical errors

Syntax errors are usually caused by simple mistakes, such as typos. They result in a "program crash"—that is, the program does not run or terminates unexpectedly, producing a (usually cryptic) error message, such as the following:

```
Traceback (most recent call last):
  File "pentatonicMelody3.py", line 21, in <module>
    for i in range(numnotes - 2):
NameError: name 'numnotes' is not defined
```

Most syntax errors are easy to fix. To do so, you should read error messages backwards, that is, starting from the bottom line. For example, the above error indicates that a word you typed in your program is not recognized (i.e., "NameError"). The message indicates that this particular word occurs on line 21 in your program. Did you type the word correctly?

Fact: Once you learn how to read error messages, syntax errors are easy to fix.

On the other hand, logical errors make a program produce wrong results or not work as intended. For example, a design error in the musical algorithm above would be to use the wrong musical scale. Although the program would run successfully (i.e., without any error messages generated by the computer), the musical outcome would be incorrect.

Design errors are hard to discover, unless we test our programs often and throughout the software development process.

Advice: Test your code sooner rather than later and as often as possible.

Seasoned programmers know that logical errors cost very little to fix, if caught early. Python makes it easy to write short excerpts of code and test them. It is advisable to develop programs incrementally. Write a little code and test it. Then write some more and test it.

4.7.4 Documentation—Good Style and Comments

Every programming language allows programmers to include comments inside their code. Comments are phrases written in human language (e.g., English) to explain a particular idea or task being performed by a part of the code. Comments are intended for humans and are ignored by the computer.

In Python comments are denoted with the "#" character—anything between "#" and the end of the line is ignored by the computer. We use this convention to capture our thoughts when they are fresh as we are coding, so that we can remember later essential ideas about our code.

Advice: If you cannot phrase it in English (or other native tongue), you cannot code it.

Fact: Seasoned programmers comment before they code.

Seasoned programmers will first write comments describing the desired goal and then translate those comments to Python code. (Some will even consider how to test their code before writing it.) Although this practice may appear counterproductive, it can actually save time in the long run.

4.8 CASE STUDY: COMPUTER-AIDED MUSIC COMPOSITION

For centuries musicians have taken existing music and created their own variations of it. Since the 1950s, researchers have been exploring how to utilize computers in this activity. One great example is David Cope's EMI (Experiments in Musical Intelligence) software, which has analyzed and produced musical works in the style of various composers (Cope 2001; 2013). Cope's fascinating research into what constitutes musical creativity and how much of it can be algorithmically encoded has produced a wide variety of musical pieces, some of which have been performed by orchestras and recorded. His latest program, called *Emily Howell*, has been used to produce two commercially available CDs.

The following program demonstrates how various Mod functions may be used to harness the computer's ability to create novel music from existing material. Although very simplistic compared to David Cope's EMI, this program demonstrates how to begin developing computer-aided composition tools to help with writer's block, or to inspire new melodic or harmonic explorations.

In this case study, we start with a theme provided by a human composer. Each musical variation algorithmically (programmatically) develops the material from the original theme. The goal here is not to create a near-perfect musical piece—that would be difficult to do with what we know so far.* Instead, we aim for a piece with reasonable musicality—a piece that can be used to explore musical possibilities, and get new ideas. To do so, we try to make the variations

a. subtle enough so that the relationship with the original is evident, but also

b. novel enough that they are interesting and surprising.

---

* Developing more intelligent music composition programs can be accomplished using more advanced methods from Artificial Intelligence and Computational Creativity, which is beyond the scope of this book.

In this case study three variations, var1, var2, and var3, are created by copying the original theme and applying Mod functions. The last variation, var4, is an exact copy of the original.

```
# themeAndVariations.py
#
# Demonstrates how to automatically develop musical material
# using a theme and transforming it through Mod functions.
#
# Here we start with the theme, continue with a few interesting
# variations, and close with the theme for recapitulation.

from music import *

# create theme
pitches = (C4, E4, G4, A4, B4, A4, B4, C5]
durations = [EN, EN, QN, SN, SN, SN, SN, QN]

theme = Phrase()
theme.addNoteList(pitches, durations)


# variation 1
# vary all pitches in theme, but keep the same
# rhythm pattern (for stability)
var1 = theme.copy()        # preserve original (make a copy)
Mod.randomize(var1, 3)     # randomize each pitch (max of +/- 3)
                           # (later we force them into the scale)


# variation 2
# slow down theme, and change two pitches using
# a random value from within the theme range
var2 = theme.copy()
Mod.elongate(var2, 2.0)  # double its time length
Mod.mutate(var2)           # change one pitch and duration value
Mod.mutate(var2)           # and another (could be the same)


# variation 3
# reverse the theme, and lower it one octave
var3 = theme.copy()
Mod.retrograde(var3)       # reverse notes
Mod.transpose(var3, -12)  # lower one octave


# recapitulation
# repeat the theme for closing (i.e., return home)
var4 = theme.copy()        # we need a copy (a phrase can be added
                           # to a part only once)


# add theme and variations in the right order (since we
# didn't specify start times, they will play in sequence)
part = Part()
part.addPhrase(theme)
```

```
part.addPhrase(var1)
part.addPhrase(var2)
part.addPhrase(var3)
part.addPhrase(var4)

# now, fix all notes to be in the C major scale,
# and with durations that line up with SN intervals
Mod.quantize(part, SN, MAJOR_SCALE, 0)

# provide alternate views
View.pianoRoll(part)
View.internal(part)
View.sketch(part)
Mod.consolidate(part) # merge phrases into one phrase, so that...
View.notation(part) #...notate() can display all the notes

# and play
Play.midi(part)
```

Notice how each variation has a distinctive character by being transformed in particular and unique ways.

Fact: By combining Mod functions in creative, yet well-designed ways, you
can create processes for transforming musical material to explore compositional spaces. These processes may involve chance, so every time you run
them, you get novel material.

In the above program, the first variation changes all the pitches in
the theme by a random interval of +/–3; however, it retains the same
note durations, thus preserving the rhythm pattern. This introduces
change, but also provides some stability, to support flow. Notice how
we make a copy of the theme before modifying it, otherwise the Mod
functions would be changing the original, which is something we do
not want.

The second variation slows down the theme and makes a couple of pitch
and duration changes. The third plays the theme backwards and one octave
lower. The fourth variation is really a recapitulation—a straight copy of
the theme, to "return home," so to speak. Recapitulations are important in
music to provide a sense of closure.

Another thing to notice is that (as stated in Chapter 3), when phrases
do not have explicit start times, they play back sequentially in the order
added to the part. If you use explicit start times, you specify precisely
when they play back. That is a simple way to create overlapping phrases.

When providing explicit start times for phrases, the order you add them to a part is inconsequential.

After the theme and variations are added into the part, we use Mod. quantize() to "straighten out" any excessive rhythmic or pitch oddities that might have been introduced by the transformation process. This compositional approach of changing (modifying) and constraining (quantizing) reflects two sides of creative processes that always need to be kept in balance, novelty and familiarity.*

4.8.1 Exercise

Starting with a theme, write various programs like the above to explore a new compositional idea. In addition to transforming melodies, explore ways to transform harmonies, for example, chord progressions. Results from this activity may evolve into a set of tools to assist you in your compositional practice. Your imagination is the limit.

4.9 SUMMARY

This chapter presented patterns of musical structure and organization. It discussed minimalism and musical canons, and showed us how to transform musical material using Mod functions, such as transpose(), elongate(), and repeat(). We learned how to recreate well-known musical works, by transcribing only a small part of the work, namely, the theme, and working from the theme to generate the remaining work, through well-defined algorithmic processes (canonic devices). This makes us realize that, although modern computers are a recent development in the history of music (and humanity), well-defined algorithmic processes have existed and benefited musical (and other) creators throughout history. J.S. Bach was one of the early masters in exploring such material and processes.

Additionally, we learned different ways to view musical material, provided by the music library, namely, as Western notation, piano roll, textual data, and sketch view. Finally, we began discussing the software development process—a high-level algorithm for designing, implementing, and evaluating (i.e., testing) a program. The objective here is to learn how to develop better programs, more efficiently, and thus better utilize our creative time.

---

* The same principle (of balancing novelty and familiarity) is applied to variation 1 (by keeping the same durations) and with repeating the opening theme at the end of the piece.

## CHAPTER 5

Iteration and Lists

Topics: Iteration, Python for loop, arpeggiators, constants, list operations, range(), frange(), FX-35 Octoplus, DNA music.

5.1 OVERVIEW

Most processes in nature and in human culture involve iteration or repetition at different levels of scale. For example, consider a tree and how it starts with a single branch (trunk), which gets split into 4–5 sub-branches at 60 (or so) degree angles, and it repeats this process until we reach the level of leaves. Another example is the human body, which starts from a single cell, which gets subdivided repeatedly, eventually giving rise to the complex organism that is reading these words.

5.2 ITERATION

Iteration or repetition is integral to making complicated and interesting things. Almost all music involves some form of repetition.* Musical repetition may occur at the high level, such as the repetition of sections in song forms like ABA and AABA. It may also happen at the middle level, such as the reuse of a melodic or rhythmic gesture that repeats with or without variation throughout a piece (e.g., the opening gesture of Beethoven's *Fifth Symphony* or the bass line in Michael Jackson's *Thriller*). It could also happen at the low level, such as a note-by-note musical pattern that repeats over and over, for example, Steve Reich's *Piano Phase*.

5.2.1 The Python for Loop

In the previous chapter, we saw the Mod function repeat(), which provides one way to do repetition of musical data. Although this works in

---

* Here we use the term "repetition" to refer to musical processes and "iteration" to refer to the mechanism for repetitive processes in Python. However, in computer science, the terms *repetition* and *iteration* are synonymous.

many cases, generally, we need something more powerful, flexible, and customizable. For this reason, all programming languages provide iteration mechanisms, such as the **for loop** and the **while loop**.

The for loop allows you to repeat a group of statements over and over. For example, consider the following code:

```
phrase = Phrase()              # create an empty phrase

pitchList = [C4, E4, G4, B4, C5]   # create a list of pitches

for pitch in pitchList:        # for each item in the list
    note = Note(pitch, QN)     # create a note
    phrase.addNote(note)       # add it to phrase

Play.midi(phrase)              # play notes
```

The first two lines create an empty phrase and a list of values — nothing new here. The next three lines are an example of a for loop. The first line (in bold below):

```
for pitch in pitchList:
    note = Note(pitch, QN)
    phrase.addNote(note)
```

is called the *loop header*. The loop header sets up the parameters for iteration, that is, how many times to iterate. The above example defines a variable called pitch.* This variable will contain values from the list, pitchList one at a time. The loop will iterate as many times as there are items in the list. At each iteration, pitch will contain the next value from pitchList, starting with the first, moving to the second, and so on, until the last value.

The next two indented lines (in bold below):

```
for pitch in pitchList:
    note = Note(pitch, QN)
    phrase.addNote(note)
```

are called the *loop body*. These two statements will be repeated as many times as the loop will iterate (see above). At each iteration, the variable pitch contains the next value from the list pitchList, in essence allowing us to use, one at a time, all the values in that list. What we do with the value in pitch depends on the Python statements we decide to use in the loop body.

---

* Note that the name of the variable could be anything (e.g., value, or x, or somePeculiarVariableName)—it does not have to be pitch. But, in this case, pitch is a good name, as it describes the variable contents.

Here the statements create a new note, using pitch as the pitch, and add it to phrase. We could do anything else we want to. A loop body may contain one or more statements.

Notice how the statements in the loop body are indented relative to the loop header. This is how Python knows that these statements belong to the loop body. In this book, we indent by three spaces. (You should avoid using tabs, since they may confuse the Python interpreter.)

To summarize, the above loop will repeatedly execute the statements in the loop body a total of five times (i.e., five repetitions)—as many as the number of items in list pitchList, namely, [C4, E4, G4, B4, C5]. At every repetition, the variable pitch is assigned to the next item in the list, starting with the first, C4. In other words, as the loop executes, the variable pitch will take the values C4, E4, G4, B4, and C5, in sequence. The loop keeps repeating until it has cycled through all the values in pitchList. When there are no more values, the loop terminates.

The final line, in the original code above, plays the constructed phrase— nothing new here either:

```
Play.midi(phrase)
```

Notice, however, that this statement is not indented. This tells Python that this statement is not in the loop. Therefore, it is executed once, after the loop terminates.

To summarize, the body of the loop is specified through indentation under the loop header. Once the loop is finished, execution continues as usual with the next statement. Indentation is important in Python. Keep this in mind; we will see it again when we explore functions and classes later in the book.

For loops iterate over Python sequences. So far we have seen two types of sequences, Python lists (e.g., [1, 2, 3, 4]) and strings (e.g., "This is a string"). We can use for and while loops to iterate through the items in a list or the characters in a string.

5.2.2 Exercises

1. What would happen if Play.midi(phrase) was indented by 3 spaces? Try it. Why is this happening? Recall that phrase is being built, one note at a time, as the loop iterates.

2. What would happen if Play.midi(phrase) was indented by 2 spaces? Why is this happening? This demonstrates how important indentation is to Python. Be careful.

5.3 CASE STUDY: ARPEGGIATORS

Arpeggiation has been used for chordal accompaniment in many Western music styles and has been a popular technique in electronic dance music for many decades. For instance, classical (and jazz) guitarists spend years practicing intricate arpeggio patterns on their instruments.

An arpeggiator is a digital instrument which creates a phrase from a set of notes (a chord) by repeatedly playing them in a series one after another. Arpeggios are often built from the notes of a chord, such as C7 (i.e., C, E, G, and B flat). Traditionally, arpeggios are repeated many times to provide rhythmic and harmonic interest in the music. There are many common arpeggio patterns; notes may be played in ascending pitch order, in descending order, in both ascending and descending order, or in some mixed order that makes sense musically and rhythmically.

Below we demonstrate a few possibilities for building arpeggiators with Python—from simple to more interesting. These programs demonstrate the power of the Python `for` loop.

5.3.1 Arpeggiator #1—Using Absolute Pitches

The first program plays an arpeggio pattern using absolute pitches. In other words, the arpeggio is fixed in a particular musical key (in this case C major).

```
# arpeggiator1.py
#
# A basic arpeggiator using absolute pitches.
#

from music import *

arpeggioPattern = [C4, E4, G4, C5, G4, E4]  # arpeggiate the C chord
duration = TN                               # duration for each note

repetitions = input("How many times to repeat arpeggio: ")

arpeggioPhrase = Phrase(0.0)   # phrase to store the arpeggio

# create arpeggiated sequence of notes
for pitch in arpeggioPattern:
    n = Note(pitch, duration)   # create note with next pitch
    arpeggioPhrase.addNote(n)   # and add it to phrase

# now, the arpeggiation pattern has been created.

# repeat it as many times requested
Mod.repeat(arpeggioPhrase, repetitions)

# add final note to complete arpeggio
lastPitch = arpeggioPattern[0] # use first pitch as last pitch
```

```
n = Note(lastPitch, duration * 2)      # using longer duration
arpeggioPhrase.addNote(n               # add it

Play.midi(arpeggioPhrase)
```

The arpeggio pattern is specified as a list of absolute pitches, stored in variable arpeggioPattern. To change the pattern (e.g., the chord or the key), we have to edit the program and change the values in this list. Try it.

Notice how the above arpeggiator asks the user to specify how many times to repeat the pattern.

5.3.2 Constants

The above example demonstrates the use of constants. For example,

```
arpeggioPattern = [C4, E4, G4, C5, G4, E4]  # arpeggiate the C chord
duration = TN                               # duration for each note
```

defines two variables, arpeggioPattern and duration.

These variables are set once at the beginning of the program and are **never** changed. These special variables are referred to as "constants."

Definition: A *constant* is a variable set once and used throughout the program (without changing it again).[*]

The purpose of constants is to hold a value commonly used throughout the program so that, if we ever need to change it, we can do so once (at the beginning of the program), as opposed to having to search throughout the program. By using this stylistic convention, we know that the behavior of the program will adjust automatically without us having to make any more changes. For example, these variables are used in the for loop (see boldface):

```
for pitch in arpeggioPattern:
    n = Note(pitch, duration)     # create note with next pitch
    arpeggioPhrase.addNote(n)     # and add it to phrase
```

to automatically build a phrase containing the appropriate notes. If we change them at the beginning of the program, this loop will adjust

---

[*] Some programming languages provide ways to "lock" variables from being changed again, i.e., turning variables into real constants. Python is not one of them. It gives you the freedom to change variables whenever you like. In Python, constants are a stylistic choice.

its behavior accordingly. That's the beauty (and convenience) of using constants.

Another example of a constant in the above program is repetitions:

```
repetitions = input("How many times to repeat arpeggio: ")
```

This variable is set once through user input and then used to adjust how many times we repeat the arpeggiated phrase:

```
# repeat it as many times as requested
Mod.repeat(arpeggioPhrase, repetitions)
```

Good Style: Use constants to store useful values at the beginning of your program. Use these throughout your program to automatically control its behavior.

Finally, notice how the two constants `arpeggioPattern` and `duration` are used at the end of the program above to create a closing note that finalizes the arpeggiation pattern. In order to create a consonant ending, this last note is the same pitch as the first note.

Since we want our program to automatically adapt if we change the values stored in arpeggioPattern, we need a way to access the first item of that list for the last note. This is done with the Python list index operation (see below). The expression

```
arpeggioPattern[0]
```

returns the first value in the `arpeggioPattern` list. Items in lists are numbered starting from 0. So the first item is at position (index) 0, the second is at position 1, and so on. Accordingly, the following code

```
lastPitch = arpeggioPattern[0]    # use first pitch as last pitch
n = Note(lastPitch, duration * 2)  # using longer duration
arpeggioPhrase.addNote(n)          # add it
```

uses the first pitch in `arpeggioPattern` list, i.e., `arpeggioPattern[0]`, to create a last note to be added to the arpeggio phrase. Also notice how we extend the duration of the last note (multiplying by 2) to give it a more final, closing feel.

The point of this section is that we can adjust the above arpeggiator code to use whatever arpeggio pattern (pitches) we wish by simply changing the two constants, `arpeggioPattern` and `duration`, at the top of the program. You should always try to write code that works this way (i.e., using constants) because it produces programs that are easy to understand, maintain, and modify.

5.3.2.1 Exercise

Modify the above to play a different arpeggio pattern and use a different rhythm value for each arpeggio note. (*Hint:* Just change the two constants, arpeggioPattern, and duration. Notice how the rest of the program automatically adjusts to generate the provided arpeggio.)

5.3.3 Interactive Processes

Using interaction (i.e., asking the user to provide information or allowing the user to modify parameters while the program is running) opens the door for the programming of user-controllable musical instruments. There is a big difference between a program that plays the same music every time you run it (like a music box) and a program that has the capacity to adjust its behavior through interaction with the user. Live interaction with a human (i.e., at runtime) extends the capacity of programs to include the intelligence and choices of a human (music performer) in real time. This forms the basis for creating flexible and intricate software instruments.[*]

Later in the book, we will see how to design a wide variety of interactive musical instruments, using our graphical user interface library. User interfaces provide the basis for enabling easy interaction with the computer music instruments you build.

5.3.4 Arpeggiator #2—Using Relative Pitches

Next, let's make the arpeggiator a little more flexible. Here we specify an arpeggio pattern using relative pitches. These pitches are added to a root note (e.g., C4) provided through user input. As we mentioned above, user input makes our program more flexible.

```
# arpeggiator2.py
#
# A basic arpeggiator using relative pitches.
#
```

---

[*] Also, by allowing users to provide input at runtime, we allow them to customize the behavior of a program without having to know how to program in Python. Input is similar to adjusting a control on a guitar effect pedal — users do not need to see the inside of the box.

```
from music import *

arpeggioPattern = [0, 4, 7, 12, 7, 4] # arpeggiate a major chord
duration = TN                          # duration for each note

rootPitch    = input("Enter root note (e.g., C4): ")
repetitions  = input("How many times to repeat arpeggio: ")

arpeggioPhrase = Phrase(0.0)           # phrase to store the arpeggio

# create arpeggiated sequence of notes
for interval in arpeggioPattern:
    pitch = rootPitch + interval       # calculate absolute pitch
    n = Note(pitch, duration)          # create note with next pitch
    arpeggioPhrase.addNote(n)          # and add it to phrase

# now, the arpeggiation pattern has been created.

# repeat it as many times requested
Mod.repeat(arpeggioPhrase, repetitions)

# add final note to complete arpeggio
lastPitch = rootPitch + arpeggioPattern[0] # close with first pitch
n = Note(lastPitch, duration * 4)          # but with longer duration
arpeggioPhrase.addNote(n                    # add it

Play.midi(arpeggioPhrase)
```

In this variation of an arpeggiator program, the arpeggio pattern is specified as a list of relative pitches (or intervals) in arpeggioPattern.

To get the corresponding absolute pitch, we simply need to add a root pitch
(see variable rootPitch) to these intervals (changes in bold):

```
for interval in arpeggioPattern:
    pitch = rootPitch + interval       # calculate absolute pitch
    n = Note(pitch, duration)          # create note with next pitch
    arpeggioPhrase.addNote(n)          # and add it to phrase
```

Variable rootPitch is set by the user:

```
rootPitch = input("Enter root note (e.g., C4): ")
```

As a result, this program can create arpeggios for different major chords
(or keys).

These lines ask the user to specify, via text input, the starting pitch and
number of repetitions for the arpeggio.

```
rootPitch = input("Enter root note (e.g., C4): ")
repetitions = input("How many times to repeat arpeggio: ")
```

Notice how the input prompt guides the user to enter a MIDI constant, such as C4.

5.3.4.1 Exercises

1. Try running this program with different roots, for example, C4, D4, and E4. Notice how versatile it is becoming, by allowing parameters to be controlled through user input.

2. Modify this program to play a different arpeggio pattern.

3. Allow the user to input the arpeggio pattern. (*Hint:* Function input() accepts any valid Python expression as input from the user. In this case, we would want the user to enter a list of intervals relative to the root (see value of arpeggioPattern above). Make sure you provide a good prompt for the user to know what is expected of him/her.)

4. Write a program called "majorScale.py", which plays the C major scale. (*Hint:* Modify the above program to create an empty phrase, and add notes to it created from the pitches in the `for` loop. If you can print out each pitch individually, you can use it to create a note, which you can then add to the phrase. After the loop, use Play. midi() to play the phrase. Remove (or comment out) all the print statements.)

5. Now write a program called "minorScale.py", which plays the A natural minor scale starting at A4, going up to A5, and returning to A4.

5.4 PYTHON LIST OPERATIONS

Python lists are created with items in square brackets separated by commas. For example,

```
>>> z = [10, 20, 30, 40, 50]
>>> z
[10, 20, 30, 40, 50]
```

List items may be numbers, strings, or any other data type supported by Python (e.g., notes, phrases, parts, etc.).

Fact: Python lists are heterogeneous, that is, they can hold values of different data types.[*]

To create a copy of a list, use the built-in function `list()`:

```
>>> w = list(z)
>>> w
[10, 20, 30, 40, 50]
```

Copies are useful if you plan to make changes but also wish to keep the original.[†]

5.4.1 Accessing List Items

Once a list has been created, we can access individual items through their index. A list item's index corresponds to the item's position in the list. The first item is stored at index 0, the second item at index 1, and so on. For example,

```
>>> z = [10, 20, 30, 40, 50]
>>> z[0]
10
>>> z[1]
20
>>> z[2]
30
>>> z[4]
50
```

Fact: List indices range from 0 to n − 1, where n is the length of the list.

If you use an index that's too large, you get an error:

```
>>> z[5]
Traceback (most recent call last):
  File "<stdin>", line 1, in <module>
IndexError: index out of range: 5
```

---

[*] Python is very flexible compared to most programming languages (e.g., Java and C/C++). The latter require that all items in a list are of the same data type. Such lists are called homogeneous.
[†] This is the reason we made copies of phrases earlier, in the "Row Your Boat" case study.

Fact: A common programming error is to use the length of the list as the index for the last item. Since list indices start at 0, the last item is at index length − 1.

Unlike other languages, Python allows use of negative indices. These access items from the right end of the list.[*] For example,

```
>>> z[-1]
50
>>> z[-2]
40
```

The index can also be a variable or expression, as long as it evaluates to an integer in the right range. For example,

```
>>> x = 1
>>> z[x]
20
>>> z[x+1]
30
```

Python also allows us to access a subset of a list, called a list *slice*. This is done using the format [start:stop], where start is the index of the first item in the sublist, and stop is the index *after* the last item. In other words, the first index (start) is included, and the second index (stop) is not included. For example,

```
>>> z
[10, 20, 30, 40, 50]
>>> z[1:3]
[20, 30]
>>> z[0:3]
[10, 20, 30]
```

5.4.2 Modifying List Items

List items can be modified by using their name as a variable. For example,

```
>>> z[0] = 1
>>> z
[1, 20, 30, 40, 50]
```

---

[*] Python allows both left and right indexing of a list. Left indexing starts at 0. Since −0 is 0, right indexing has to start at −1. Either way, if you go too far, you get an out-index error.

5.4.3 List Functions

Python provides a rich set of functions for lists. Here are the most common. Assuming that `someList` is a list:

Function	Description
`len(someList)`	Returns the number of items in the list.
`sum(someList)`	Returns the summation of items in the list (only for numeric lists).
`min(someList)`	Returns the smallest item in the list.
`max(someList)`	Returns the largest item in the list.
`list(someList)`	Returns a copy of the list (useful if you plan to modify the list, but wish to also keep the original).
`someList.append(x)`	Appends item x to the end of the list.
`someList.insert(index, x)`	Insert item x in the list at position index (e.g., `insert(0, x)` inserts at beginning).
`someList.extend(otherList)`	Concatenate `otherList` to the end of `someList`.
`del someList(index)`	Delete (remove) item at position `index`.
`someList.pop(index)`	Same as above, but also returns the item (to be used in an expression).
`someList.remove(x)`	Removes the first (leftmost) occurrence of item x from `someList`.
`someList.index(x)`	Returns the index of the first (leftmost) occurrence of item x.
`someList.count(x)`	Returns how many times x occurs in `someList`.
`someList.sort()`	Sorts the list in increasing order.
`someList.reverse()`	Reverses the order of items in `someList`.
`x in someList`	Checks if x appears in `someList`.

Fact: Most of the above operations are "destructive," that is, they modify the list. These are append(), insert(), extend(), del(), pop(), remove(), sort(), and reverse().

If you wish to keep a copy of the original, use the list() operation.

5.4.4 Case Study: Scale Tutor

This case study shows how easy it is to build useful little tools, given what we know so far. This particular program teaches music students the notes (pitches) in particular scales.

```
# scaleTutor.py
#
# Outputs the pitches of a scale, starting at a given root.
```

```
#
# Also demonstrates the reverse look-up of MIDI constants using music
# library's MIDI_PITCHES list.

from music import *

print "This program outputs the pitches of a scale."

# get which scale they want
scale = input("Enter scale (e.g., MAJOR_SCALE): ")

# get root pitch
root = input("Enter root note (e.g., C4): ")

# output the pitches in this scale
print "The notes in this scale are", # print prefix (no newline)

# iterate through every interval in the chosen scale
for interval in scale:
    pitch = root + interval     # add interval to root
    print MIDI_PITCHES[pitch],  # print pitch name (no newline)

# after the loop
print "."   # done, so output newline
```

Notice the use of MIDI_PITCHES, a list defined by the music library, which contains names of MIDI pitches as strings. By using the pitch integer as an index into it, we get the corresponding pitch name—very useful. (A similar list for MIDI instruments is presented in the next case study.)

Also notice the use of the comma "," operator at the end of print statements to prevent output of a new line. This allows the building of more complex output lines.

This program demonstrates that we know enough to start developing useful computer-based tools. What other possibilities can you think of? Some more possibilities are illustrated in the remaining case studies.

5.4.5 Case Study: Interactive PianoRoll Generator

Imagine having access to a special printer that can print large piano rolls for player pianos. This case study explores a program for driving such a printer, that is, an interactive pianoroll generator. This program also allows us to specify which MIDI instrument to use (imagine that the player pianos are fitted with a modern synthesizer module, so they can handle that).

For now, all user input is done via the keyboard (as typed text).* As is, the program accepts only one melodic line. We leave, as an exercise, making this more flexible. As seen below, this application presents several opportunities for iteration with lists.

```
# pianoRollGenerator.py
#
# Task: An interactive pianoRoll generator.
#
# Input: User selects MIDI instrument and enters pitches
#        one at a time.
#
# Output: Program generates a pianoRoll for the entered pitches,
#         and plays the corresponding notes (for verification).
#
# Limitation: Currently, all notes have QN durations.

from music import *

# ask user to select a MIDI instrument
instrument = input("Select MIDI instrument (0-127): ")

# output the name of the selected instrument
print "You picked", MIDI_INSTRUMENTS[instrument]

howMany = input("How many notes to play (0 or more): ")

pitches = []  # to be populated via user input

for i in range(howMany):                       # loop this many times
    pitch = input("Enter note (e.g., C4): ")   # get next pitch
    pitches.append( pitch )                     # append to pitch list

# now, all pitches have been entered by user and stored in pitches

# create notes
phrase = Phrase()                          # create empty phrase
phrase.setInstrument( instrument )         # use selected instrument

for pitch in pitches:                      # for each pitch in the list
    note = Note( pitch, QN )               # create next note
    phrase.addNote( note )                 # and add it to phrase

# now, all notes have been created and stored in phrase

# generate pianoRoll and play notes
View.pianoRoll( phrase )
```

---

* In a later chapter we will study how to create graphical user interfaces.

First, notice the documentation at the top of the program listing. It describes the task performed by the program, the user input expected, the output generated, and the program limitations. Consider this an example of a completely documented program. Also, notice how, for each input prompt, we provide the user with examples. This type of documentation makes our programs more user friendly and facilitates robustness. Although it is still up to the user to enter the correct input, we have made an effort to facilitate error-free interaction and to minimize user confusion (or frustration).*

Notice how we give the user a chance to select a MIDI instrument. Then the user input (an integer from 0 to 127) becomes an index into a list provided by the music library, MIDI_INSTRUMENTS. This provides a string representation of the selected instrument. In addition to the main objective, this capability makes this program a tool for exploring available MIDI sounds (timbres).

In the first loop, we use the list function `append()` to accumulate all the pitches entered by the user, one by one. Also, the first loop introduces a new function, `range()`. This function will be discussed in the next section. For now, think of it as a way to tell the loop how many times to iterate. For example,

```
for i in range(4):
```

will iterate 4 times, whereas

```
for i in range(15):
```

will iterate 15 times.

The second loop iterates through each item in the pitches list, creates a quarter note using this pitch, and stores it in a `Phrase` object.

Finally, we generate the `pianoRoll` representation and play the musical phrase consisting of all the pitches entered by the user.

5.4.5.1 Exercises

1. Modify the above program to allow the user to also enter the rhythm value for each note. (*Hint:* Create a second list to store these values. Then, for each pitch value, also ask the user to enter the corresponding duration.)

2. Modify the program to allow the user to enter a separate instrument for each note. (*Hint:* Create a third list to store these values.

---

* In the next chapter, we will see the if statement, which allows us to write code to perform error-checking against user input. This will make our programs even more robust and user friendly.

Then create a separate phrase for each note. Store these phrases into yet another list. In the end, create a part and add all the generated phrases into it.)

3. Modify these programs to use only one loop. (*Hint:* Start with the original case study. How can you consolidate the bodies of the two loops? Notice that they loop the same number of times. Can you simplify the code by reducing unnecessary variables?*)

4. Which way do you think is better, one or several loops? Why?

5. Which way is more efficient for the computer? Why?

6. Which way is more likely to be understandable (and help in maintaining this program) by future programmers? Why?

5.4.6 The range() Function

Python has two special list functions, `range()` and `frange()`.[†] These functions generate lists of integers and floats, respectively. They are used mainly with `for` loops.

The `range()` function returns a list of integers. For example,

```
>>> range(10)
[0, 1, 2, 3, 4, 5, 6, 7, 8, 9]
>>> range(1)
[0]
>>> range(3)
[0, 1, 2]
```

The `range()` function returns a list of as many integers as specified in its argument, i.e., if the argument is 10, it will return a list of 10 integers. The first integer is 0, so the last is the argument minus 1.

The `range()` function has two additional forms:

```
>>> range(2, 10)
[2, 3, 4, 5, 6, 7, 8, 9]
```

Here the two arguments specify the starting value and ending value of the sequence, respectively. The starting value (e.g., 2) is included in the sequence, whereas the ending value (e.g., 10) is not included.

---

[*] This demonstrates that, when tasks become more interesting, there may be several ways to accomplish them. Some of them are cleaner and easier to understand or implement than others.

[†] Actually, *frange()* is not in standard Python; it is included in the music library because it is useful.

The last form of the `range()` function takes three arguments:

```
>>> range(1, 10, 2)
[1, 3, 5, 7, 9]
>>>
```

The first two arguments are, again, the starting and ending value of the sequence. The third argument is the increment (or step), i.e., how much to add to one value to get to the next in the sequence. If the third argument is omitted, it defaults to 1. You could use a negative increment, for example,

```
>>> range(10, 1, -2)
[10, 8, 6, 4, 2]
```

Here the starting value is 10 (included), the ending value (excluded) is 1, and the increment is −2.

Notice that, if you use a negative increment, as in range(10, 1, −2), the starting value, 10, has to be larger than the ending value, 1. Otherwise, range returns an empty list, [], indicating that there is no way to get from the starting value to the ending value by adding −2.

A similar thing applies to positive increments.

5.4.6.1 Exercises

1. What do you expect is the output of the following?

   ```
   >>> range(10, 1, -1)
   >>> range(1, 10, -1)
   >>> range(10, 1, 1)
   >>> range(1, 10, 1)
   ```

2. Modify the above program to allow the user to also enter the rhythm value for each note. (*Hint:* Create a second list to store these values. Then, for each pitch value, also ask the user to enter the corresponding duration.)

5.4.7 The frange() Function

The `frange()` function behaves exactly as `range()`, except it generates a list of floats, as opposed to integers. Unlike `range()` above, `frange()` is not available in standard Python — it is provided in the music library. For example,

```
>>> from music import *

>>> frange(2.0, 3.0, 0.1)
[2.0, 2.1, 2.2, 2.3, 2.4, 2.5, 2.6, 2.7, 2.8, 2.9]
```

```
>>> frange(2.0, -3.0, -1.0)
[2.0, 1.0, 0.0, -1.0, -2.0]

>>> frange(2.0, 3.0, 0.192)
[2.0, 2.192, 2.384, 2.576, 2.768, 2.96]
```

The first two arguments are the starting and ending value of the sequence, and the third argument is the increment or step, that is, what value to add to one item in order to get the next. The number of decimal places in the step argument is used to determine the accuracy of the list's elements. This is very useful.

5.4.8 Iterating with Lists

To summarize, `for` loops work hand-in-hand with lists. *For* loops allow you to iterate over every item in a list.

For example, assuming that x = [1, 2, 3, 4],

```
for item in x:
    print item
```

generates this output:

```
1
2
3
4
```

At each iteration the `for` loop assigns the variable item to the next item in the list x. In the first iteration, item is set to 1 (first item of x); in the second iteration, item is set to 2, and so on.

Fact: We can use the range() function to specify how many times we wish a for loop to iterate.

For example, the following code

```
for i in range(10):
<some Python statements>
```

will loop 10 times. This is because `range(10)` generates the list [0, 1, 2, 3, 4, 5, 6, 7, 8, 9]. The for loop will iterate over that list, as described above. Here is another example that demonstrates this point:

```
for i in range(4):
    print i
```

will loop 4 times, and output the following:

```
0
1
2
3
```

Fact: The statement, for i in range(n), iterates n times (where n is a positive integer).

Now that we have seen range used with a for loop, here is another way to iterate over all the items of an arbitrary list:

```
for i in range(len(x)):
    print x[i]
```

Here the for loop iterates over the list generated by range(). The argument provided to range() is the length of x, i.e., 4. So range() generates the list [0, 1, 2, 3]. At each iteration, i gets the next value from that list, starting with 0. It so happens that these values work as indices to x. We use this in the body of the loop to access and print each item in x.

5.5 ITERATIVE MUSICAL PROCESSES

Since music is temporal, most musical transformations require performing an operation over and over on sequential data (e.g., notes). This iterative process is most naturally implemented by using a for loop to iterate over a list (e.g., a phrase).

Below are a few examples that implement common musical transformations, some of which are already provided as Mod functions. By showing the internals of these operations (i.e., how to design these processes), you gain necessary experience to develop your own, innovative processes for musical transformation. This allows you to cross the threshold from being a user of other people's code to being a developer of your own code—a most important moment in learning to program.

5.5.1 Case Study: Mod Retrograde

As seen in the previous chapter Mod.retrograde() reverses the notes in a phrase. Let's see how we can implement this with what we have learned so far using iteration with lists.

```
# retrograde.py
#
# Demonstrates one way to reverse the notes in a phrase.
#

from music import *

# create a phrase, add some notes to it, and save it (for comparison)
pitches = [C4, D4, E4, F4, G4, A4, B4, C5]        # the C major scale
rhythms = [WN, HN, QN, EN, SN, TN, TN/2, TN/4]    # increasing tempo

phrase = Phrase()
phrase.addNoteList( pitches, rhythms )
Write.midi(phrase, "retrogradeBefore.mid")

# now, create the retrograde phrase, and save it
pitches.reverse() # reverse, using the reverse() list operation
rhythms.reverse()

retrograde = Phrase()
retrograde.addNoteList( pitches, rhythms )
Write.midi(retrograde, "retrogradeAfter.mid")
```

In reality (i.e., if we were implementing the functionality of Mod.retrograde() precisely), we would be working with an arbitrary phrase already constructed. That is, we would not be constructing the phrase ourselves from a list of pitches and rhythms, as above. In this general case, we would need to ask the phrase to give us its notes, as follows.[*]

```
# now, a more general way...

# get the notes from the phrase
noteList = phrase.getNoteList()  # this gives us a list

pitches   = [] # create empty lists of pitches
durations = [] # ...and durations

# iterate through every note in the note list
for note in noteList:
    # for each note, get its pitch and duration value
    pitches.append( note.getPitch() )       # append this pitch
    durations.append( note.getDuration() )  # append this duration

# now, create the retrograde phrase, and save it
pitches.reverse()   # reverse, using the reverse() list operation
durations.reverse()
```

---

[*] Usually, there is more than one way to implement a particular operation. It is important to stop, think, and weigh alternatives; some them may be more efficient or more elegant. As mentioned earlier, "2 hours of design can save you 20 hours of coding." We have reached the point where this is beginning to apply, so pay particular attention to this guideline.

```
retrograde = Phrase()
retrograde.addNoteList( pitches, durations )
Write.midi(retrograde, "retrogradeAfter.mid")
```

Yet another way would be to iterate through the note list in reverse order directly, thus saving the computational effort of reversing the two lists, pitches and durations, after the loop. The idea is, since we are iterating through the note list already, why not iterate backwards, i.e., starting from the last note and moving toward the first one. The following code demonstrates this[*]:

```
# a more economical way...

# get the notes from the phrase
noteList = phrase.getNoteList() # this gives us a list

pitches  = [] # create empty lists of pitches
durations = [] #...and durations

# iterate *backwards* through every note in the note list
numNotes = len( noteList ) # get the number of notes in the list

# iterate through all the notes in reverse order
for i in range( numNotes ):

    # calculate index from the other end of the list
    reverseIndex = numNotes - i - 1

    note = noteList[reverseIndex] # get corresponding note

    pitches.append( note.getPitch() )       # append this pitch
    durations.append( note.getDuration() )  # append this duration

# now, create the retrograde phrase, and save it
retrograde = Phrase()
retrograde.addNoteList( pitches, durations )
Write.midi(retrograde, "retrogradeAfter.mid")
```

The "trick" here lies in this statement:

```
reverseIndex = numNotes - i - 1
```

---

[*] As with most of the examples in the chapter, this provides standard code patterns for doing common list manipulations. Study these examples well. They will come in handy in various unexpected ways in your music programming. Remember, music consists of sequences (i.e., lists) of events.

For example, consider a list with 4 items (i.e. `numNotes` is 4). In this case, `range(numNotes)` generates the list [0, 1, 2, 3]. The for loop will assign `i` to 0, then to 1, and so on. Given the above calculation, when `i` is 0, `reverseIndex` is 3; when `i` is 1, `reverseIndex` is 2, and so on, that is, `reverseIndex` is "moving" in the opposite direction from `i`, which is what we wanted.[*]

5.5.2 Exercises

Write code to perform the following tasks:

1. Given a phrase, create a copy of that phrase. (*Hint:* Get the notes from the first phrase, build a list of pitches and durations, and then use them to populate another phrase.) This is the same as `phrase.copy()`.

2. Given two phrases, `phrase1` and `phrase2`, append the notes of `phrase2` to `phrase1`. (*Hint:* Iterate over `phrase2` notes and `addNote()` them to `phrase1`.) This is the same as `Mod.append()`.

3. Given a phrase and a factor (a real number), stretch the duration of each note in the phrase by that factor. (*Hint:* Iterate through the notes of the phrase, and for each note multiply its duration by the factor. You will need note functions `getDuration()` and `setDuration()`.[†]) This is the same as `Mod.elongate()`.

4. Pick some additional Mod functions from the previous chapter and explore how you could implement them, using what you have learned so far.

Iterating over a list, as shown above, is one of the most common list operations. Using this technique, you can perform a specific operation to every item in a list. The following case study demonstrates how to use this technique to create a real musical effect.[‡]

---

[*] Don't be discouraged by the inventiveness of this calculation. It is a standard "trick" that most computer scientists learn from a book (like this one). To do music computing you are not expected to invent such techniques—just to understand them and use them.

[†] Also consider the function `Note.setDurationAndLength(double newDuration)` that changes both the notated duration and the performed length of the note in one step. This saves having to do two functions, one to set duration and the other to set length.

[‡] Soon we will learn how to "package" these operations inside functions. This way, other programmers will be able use your operations the same way you use Mod functions, such as `Mod.repeat()`, that is, by calling the function (without having to type all this extra code).

5.5.3 Case Study: Guitar Effect, FX-35 Octoplus

The following code shows how to implement a more advanced version of the classic guitar effect box, DOD FX-35 Octoplus. The FX-35 Octoplus generates a note one octave below the original note, allowing guitarists to add "body" to their sound or play bass lines on regular guitars. Here we generate one octave below plus a fifth below, creating a "power chord" effect from a single melodic line.

This example demonstrates how list operations allow us to transform musical data.

```
# octoplus.py
#
# A music effect based loosely on the DOD FX-35 guitar pedal.
# It takes a music phrase and generates another phrase containing
# the original notes plus an octave lower, and a fifth lower.

from music import *

# program constants
instrument = STEEL_GUITAR
tempo = 110

# test melody - riff from Deep Purple's "Smoke on the Water"
pitches = [G2, AS2, C3,  G2, AS2, CS3, C3, G2, AS2, C3,  AS2, G2]
durs    = [QN, QN,  DQN, QN, QN,  EN,  HN, QN, QN,  DQN, QN,  DHN+EN]

#################
# create original melody
originalPhrase = Phrase()

# set parameters
originalPhrase.setInstrument( instrument )
originalPhrase.setTempo( tempo )

originalPhrase.addNoteList(pitches, durs)

#################
# create effect melody (original + octave lower + fifth lower)
octoplusPhrase = Phrase()

# set parameters
octoplusPhrase.setInstrument( instrument )
octoplusPhrase.setTempo( tempo )

# for every note in original, create effect notes
for note in originalPhrase.getNoteList():

   pitch = note.getPitch()          # get this note's pitch
   duration = note.getDuration()    # and duration
```

```
# build list of effect pitches, for given note
chordPitches = []                    # create list to store pitches
chordPitches.append(pitch)       # add original pitch
chordPitches.append(pitch - 12)# add octave below
chordPitches.append(pitch - 5) # add fifth below
# now, list of concurrent pitches if ready, so...


# add effect pitches (a chord) and associated duration to phrase
octoplusPhrase.addChord( chordPitches, duration )

# now, we have looped through all pitches, and effect phrase is built


##################
# save both versions (for comparison purposes)
Write.midi(originalPhrase, "octoplusOriginal.mid")
Write.midi(octoplusPhrase, "octoplusEffect.mid")
```

Notice the use of constants at the beginning of the program. Since both phrases (original and octoplus) need to set the same instrument and tempo, by using constants we ensure that, if any changes are made, they are always consistent.

Also notice the use of a list in the body of the loop, to hold the three effect pitches created based on the original pitch. The loop repeats this operation for every note in the original phrase.

Finally, notice the use of Phrase and Note functions from Appendix B.

5.5.4 Exercises

1. Using list operations, create a more elaborate arpeggiator. It should take a basic arpeggio pattern (see two arpeggiator examples earlier) and create more intricate arpeggios. For instance, it plays the arpeggio in both ascending and descending order, or in some mixed order that makes sense musically and rhythmically. Explore the possibilities.

2. More advanced: Using the FX 35 Octoplus guitar effect as a model, implement a voice harmonizer. This effect should take a single melodic line and add one voice a 3rd apart following the major scale. Explore adding additional harmonic voices—what intervals work best? What intervals create more "exotic" sounds?

3. Apply the functions seen earlier in this book to make the arpeggiator program play the arpeggio motif in different keys, following a chord progression. For example, try to make it repeat the motif 8 times in C then 4 times in F, 4 times in G, and another 8 in C. To do this you

will need to make copies of phrases, transpose them as required, set their start times to the right locations, and add all phrases to a part for writing and playing.

5.6 DNA MUSIC

Hofstadter, in his seminal work "Gödel, Escher, Bach," explores repetition and similarities between genetic material (e.g., DNA) and music. In particular, he states:

> Imagine the mRNA to be like a long piece of magnetic recording tape, and the ribosome to be like a tape recorder. As the tape passes through the playing head of the recorder, it is "read" and converted into music, or other sounds ... When a "tape" of mRNA passes through the "playing head" of a ribosome, the "notes" produced are amino acids and the pieces of music they make up are proteins (Hofstadter 1979, p. 159)

In addition to the above, Clark (2005) provides an extensive list of researchers who have explored ways to convert human proteins into music. This conversion (also known as sonification) allows people to better understand and study the complicated structures and interdependencies present in genetic material.[*]

5.6.1 Case Study: Protein Music—Human Thymidylate Synthase A

In one study, Takahashi and Miller made music from the amino acid sequence in the human thymidylate synthase A protein (Takahashi and Miller 2007). They converted protein sequences into a range of 13 notes. In particular, they grouped similar amino acids together and then used a unique chord to represent each group (e.g., a D minor chord). To identify differences between similar amino acid groups, they used chord variations (i.e., chord inversions).[†] Also, they used additional genetic characteristics of each amino acid (i.e., codon distribution) to generate unique rhythm values.

---

[*] The concept of sonification is explored further in a later chapter.

[†] Chord inversions are a music theory topic beyond the scope of this book. In summary, by moving each chord note, in turn, to a higher (or lower) register, you create chord variations (i.e., inversions), which maintain the same overall chord quality, but sound slightly different from one another.

Takahashi and Miller's objective was to make genetic material more approachable to "the general public, young children, and vision-impaired scientists" (ibid, p. 405). Also, they considered the possibility that the theme provided by each protein could be used as inspiration for improvisation and elaboration, "which would allow the investigator/author to contribute an artistic component to the original melody" (ibid).

The program below explores this possibility to elaborate on the musical theme generated from the human thymidylate synthase A protein, using the musical encoding of Takahashi and Miller. Here we explore how repetition can represent the gradual unfolding and growth of genetic material in life.

This program makes use of Python list slices, which allow it to capture subsets (i.e., slices) of the protein sequence encoding and incrementally build a larger musical phrase. Given the limited space, here we concentrate only on the first 13 amino acids (i.e., chords) of the protein.

```
# proteinMusic.py
#
# Demonstrates how to utilize list operations to build an unfolding
# piece of music based on the first 13 amino acids in the human
# thymidylate synthase A (ThyA) protein.
#
# The piece starts with the 1st amino acid, continues with the 1st
# and 2nd amino acids, then with the 1st, 2nd, and 3rd amino acids,
# and so on, until all 13 amino acids have been included.
#
# See: Takahashi, R. and Miller, J.H. (2007), "Conversion of
# Amino-Acid Sequence in Proteins to Classical Music: Search for
# Auditory Patterns", Genome Biology, 8(5), p. 405.

from music import *

# set of pitches/rhythms to use for building incremental piece
pitches  = [[D3, F3, A3], [E3, G3, B3], [B3, D4, F4], [D4, F4, B4],
            [D4, F4, A4], [G4, B4, E5], [G4, B4, D5], [A4, C4, E4],
            [B3, G3, E3], [A4, C5, E5], [A4, C5, E5],
            [E3, G3, B3], [A3, C4, E4]]
durations = [HN,          QN,          HN,          QN,
             HN,          EN,          WN,          WN,
             EN,          QN,          QN,
             QN,          QN]

# we will store each incremental portion in a separate phrase
phraseList = []   # holds incremental phrases
```

```
# iterate through every index of the pitch/duration set
for i in range( len(pitches) ):

   # get next incremental slice of pitches/durations
   growingPitches = pitches[0:i+1]
   growingDurations = durations[0:i+1]

   # build next incremental phrase (no start time - sequential play)
   phrase = Phrase()               # create empty phrase
   phrase.addNoteList( growingPitches, growingDurations )
   silenceGap = Note(REST, HN)   # add separator at end of phrase...
   phrase.addNote( silenceGap )  # ...to distinguish from next phrases

   # remember this phrase
   phraseList.append( phrase )

# now, phraseList contains incremental phrases from pitches/durations

# add incremental phrases to a part
part = Part()
for phrase in phraseList:
  part.addPhrase( phrase )
# now, all phrases have been added to the part

# set the tempo
part.setTempo(220)

# view part and play
View.sketch( part )
Play.midi( part )
```

As mentioned above, the program builds the piece incrementally by adding one new amino acid every iteration. It first creates a phrase consisting of the chord associated with the first amino acid, followed by some silence. Next it creates a phrase with the chords associated with the first and second amino acids, plus some silence. Then comes a phrase with the chords for first, second, and third amino acids, and so on. These phrases are stored in a list. After the list is completed, we use it in the second loop to populate a Part object. The part is then displayed and performed on the computer's MIDI synthesizer.

Notice the use of list slices in the first loop, i.e.,

```
# iterate through every index of the pitch/rhythm set
for i in range(len(pitches)):
```

```
# get next incremental slice of pitches/rhythms
    growingPitches = pitches[0:i+1]
    growingRhythms = durations[0:i+1]
```

The slice [0:i+1] refers to the sublist starting from index 0 to the index i+1. For instance,

- In the first iteration i is 0, so the list slices are

  - pitches[0:1], which evaluates to [[D3, F3, A3]]

  - durations[0:1], which evaluates to [HN].

- In the second iteration i is 1, so the list slices are

  - pitches[0:2], which evaluates to [[D3, F3, A3], [E3, G3, B3]]

  - durations[0:2], which evaluates to [HN, QN].

- In the third iteration i is 2, so the list slices are

  - pitches[0:3], which evaluates to [[D3, F3, A3], [E3, G3, B3], [B3, D4, F4]]

  - durations[0:3], which evaluates to [HN, QN, HN], and so on.

Finally, notice how the piece evolves, both harmonically and rhythmically, sometimes maintaining a regular beat, other times changing to an off-beat pattern, and at the very end reaching a subtle, yet closing point. After you listen to it a few times, you may realize that this could be included verbatim (or with slight modifications) into a larger musical work, for example, a jazz piece or an avant garde composition.

5.6.1.1 Exercises

1. Add more pitches and durations to the above program from the Takahashi and Miller study (see Takahashi and Miller 2007, Figure 2). Explore adding a melody to this music. Where would you get your inspiration?

2. Modify the above program to render other sonifications of proteins, for example, see Clark 2005.

3. Extend this program by adding Mod functions to transform the musical material in various ways to create a larger, more complete piece (e.g., see the Theme and Variations case study in Chapter 4).

5.7 SUMMARY

This chapter addressed modeling of iterative musical (and other) processes using the Python for loop. *For* loops are very convenient for iterating over sequences. One possibility is to iterate over sequences of musical material. This is useful for creating arbitrary musical effects and transformations that need to be applied on every element of the musical material. We also learned about the *list* data type and its various operations in more detail. Through the various examples presented, we saw how versatile Python lists are for storing and manipulating sequences of data as is often required for musical material. These important concepts and techniques will be used throughout the remainder of the book.

## CHAPTER 6

Randomness and Choices

Topics: Randomness and creativity, Mozart, indeterminism, serialism, Python random functions, stochastic music, Iannis Xenakis, probabilities, wind chimes, melody generator, selection, Python if statement, flipping a coin, Russian roulette, throwing dice, realistic drums, relational and logical operators, generative music.

6.1 OVERVIEW

Creative acts start with a flash of inspiration—a new idea, a new something—which gets channeled through our knowledge, experiences, and aesthetics to be transformed (hopefully) into something structured yet appealing and that is inspiring to others. In a way, creativity is perhaps the act of taking untamed, open possibilities and shaping them into an artifact (e.g., a poem, a musical piece, a research paper, a program) which is hopefully successful in creating an aesthetic experience for our audience.*

6.2 RANDOMNESS AND CREATIVITY

Computers, through human ingenuity, are provided with a source of untamed possibilities in the form of a *random number generator*. Mathematically, a random number generator is simply an algorithm (a program) written to generate a sequence of numbers which, to our eyes, appear unpredictable; in other words, we cannot work out, having seen the numbers generated so far, what numbers may appear next. Philosophically a random number generator may serve as that flash of inspiration, as an

---

* Interestingly, our scientists tell us that the most creative act of all (from our perspective), namely, "Life, the Universe and Everything" to quote Douglas Adams, started with a big bang of untamed creativity, that is, a moment when all possibilities materialized through a flow of energy and potential, which somehow produced our universe, our laws of physics, our space and time, the structures that we see around us.

element of surprise* that gives our programs the capability, if channeled well, to generate outcomes that might be perceived as creative or intriguing by others.

Many composers have experimented with randomness (chance or indeterminism) in their compositions, including Wolfgang Amadeus Mozart, John Cage, and Iannis Xenakis. Ultimately though it is the interplay of, or balance between, unpredictability and predictability that creates aesthetically pleasing artifacts (Arnheim 1971). If an artifact is too chaotic it may be difficult to comprehend and appreciate. At the other extreme, if the artifact is too regular it may be uninteresting and boring.

Randomness is a source of surprise, and it is our choices about how to channel (or sieve) that randomness that provide opportunities to balance stability and unpredictability.

In this chapter we explore the tools provided by Python to give our programs access to randomness and how to harness it. Using these tools you may produce balanced, interesting, and aesthetically pleasing musical results.

Fact: Computers, by definition, cannot be creative. Computer-generated artifacts are actually produced by programs (models of process) that are written by humans.

Ultimately, a computer serves as an extension of human ingenuity, or as a *cognitive amplifier*. For the music composer, in particular, a computer is a tool for trying out new ideas, exploring compositional spaces, performing sonic experiments, and achieving musical tasks efficiently. In the end, computer "creativity" is limited by the ingenuity, creativity, inspiration, and effort of its human programmers. Randomness is a useful technique when used creatively.

An early, but quite ingenious example of using randomness to generate music is Mozart's dice game.

6.2.1 Case Study: Mozart—"Musikalisches Würfelspiel"

In 1787, Wolfgang Amadeus Mozart wrote "Musikalisches Würfelspiel," a musical process involving randomness (by way of rolling dice) for

---

* Extreme surprise, actually, since there are no apparent patterns, no predictability whatsoever, from the human observer's perspective, in the numeric sequence generated by a good random number generator.

generating a 16-measure waltz. In this process, each measure is selected from a set of 11 precomposed chunks of music (the number of possible outcomes of throwing two dice). These selections are concatenated to generate the whole piece.

The sets of precomposed measures were constructed so that any measure from one set could connect with any measure from the next set, and so on. In one particular implementation of this musical "game," there are $11^{16} =$ 45,949,729,863,572,161 different waltzes possible (Zbikowski 2005, p. 148).

In Mozart's case, by harnessing randomness and forcing it to select among choices that flow nicely from one to another, there is actually very little left to chance. Chance is simply producing unique combinations of things that already go well together. Clearly, randomness plays "little part in the success of the music produced by such games. Instead, what was required of the [composers] ... [is] a little knowledge about how to put the game together and an understanding of the formal design of [music]" (ibid., pp. 142–143).

For the sake of space, the code below implements a simplified version of Mozart's musical game, as the comments indicate. Adding more possibilities is left as an exercise.

```
# Mozart.MusikalischesWurfelspiel.py
#
# This program generates an excerpt of Mozart's "Musikalisches
# Wurfelspiel" (aka Mozart's Dice Game). It demonstrates how
# randomness may be sieved (harnessed) to produce aesthetic results.
#
# See Schwanauer, S, and D Levitt. 1993. Appendix
# pp. 533-538.
#
# The original has 16 measures with 11 choices per measure.
# This excerpt is a simplified form. In this excerpt,
# musical material is selected from this matrix:
#
# I II III IV
# 96 6 141 30
# 32 17 158 5
# 40
#
# Columns represent alternatives for a measure. The composer throws
# dice to select an alternative (choice) from first column.
# Then, connects it with the choice from second column, and so on.

from music import *
from random import *

# musical data structure
walzerteil = Part()      # contains a four-measure motif generated
                         # randomly from the matrix above
```

```
# measure 1 - create alternatives
# choice 96
pitches96      = [[C3, E5], C5, G4]
durations96    = [EN, EN, EN]
choice96 = Phrase()
choice96.addNoteList(pitches96, durations96)


# choice 32
pitches32      = [[C3, E3, G4], C5, E5]
durations32    = [EN, EN,      EN]
choice32 = Phrase()
choice32.addNoteList(pitches32, durations32)


# choice 40
pitches40      = [[C3, E3, C5], B4, C5, E5, G4, C5]
durations40    = [SN,  SN, SN, SN, SN, SN]
choice40 = Phrase()
choice40.addNoteList(pitches40, durations40)


# measure 2 - create alternatives
# choice 6 (same as choice 32)
choice6 = Phrase()
choice6.addNoteList(pitches32, durations32)


# choice 17
pitches17      = [[E3, G3, C5], G4, C5, E5, G4, C5]
durations17    = [SN,  SN, SN, SN, SN, SN]
choice17 = Phrase()
choice17.addNoteList(pitches17, durations17)


# measure 3 - create alternatives
# choice 141
pitches141     = [[B2, G3, D5], E5, F5, D5, [G2, C5], B4]
durations141 = [SN,     SN, SN, SN, SN, SN]
choice141 = Phrase()
choice141.addNoteList(pitches141, durations141)


# choice 158
pitches158      = [[G2, B4], D5, B4, A4, G4]
durations158 = [EN, SN, SN, SN, SN]
choice158 = Phrase()
choice158.addNoteList(pitches158, durations158)


# measure 4 - create alternatives
# choice 30
pitches30 = [[C5, G4, E4, C4, C2]]
durations30 = [DQN]
choice30 = Phrase()
choice30.addNoteList(pitches30, durations30)


# choice 5
pitches5       = [[C2, C5, G4, E4, C4], [G2, B4], [C2, E4, C5]]
durations5     = [SN,           SN, QN]
choice5 = Phrase()
choice5.addNoteList(pitches5, durations5)
```

```
# roll the dice!!!
measure1 = choice([choice96, choice32, choice40])
measure2 = choice([choice6, choice17])
measure3 = choice([choice141, choice158])
measure4 = choice([choice30, choice5])

# connect the random measures into a waltz excerpt
walzerteil.addPhrase(measure1)
walzerteil.addPhrase(measure2)
walzerteil.addPhrase(measure3)
walzerteil.addPhrase(measure4)

# view and play randomly generated waltz excerpt
View.sketch(walzerteil)
Play.midi(walzerteil)
```

The first thing that implies randomness in the above code is the inclusion of the `random` library, at the top. As we will discuss in more detail below, the random library makes available various functions that provide random outcomes. The program above makes use of the `choice()` function, which randomly picks an item from a list. Each list item has an equal probability of being selected, so every time you run the program you will most likely get a different outcome. In this reduced version of Mozart's dice game, there are $3 \times 2 \times 2 \times 2 = 24$ possible outcomes. Go ahead, hear a few of them. See how Mozart weaves the different possibilities together.

As you work through this chapter it will become more clear how to leverage algorithmic processes to shift the balance of effort (and creativity) to produce vast quantities of interesting music with modest amounts of code. Shifting the burden of many choices to Python's random functions is an interesting first step toward exploring the generative possibilities available through computers. This is similar to the use of randomness in computer games, where a computerized opponent makes unpredictable, yet intelligent, choices that add variation, interest, and excitement to the game playing process.

6.2.1.1 Exercise

Add a few more possibilities. You have two options: either find a score of the original and transcribe a few more of Mozart's measures, or put together (compose) your own measures. Either way, adding just one more measure for each of the II, III, and IV columns (see program documentation above) increases the possible outcomes from 24 to 81 (i.e., $3 \times 3 \times 3 \times 3$). That is not a bad outcome for a little extra work.

6.3 INDETERMINISM AND SERIALISM

An interesting way of applying randomness in music is in a style referred to as *chance music*. The use of chance in music composition (or performance) stretches as far back as Mozart's Dice Game and became established in Western music in the 20th century, particularly in the works of John Cage and Iannis Xenakis.

Definition: *Chance music,* also known as *aleatoric music,* is a compositional technique that introduces elements of randomness (or indeterminism) into the compositional process.

Indeterminism was explored extensively in the middle of the 20th century. John Cage, among other composers, is particularly well known for his indeterministic compositions, where events and their timing were chosen by rolling dice, selecting cards from a deck, and by other activities that would vary from performance to performance.

Aleatoric techniques are sometimes compared to serialism.

Definition: *Serialism* involves using deterministic rules to control choices within the compositional process.

A common form of serialism is *atonal* music, where every pitch in the chromatic scale has to be used before any pitch can be repeated.* This is designed to create music that does not have any obvious tonal center. This method of atonal music was developed in the early 20th century by Arnold Schoenberg and employed by other composers, including Alban Berg and Anton Webern.

From their definitions, it is clear that aleatoric and serial techniques are compositionally opposite each other. One is relinquishing control to randomness. The other is applying maximum control of choices through well-defined rules. However, the music outcomes can appear to be very similar. Also, both involve the human composer for meta-level decision

---

* A "row" consists of the 12 chromatic pitches in some order devised by the composer. Rows are cycled through with a forward motion. Within a row, a pitch may be repeated several times in sequence, but cannot be repeated otherwise until all others have sounded. There can be more than one row going simultaneously in different phrases.

making (defining rules and processes) and defer compositional details to the outcomes of those processes. Because these compositional techniques include defined rules and processes, they are obvious candidates for use in computer music programs.

The similarity of aleatoric and serial compositions from the 1950s, for example, can be observed in these two pieces:

- John Cage, "Music of Changes, Book I" (1951), and

- Pierre Boulez, "Structures I for Two Pianos" (1951–2)

The following case study capitalizes on this similarity to create a program where it is impossible to determine, simply by listening to it, if the compositional approach was aleatoric or serialist.

6.3.1 Case Study: Pierre Cage—"Structures pour deux Chances"

The following Python program creates music by Pierre Cage. Pierre Cage is a fictitious composer.* The piece "Structures pour deux Chances" is generated through randomness and is inspired by the two pieces above. The program generates two parallel melodic lines (phrases) consisting of notes with random pitch, duration, and dynamics, and combines them into one Part object.

```
# PierreCage.StructuresPourDeuxChances.py
#
# This program (re)creates pieces similar to:
#
# Pierre Boulez, "Structures I for two pianos", and
# John Cage, "Music of Changes, Book I".
#
# The piece generated consists of two parallel phrases containing
# notes with random pitch and duration.
#

from music import *
from random import *     # import random number generator


numberOfNotes = 100      # how many notes in each parallel phrase


##### define the data structure
part = Part()            # create an empty part
melody1 = Phrase(0.0)    # create phrase (at beginning of piece)
melody2 = Phrase(0.0)    # create phrase (at beginning of piece)
```

---

* Pierre Cage is a remix of the names Pierre Boulez and John Cage.

```
##### create musical data
# create random notes for first melody
for i in range(numberOfNotes):
    pitch = randint(C1, C7)      # get random pitch between C1 and C6
    duration = random() * 1.0    # get random duration (0.0 to 2.0)
    dynamic = randint(PP, FFF)   # get random dynamic between P and FF
    note = Note(pitch, duration, dynamics)   # create note
    melody1.addNote(note)        # and add it to the phrase
# now, melody1 has been created

# create random notes for second melody
for i in range(numberOfNotes):
    pitch = randint(C1, C7)      # get random pitch between C1 and C6
    duration = random() * 1.0    # get random duration (0.0 to 2.0)
    dynamic = randint(PP, FFF)   # get random dynamic between P and FF
    note = Note(pitch, duration, dynamics)   # create note
    melody2.addNote(note)        # and add it to the phrase
# now, melody2 has been created

##### combine musical material
part.addPhrase(melody1)
part.addPhrase(melody2)

##### play and write part to a MIDI file
Play.midi(part)
Write.midi(part, "Pierre Cage.Structures pour deux chances.mid")
```

Notice the two *for* loops, each of which creates random notes for the first and second phrase, respectively. During each iteration, a new note is created with random pitch, random duration, and random dynamics. The random functions used, that is, random() and randint(), are presented next.

6.3.1.1 Exercises

1. Try varying the program by changing the parameter values. For example, change the pitch range from C1–C6 to C4–C5, or the duration range multiplier from 2.0 to 1.0. What musical effect do these changes have?

2. How could you modify this program to also create random chords (i.e., chords with random pitches and durations)?

3. Also, how would you modify the above program to merge the two *for* loops into one? Would this result in a program that is easier or harder to understand?

6.4 PYTHON RANDOM FUNCTIONS

The Python *random* library contains numerous functions. Of these, random(), randint(), and choice() are the most commonly used ones. They are described below.

```
random()
```

The function random() returns a random float number between 0.0 and 0.9999999... (according to the computer's accuracy in decimal digits). Since this function returns a float, it may be used in places where a float value is used, such as note durations, phrase start times, and note panning values.

It is important to remember that each value in this range (0.0 to 0.9999999 ...) has an equal probability of appearing. For example,

```
>>> from random import *
>>> random()
0.64942495860364424
>>> random()
0.3450751019588989
>>> random()
0.98337237477515838
```

When needed for float values that may have wider ranges (e.g., 4.0 to 0.0), you may use simple *multiplication* to expand the range of the values returned by random(). For example,

```
>>> random() * 4.0
2.5486445467247525
>>> random() * 4.0
0.41445155125611377
>>> random() * 4.0
1.3455937087811325
>>> random() * 4.0
0.9648149453199566
>>> random() * 4.0
2.9250401899846152
>>> random() * 4.0
1.5926868696828085
>>> random() * 4.0
3.8686865532078105
```

The values now range from 0.0 to 3.9999999 ... Why? The answer is straightforward. If random() returns 0.0, then multiplying it by 4.0 remains 0.0; so the new range starts at 0.0. If random() returns the largest possible value (i.e., 0.9999999 ...), then multiplying it by 4.0 produces 3.9999999 ...; so the new range ends there.

Adding a number to random() simply shifts the range, i.e.,

```
>>> random() + 50
50.287401757617445
>>> random() + 50
50.021080369457415
>>> random() + 50
50.245497900440455
>>> random() + 50
50.54759454925399
>>> random() + 50
50.99158613906523
```

Here, by adding 50, the random range became 50.0 to 50.9999999 ...

We can create any desired range of random float numbers through both multiplication (stretching of the range) and addition (shifting of the range).

6.4.1 Exercise

How would you create the following ranges using random(), addition, and/or multiplication?

- 1.0 to 1.9999999 ...

- −1.0 to −0.0000001 ...

- 0.0 to 8.9999999 ...

- 1.0 to 9.9999999 ...

- 25.0 to 27.9999999 ...

6.4.2 randint()

The function randint(x, y) returns a random integer between x and y (including x and y). Since this function returns an integer, it is ideal to use for generating random note pitches (among others). Each value in the range between x and y has an equal probability of appearing.

For example (recall that C4 is MIDI pitch 60, whereas C6 is 84):

```
>>> from random import *
>>> from music import * # needed to use C4 and C6
>>> randint(C4, C6)
78
```

```
>>> randint(C4, C6)
63
>>> randint(C4, C6)
83
>>> randint(C4, C6)
67
>>> randint(C4, C6)
76
>>> randint(C4, C6)
64
```

In other words, randint (C4, C6) generates random integers in the range 60 to 84 (inclusive).

6.4.3 choice()

Another useful random function is choice(). This function takes a list as its argument and returns a random item chosen from this list. Every item has an equal probability of appearing. For example,

```
>>> from random import *
>>> choice([60, 62, 64, 67, 69])
64
>>> choice([60, 62, 64, 67, 69])
60
>>> choice([60, 62, 64, 67, 69])
69
>>> choice([60, 62, 64, 67, 69])
64
>>> choice([60, 62, 64, 67, 69])
67
```

This function is used when we have a list of elements, e.g., [C4, D4, E4, G4, A4], and we wish to randomly pick one of them, for example, when picking random pitches from a scale.

6.5 STOCHASTIC MUSIC

Stochastic music is a compositional method employed by Iannis Xenakis as a reaction to the abstractness and complexity of music from the Serialist movement. Although serialists employ certain rules (e.g., the 12-tone technique requires that all 12 pitches have to be used before they can be repeated), Xenakis points out that these rules (or "causality") are not readily accessible, nor can they be easily appreciated by the listener. As a result, the listener may be aesthetically overwhelmed by the complexity of the music. Xenakis proposed that the mathematics of probability could be the basis of a more general and manageable compositional technique (Xenakis 1971).

Xenakis proposed a new idea, to create music that follows a stochos ("στόχος") or target, that is, music that evolves over time within certain statistical tendencies and has points of origin and destination.

Definition: *Stochastic music* refers to music whose various aspects are guided by probability.

Stochastic music is made with the help of mathematics and statistics (e.g., probability functions) that may include random elements. The numeric results of these formulas (or functions) are mapped to certain attributes of music, such as pitch, duration, dynamic, pan position, etc. Although the music may appear random at the local level, when seen as a whole (at the macroscopic level), it has tendencies and direction.

In particular, Xenakis' work constitutes one of the most extensive attempts to produce music from a standpoint which is not only mechanistic but expresses a technological aesthetic. Xenakis wrote his early music in the 1950s for acoustic instruments, calculating the probabilities by hand, using math to determine elements of pitch, duration, dynamic, and structure. Beginning in the 1960s, he turned to the computer to assist in both note element determination and sound synthesis. Xenakis applied his stochastic perspective to music at all levels, from the micro-structure of sound to the macro-structure of form.

6.5.1 Case Study: Iannis Xenakis—"Concret PH"

Concret PH is a very influential piece of stochastic music. It was created by Xenakis to be played inside the Philips Pavilion in the 1958 World's Fair in Brussels (Valle, et al. 2010). This building was designed by architect Le Corbusier, who employed Xenakis as an architect and mathematician at the time.

The following program creates a piece inspired by Concret PH. It uses randomness to handle start time, pitch, duration, and pan position of sound events. In the original piece, Xenakis used spliced tape of sounds made by burning charcoal. Here we mimic the sound using the MIDI instrument BREATHNOISE, which at short "bursts" (notes with short duration) sounds much like Xenakis' original sound elements. Certain constants have been defined at the top of the program to control the texture and organization of the piece. Notice how these constants are used throughout the piece to avoid using literal values (or "magic numbers," as such literal values are called by expert programmers).

Good Style: Refrain from using "magic numbers" in your code.

```
# ConcretPH_Xenakis.py
#
# A short example which generates a random cloud texture
# inspired by Iannis Xenakis's 'Concret PH' composition
#
# see http://en.wikipedia.org/wiki/Concret_PH

from music import *
from random import *


constants for controlling musical parameters
cloudWidth    = 64      # length of piece (in quarter notes)
cloudDensity = 23.44    # how dense the cloud may be
particleDuration = 0.2 # how long each sound particle may be
numParticles = int(cloudDensity * cloudWidth) # how many particles


part = Part(BREATHNOISE)

# make particles (notes) and add them to cloud (part)
for i in range(numParticles):

    # create note with random attributes
    pitch = randint(0, 127)     # pick from 0 to 127
    duration = random() * particleDuration  # 0 to particleDuration
    dynamic = randint(0, 127)   # pick from silent to loud
    panning = random()          # pick from left to right
    note = Note(pitch, duration, dynamic, panning)   # create note

    # now, place it somewhere in the cloud (time continuum)
    startTime = random() * cloudWidth # pick from 0 to end of piece
    phrase = Phrase(startTime)  # create phrase with this start time
    phrase.addNote(note)        # add the above note
    part.addPhrase(phrase)      # and add both to the part
# now, all notes have been created

# add some elegance to the end
Mod.fadeOut(part, 20)

View.show(part)
Write.midi(part, 'ConcretPh.mid')
```

Notice how we multiply `random()` by different parameters to extend (stretch) the range of random values created. For example, `duration = random() * particleDuration` generates a random value between 0 and `particleDuration`. Similarly, `startTime = random() * cloudWidth` generates a float value between 0 and `cloudWidth`.

Notice how `numParticles` is defined in terms of earlier parameters. When a piece like this is designed, some numerical experimentation may take place. Once the numerical parameters have been established and the composer (you) has a better understanding of the process being created, it then pays off to formalize (crystallize) these thoughts by selecting appropriate variable names and by making sure that these variables are used as constants throughout the program to control its behavior. This makes your program easier to maintain and understand by future programmers/composers (or even you, six months down the road).

Good Style: Pick named constants to control important aspects of your program's behavior.

As discussed in Chapter 1, Iannis Xenakis was a pioneer in using computers to create music. Stochastic music was made feasible through the use of early computers. Through the skills and techniques we have been exploring, it is now also accessible to you.

6.6 HARNESSING (OR SIEVING) RANDOMNESS

Random numbers, like any other numbers, can be mapped to musical pitches, note durations, and other kinds of musical attributes — after all, this book's underlying theme has been that music is numbers and that numbers are music. Moreover, an interesting thing happens if you do not leave everything to chance.

If you take a purely random process and sieve it (filter it) through a set of predetermined, aesthetically pleasing choices, you can end up with artifacts that exhibit organization, coordination, and (some might argue) elements of interest and beauty. This, of course, does not necessarily come close to the creative and aesthetic potential of someone, like say, J. S. Bach. However, in the hands of a creative composer (who programs), randomness can be harnessed to contribute creative and unexpected outcomes to a musical process.

Definition: A *sieve* is a mechanism applied to data which ensures that certain values are eliminated and/or mapped to particular output values.

In the rest of the chapter, we see various examples of applying sieves to randomness. Let's begin.

6.6.1 Case Study: Wind Chimes

As mentioned above, a way to generate artifacts that are aesthetically pleasing, starting with pure randomness, is to filter a random process through a sieve. For example, consider wind chimes. Wind chimes take the random displacements of air molecules in the atmosphere and map them into melodic sequences. In other words, wind chimes capture random movements of air and force them onto a narrow set of aesthetic possibilities, typically limiting sounding options to tubes that are already tuned to a particular music scale, for instance.

```
# windChimes.py
#
# Simulates a 4-tube wind chime.
#
# Demonstrates how we may sieve (harness) randomness to generate
# aesthetically pleasing musical outcomes.

from music import *
from random import *

# program parameters
cycles = 24        # how many times striker hits all four tubes
duration = 8.0     # tubes sounds last from 0 to this time units
minVol = 80        # low and high limit for random volume
maxVol = 100

# tube tuning
tube1 = C5
tube2 = F5
tube3 = G4
tube4 = D6

# wind chime part
windChimePart = Part(BELLS)

# wind chime consists of four tubes
tube1Phrase = Phrase(0.0) # first tube starts at 0.0 time
tube2Phrase = Phrase(1.0) # second tube starts at 1.0 time,...
tube3Phrase = Phrase(3.0) #... and so on.
tube4Phrase = Phrase(5.0)
```

```
# generate wind chime notes
for i in range(cycles):

    # create random tube strikes (notes)
    note1 = Note(tube1, random() * duration, randint(minVol, maxVol))
    note2 = Note(tube2, random() * duration, randint(minVol, maxVol))
    note3 = Note(tube3, random() * duration, randint(minVol, maxVol))
    note4 = Note(tube4, random() * duration, randint(minVol, maxVol))

    # accumulate notes in parallel sequences
    tube1Phrase.addNote(note1)
    tube2Phrase.addNote(note2)
    tube3Phrase.addNote(note3)
    tube4Phrase.addNote(note4)

# add note sequences to wind chime part
windChimePart.addPhrase(tube1Phrase)
windChimePart.addPhrase(tube2Phrase)
windChimePart.addPhrase(tube3Phrase)
windChimePart.addPhrase(tube4Phrase)

# view and play wind chimes
View.sketch(windChimePart)
Play.midi(windChimePart)
```

As in the previous case study, notice how we multiply by duration to stretch the range of values generated by `random()` above. Also, by using pitches that sound harmonious together (e.g., taken from a chord), we make sure that generated notes will sound good together, regardless of the order and combinations produced.

6.6.1.1 Exercises

1. Add random panning to each of the notes to provide a greater sense of sound spatialization.

2. Try varying the pitches used to create wind chimes with different moods.

6.6.2 Case Study: Pentatonic Melody Generator

In this case study, we explore how to harness randomness to create a melodic line within a particular scale.

```
# pentatonicMelody.py
# Generate a random pentatonic melody. It begins and ends
# with the root note.

from music import *     # import music library
from random import *    # import random library
```

```
pentatonicScale = [C4, D4, E4, G4, A4]  # which notes to use
durations = [QN, DEN, EN, SN]           # which durations to use

# pick a random number of notes to create (between 12 and 18)
numNotes = randint(12, 18)              # number of notes to create

phrase = Phrase()                       # create an empty phrase

# first note should be root
note = Note(C4, QN)                     # create root note
phrase.addNote(note)                    # add note to phrase

# generate enough random notes (minus starting and ending note)
for i in range(numNotes - 2):
    pitch = choice(pentatonicScale)     # select next pitch
    duration = choice(durations)        # select next duration
    dynamic = randint(80, 120)          # randomly vary the volume
    panning = random()                  # place in stereo field
    note = Note(pitch, duration, dynamic, panning) # create a note
    phrase.addNote(note)                # add note to phrase

# last note should be root also (a half note, to signify end)
note = Note(C4, HN)                     # create root note
phrase.addNote(note)                    # add note to phrase

Play.midi(phrase)                       # play the melody
```

Notice how the program generates a random number of notes, ranging between 12 and 18 (inclusive). This facilitates exploration of different melodic phrase possibilities. Also, notice how we "anchor" the melody by attaching a fixed (nonrandom) note at the beginning of the phrase (the root note) and also at the end of the phrase (again, the root). Even though the program will meander through the pentatonic scale in the middle of the phrase, we are guaranteed that the phrase will begin and end in the root note.* Accordingly, we make the for loop iterate two fewer times, since we hand-pick the beginning and ending notes. Finally, we provide random variation of note volume (between 80 and 120) and stereo-field placement (panning) anywhere from extreme left to extreme right.

6.6.3 Weighted Probabilities

Although choice() has an equal probability of returning any item in a list, we can add more weight to an item simply by making it appear more than once in the list. In other words, if we wish for some values to be randomly

---

* This trick brings some balance between unpredictability and predictability, as discussed above.

generated more frequently than others (such as the tonic or the fifth of a scale), we can add more copies of those values in the list. This is called a *weighted list.*

For example, if we wish C4 to be randomly generated 75% of the time, and, say, A4 to be randomly generated 25% of the time, we could use this weighted list:

```
[C4, C4, C4, A4]
```

This list, given as an argument to `choice()`, has a 75% (3 out of 4, or ¾, or 0.75) chance of returning C4, and only a 25% (1 out of 4, or ¼, 0.25) chance of returning A4.

This is nice trick to create music that sounds more "natural," that is, with some bias toward some values over others. This should be contrasted to completely random and unpredictable music generated by the "Structures pour deux Chances" case study above.

List repetition (using "*") and concatenation (using "+") provide efficient ways to generate weighted lists for use with `choice()`. For example, the following

```
durations = [WN] + [HN]*2 + [QN]*4 + [EN]*8 + [SN]*16
```

generates the following list:

```
>>> durations
[4.0, 2.0, 2.0, 1.0, 1.0, 1.0, 1.0, 0.5, 0.5, 0.5, 0.5, 0.5, 0.5, 0.5,
0.5, 0.25, 0.25, 0.25, 0.25, 0.25, 0.25, 0.25, 0.25, 0.25, 0.25, 0.25,
0.25, 0.25, 0.25, 0.25, 0.25]
```

Clearly, the first expression is more readable and less error-prone to create, as opposed to the second list (i.e., hard to type and verify). Either way, you get a list that will favor SN (i.e., 0.25) notes 16 out of 31 times (i.e., 51%), EN notes 8 out of 31 times (i.e., 26%), QN notes 4 out of 31 times (i.e., 13%), HN notes 2 out of 31 times (i.e., 7%), and WN notes 1 out of 31 times (i.e., 3%).

6.7 SELECTION

Selection statements allow us to specify that different circumstances execute different code. In other words, selection allows a computer to adapt to particular situations. Of course, these situations (conditions) and their outcomes have to be specified by the programmer. We are, as programmers, in the business of modeling process. Through selection, we may model processes like the ones we find in nature where entities are continuously adjusting their behavior to a changing environment.

In our programs, these changing conditions are specified in terms of information from the outside world (e.g., through randomness, reading external data from a file, input from a MIDI instrument, etc.).* Then, as data is flowing in, the program can "decide" what to do based on the alternatives specified in a conditional statement, such as the if statement.

6.7.1 Python if Statement

In general, if statements are useful for any situation where we want the computer to decide what to do next, based on what has already happened. Keep in mind, initial conditions may change every time we run a program. For example, if a program uses external sensors, the outside temperature may have changed, or today may be a rainy day (as opposed to yesterday, which was a sunny day). Similarly, if a program works with user input, the user may enter something different—even something erroneous.† Finally, if our program utilizes randomness, we may wish to react differently given different randomly generated values. For all these situations, if statements provide a mechanism to decide what behavior the program should exhibit under different conditions. Let's see how if statements work.

If statements ask a question. For example, the following statement:

```
if number > 1:
   note = Note(C4, HN)
else:
   note = Note(G4, HN)
```

asks if the value stored in variable number is larger than 1. If indeed number contains such a value, then the computer executes the indented statement below the if:

```
if number > 1:
   note = Note(C4, HN)
else:
   note = Note(G4, HN)
```

Otherwise, if number does NOT contain a value larger than 1, then the computer executes the indented statement below the else, i.e.,

---

* Reading data from a file and getting input from MIDI instruments will be explored later. For now, we want to learn how to construct selection statements using the Python if statement.

† If statements may be used to implement code that checks against user error (i.e., did the user provide input in the proper numerical range).

```
if number > 1:
   note = Note(C4, HN)
else:
   note = Note(G4, HN)
```

This is key. Look at the code above one more time. There is much hiding in it. Although the statement appears sequential (i.e., there are four lines placed in sequence), the statement actually does not execute sequentially. The two indented statements (under the if part and under the else parts) will *never* be executed together. It will always be one or the other.

To summarize, if number contains a value larger than 1, we create a C4 note. Otherwise, we create a G4 note.

Actually, in if statements, the else part is optional. If no else part is present, then the indented statement(s) under the if part are executed conditionally (i.e., if the answer to the question is yes or true). Otherwise, the indented statement(s) are skipped. It's as simple as that.

An if statement defines a fork in the road (a crossroads in the flow of execution), where the computer will decide which statements to execute based simply on the answer to the question we are posing, for example, is number > 1?

Just as with for loops, indentation is important in if statements. The body of the if and else parts are identified through indentation. In this book, we use three spaces for indentation. Do not use tabs as they may confuse the Python interpreter.

6.7.1.1 Many Cases
If statements can have more than one alternative, as follows:

```
if <some condition>:
   ...
elif <some other condition>:
   ...
elif <yet another condition>:
   ...
   ...
else:
   ...
```

There is no limit as to the number of elif alternatives you can have. As soon as one condition evaluates to true, the corresponding indented statements will be executed. Also, all remaining conditions will be skipped. For example,

```
if x = = 1:
   note = Note(C4, QN) # C4 note
```

```
elif x = = 2:
   note = Note(D4, QN) # D4 note
elif x = = 3:
   note = Note(E4, QN) # E4 note
else:             # otherwise (i.e., tails)
   note = Note(G4, HN) # G4 note
```

If variable x contains 1, then the above creates a C4 note. If variable x contains 2, it creates a D4 note. If x is 3, it creates a E4 note. For all other values of x (i.e., the else part), it creates a G4 note.

Multi-case if statements allow us to model mutually exclusive outcomes in our code.

6.7.2 Case Study: Flipping a Coin

If statements allow us to harness randomness. For example, the following program mimics the flipping of a coin (i.e., a 50–50 chance).

```
from music import *
from random import *

heads = random() < 0.5              # a 50-50 chance to be True

if heads:                           # if we got heads,
   note = Note(C4, HN)                # create a C4 note
else:                               # otherwise (i.e., tails)
   note = Note(G4, HN)                # create a G4 note

Play.midi(note)                     # and play it!
```

Notice how this program takes a random value between 0.0 and 1.0 and sieves it into two choices:

1. either less than 0.5, or

2. larger than (or equal to) 0.5.

There is a 50–50 chance that each of the two outcomes will happen. Why? The answer depends on the fact that random() is uniformly distributed. This means that the number it returns may be anywhere between 0.0 and 1.0 with equal probability (no particular region is more preferred than another).

As a result, over time, about 50% of the values will be less than 0.5, and 50% of the values will be larger than (or equal) to 0.5. That's the same as flipping a coin. So, 50% of the time this program will play a C4 note, and the other 50% of the time it will play a G4 note. It is possible that we would get a sequence of more and more, say, C4 notes, but that would become increasingly improbable — similar to throwing a coin.

Notice how the `Play.midi()` statement is not indented. It is *always* executed after the selection statement. If `Play.midi()` were indented, it would be part of the `else` part. As a result, only the G4 note would be played, when it was selected via the "coin toss," but not the C4 note. Why?

Clearly, this program is very limited musically. However, it demonstrates the power of `if` statements and selection. They allow us to control the flow of execution and thus give us an additional way to shape the process modeled by our programs.

6.7.3 Case Study: Russian Roulette

Russian roulette involves distributing randomness between six equally probable outcomes (i.e., stemming from revolvers with 6 chambers). This translates to a 1-in-6 (17%) probability for each outcome. Besides its unpalatable origins, the idea is very useful when applied to music generation.

For example, consider the following program (a slightly modified version from above), which presents a 1-in-6 (17%) chance to play a C4, and a 5-in-6 (83%) chance to play a G4.

```
from music import *
from random import *

oneInSixChance = random() < 0.17      # a 17% chance to be True

if oneInSixChance:                    # 17% of the time,
    note = Note(C4, HN)               #   we create a C4 note
else:                                 # otherwise (83% of the time)
    note = Note(G4, HN)               #   we create a G4 note

Play.midi(note)                       # and play it!
```

Notice how this program again receives a random value between 0.0 and 1.0 and sieves it into two choices with biased probabilities (17% and 83%, respectively).

6.7.3.1 Exercise

Change the above program to implement a different probability, for example, 33–67%. How would you do that?

What if you wanted to have more choices? The following program demonstrates one way of doing it. We will see a more general way later in the chapter.

6.7.4 Case Study: Throwing Dice

In this case study, we generalize the distribution of randomness to several choices. This can be done in various ways. Here we use the randint() function together with the general form of the if statement. This is useful when we are faced with many alternatives (i.e., what to do under different cases).

```python
# throwingDice.py
#
# Demonstrates the distribution of randomness to several choices.

from music import *
from random import *

numNotes = 14          # how many random notes to play

phrase = Phrase()      # create an empty phrase

for i in range(numNotes):

    dice = randint(1, 6) # throw dice (1 and 6 inclusive)

    # determine which dice face came up
    if dice == 1:
      note = Note(C4, QN)    # C4 note
    elif dice == 2:
      note = Note(D4, QN)    # D4 note
    elif dice == 3:
      note = Note(E4, QN)    # E4 note
    elif dice == 4:
      note = Note(F4, QN)    # F4 note
    elif dice == 5:
      note = Note(G4, QN)    # G4 note
    elif dice == 6:
      note = Note(A4, QN)    # A4 note
    else:
      print "Something unexpected happened... dice =", dice

    phrase.addNote(note)   # add this random note to phrase

# now, all random notes have been created

# so, play them
Play.midi(phrase)
```

Notice how the alternatives we handle above are dictated by the randint() function. Actually, we use the same approach to handle input notes from a MIDI instrument or to handle data from other user input.

Using a multicase `if` statement allows us to handle complex situations, by specifying particular behavior (Python code) to handle them. This is very useful in musical and other tasks.

Also, notice how we included a final `else` statement to handle an unexpected situation. Given the above program, the `else` statement would never be executed (since the result of `randint()` is guaranteed to be between 1 and 6). However, if the program was ever modified, we will be happy to have this catch-all `else` statement.*

In the case of user input (i.e., using the `input()` function instead of `randint()`), the else part provides a way to handle erroneous input. Just because we give user instructions through the prompt, it does not mean that users will not make mistakes. In such cases, we want our programs to behave graciously/robustly, and issue useful error messages, to help users cope with and hopefully recover from the errors. This kind of error handling is a good software development strategy.

6.7.4.1 Nesting "If" Statements

The above program demonstrates that it is possible to put an `if` statement inside a loop. Actually, both `for` loops and `if` statements allow any Python statement to be included in their body, as long as it is properly indented. This may include other `for` loops or `if` statements. The combinations are endless, so they need to be guided by the type of logic (process) you are trying to model. One has to be careful when nesting statements, as it is easy to make mistakes. Seasoned programmers will draw the options on paper first, before implementing them in Python, to make sure that things make sense and thus to avoid errors. Also, you should pay special attention to indentation.

6.7.4.2 Exercise

Change this program to get a choice from the user. Provide a user-friendly prompt (i.e., tell them what choices are legal). Also, update the error message to something more useful. This will help you get into the habit of providing user-friendly messages to make your programs more usable and robust.

---

* For instance, someone could change the parameters to `randint()`, for example, `randint(1,8)`, and forget to add more cases in the `if` statement.

6.7.5 Case Study: Let the Drums Come Alive

Earlier in the book, in Chapter 3, we saw how to introduce drums and percussion sounds into our programs. As you may have noticed, these drum patterns are very mechanical and repetitious—a problem also prevalent with patterns generated by drum machines. What sets human drummers apart is that they add variation to what they play.

Here we add some random elements to drum machine patterns and generate drum parts that have a "human" element to them, that is, they sound a little less predictable, less mechanical, and hopefully more "alive."

In the following program, every now and then (randomly, 35% of the time) we interject an open hi-hat sound to the sequence of closed hi-hat sounds. It also randomly varies the loudness (dynamic level) of the notes. Clearly, these are just some of many possibilities.

```
# drumsComeAlive.py
#
# Demonstrates how to uses randomness to make a drum pattern come
# "alive", i.e., to sound more natural, more human-like.
# In this example, every now and then (randomly, 35% of the time),
# we play an open hi-hat sound (as opposed to a closed one).
#

from music import *
from random import *


##### musical parameters
# 35% of the time we try something different
comeAlive = 0.35

# how many measures to play
measures = 8

##### define the data structure
score = Score("Drums Come Alive", 125.0) # tempo is 125 bpm

drumsPart = Part("Drums", 0, 9)  # using MIDI channel 9 (percussion)

bassDrumPhrase = Phrase(0.0)      # create phrase for each drum sound
snareDrumPhrase = Phrase(0.0)
hiHatPhrase = Phrase(0.0)


##### create musical data
# kick
# bass drum pattern (one bass 1/2 note) x 2 = 1 measure
# (we repeat this x 'measures')
```

```
for i in range(2 * measures):

    dynamics = randint(80, 110)    # add some variation in dynamics
    n = Note(ACOUSTIC_BASS_DRUM, HN, dynamics)
    bassDrumPhrase.addNote(n)

# snare
# snare drum pattern (one rest + one snare 1/4 notes) x 2 = 1 measure
# (we repeat this x 'measures')
for i in range(2 * measures):

    r = Note(REST, QN)
    snareDrumPhrase.addNote(r)

    dynamics = randint(80, 110) # add some variation in dynamics
    n = Note(SNARE, QN, dynamics)
    snareDrumPhrase.addNote(n)

# hats
# a hi-hat pattern (one hi-hat + one rest 1/16 note) x 8 = 1 measure
# (we repeat this x 'measures')
for i in range(8 * measures):

    # if the modulo of i divided by 2 is 1, we are at an odd hit
    # (if it is 0, we are at an even hit)
    oddHit = i%2 == 1

    # time to come alive?
    doItNow = random() < comeAlive

    # let's give some life to the hi-hats
    if oddHit and doItNow:      # on odd hits, if it's time to do it,
      pitch = OPEN_HI_HAT       # let's open the hit-hat
    else:                       # otherwise,
      pitch = CLOSED_HI_HAT     # keep it closed

    # also add some variation in dynamics
    dynamics = randint(80, 110)

    # create hi-hat note
    n = Note(pitch, SN, dynamics)
    hiHatPhrase.addNote(n)

    # now, create rest
    r = Note(REST, SN)
    hiHatPhrase.addNote(r)


##### combine musical material
drumsPart.addPhrase(bassDrumPhrase)
drumsPart.addPhrase(snareDrumPhrase)
drumsPart.addPhrase(hiHatPhrase)
score.addPart(drumsPart)
```

```
##### view and play
View.sketch(score)
Play.midi(score)
```

Earlier we have seen that `Mod.repeat()` and `for` loops have a similar effect; they can be used to create repetitions of musical material. In many cases, they can be used interchangeably. Here, however, they cannot. Why? (Hint: What is special about the repetition of musical material above? Is the repeated musical material exactly identical?)

This is yet another example of how to use randomness. Here it is happening at the note level. Nothing prevents us from adding randomness and choices at the macro level too, similar to Mozart's Dice Game. This way we can construct more versatile (creative, intelligent?) drum machines. Or, at least, drum machine processes that could inspire a human drummer to explore new patterns and new combinations of percussive sounds.

6.7.5.1 Exercises

1. Explore more possibilities for making this drum machine come alive. Try adding variations to volume and introducing variety through other drum sounds (e.g., a cowbell or tom sound every now and then) instead of a snare.

2. Develop a drum-machine improviser that selects randomly between patterns. (Hint: See Mozart's Musical Dice Game above.) Consider using Mod functions, such as elongation, retrograde, transposition (for different drum sounds) to create variations from a basic pattern. Also see the "Computer-Aided Music Composition" case study in Chapter 4 for more ideas.

6.8 PYTHON RELATIONAL OPERATORS

As we saw in the previous section, there are two important considerations when using selection:

a. The type of question(s) you are asking and

b. What statement(s) to include under each condition.

The latter you already know how to do. Any Python statement or statements can be used.[*]

---

[*] As long as they are consistently intended.

Below, we explore how to build `if` questions, or conditions (as they are called formally).

Definition: A *condition* determines which statements will be executed from a set of alternatives.

Conditions are built using relational and logical operators. The next two sections describe how to use them.

The simplest way to construct conditions (or questions to be answered by the computer) involves relational expressions.

Definition: A *relational expression* consists of values and relational operators.

These ask the computer to determine the relationship between different values or variables. Relational expressions are built using relational operators such as "equal to" or "greater than" whose answer is either yes or no (in Python, True or False). The relational operators are demonstrated below.

Python supports the following relational operators:

Operator	Description	Example
==	Checks for equality (i.e., are these two values equal?)	x == 3
!=	Checks for nonequality (i.e., are these two values NOT equal?)	x != 3
>	Greater than	x > 3
>=	Greater than or equal	x >= 3
<	Less than	x < 3
<=	Less than or equal	x <= 3
in	Membership (i.e., is value in a sequence?)	x in [1, 2, 3]
not in	Nonmembership (i.e., is value NOT in a sequence?)	x not in [1, 2, 3]

Below are some examples.

The "equals" operator, ==, asks if two values are equal. In practice, we use this with variables (since we may need to figure out what values they have, e.g., from user input, and instruct the computer what to do in different cases). For example,

```
>>> x = 3        # let's put 3 in variable x
>>> x            # verifying the value of x
```

```
3
>>> x == 3        # let's ask if x is equal to 3?
True              # yes, it is
>>> x == 4        # is x equal to 4?
False             # no, it is not
```

Be careful to not confuse assignment, =, and equality, ==, operators. For example,

```
>>> x = 4         # assignment (store the value 4 in variable x)
>>> x             # verifying the value of x
4
>>> x == 5        # is x equal to 5?
False   # no, it is not (x is NOT altered)
>>> x
4                 # x is still 4
>>> x = 5         # assignment!!!
>>> x             # x is now 5
5
```

The "not equal" operator, ! =, asks if two values are NOT equal. For example (assume x is 3),

```
>>> x != 4        # is x NOT equal to 4?
True              # yes, it is NOT equal to 4
>>> x != 3        # is x NOT equal to 3?
False             # no, x is 3
```

The "greater than" operator, >, asks if the left value is larger than the second value. For example (assume x is 3):

```
>>> x > 2         # is x greater than 2?
True
>>> x > 4         # is x greater than 4?
False
```

The "less than" operator, <, asks if the left value is smaller than the second value. For example (assume x is 3):

```
>>> x < 4         # is x less than 4?
True
>>> x < 2         # is x less than 2?
False
```

The "less than or equal to" and "greater than or equal to" operators. For example (assume x is 3):

```
>>> x >= 3        # is x greater than or equal to 3?
True
```

```
>>> x <= 4       # is x less than or equal to 4?
True
```

Do not put a space between > and = (or between < and =). This will cause an error. Also, the order is fixed, i.e., >=.

The "membership" operator, in. For example (assume x is 3):

```
>>> x in [1, 2, 3]
True
>>> x in [2, 4, 6]
False
>>> 45 in [25, 3, 45]
True
>>> 'Y' in 'New York'
True
'y' in 'New York'
False
>>> 'a' in 'New York'
False
>>> 'a' not in 'New York'
True
```

Notice that you may combine relational operators, as you do in math. This is very helpful. For example (assume x is 3):

```
>>> 2 < x < 5
True
>>> 3 < x < 5
False
```

In the last example, the answer is False because x is exactly 3. If the condition was 3 <= x < 5, then the answer would be True.

6.9 PYTHON BOOLEAN VALUES

Boolean (or bool) is another data type recognized by Python. It has only two values, True and False:

```
>>> True
True
>>> False
False
```

Similarly to other data types we saw in Chapter 3 (e.g., integers, floats, strings, Scores, etc.), boolean values are like any other values, that is, they can be stored in variables for later use. For example,

```
>>> x = 3 < 4
>>> x
True
```

Mechanics: As always with an assignment statement, first the expression on the right-hand side of the assignment operator (=) is evaluated, in this case, to True, and subsequently that value is stored in the variable on the left-hand side of operator =.

```
>>> type(x)
<type 'bool'>
```

Function `type()` was introduced in Chapter 3 to help us determine the datatype of values, if needed. Here it demonstrates that Python indeed recognizes variable x as a boolean, since we stored a boolean value to it.

6.10 PYTHON LOGICAL OPERATORS

If we consider everyday life and the types of complex conditions we are faced with, we realize that relational operators are somewhat simplistic and may not be adequate to handle (or help model) all realistic processes. For example, if it is a work day and it is raining, then I need bring the umbrella; otherwise, if it is Sunday and it is raining I might just stay in bed.

Since Python tries to help us model realistic processes, it also provides operators to combine simpler conditions to build more complicated ones. These Python operators, i.e., and, or, and not, have very similar meanings to those in everyday English—they are called *logical operations*.

Operator	Description	Example
and	True if both sides are true	`x == 3 and y == 4`
or	True if at least one side is true	`x == 3 or y == 4`
not	Reverses truth value	`not x > 3`

Below are some examples (assume that x is 3 and y is 4).

```
>>> x > 2 and y < 5
True
```

This evaluates to True since both sides are True.

```
>>> x > 2 and y < 3
False
```

This evaluates to False since one of them is False.

```
>>> x <= 2 and y < 5
False
```

This evaluates to False since one of them is False.

```
>>> x <= 2 or y < 5
True
```

This evaluates to True since one of them is True.

```
>>> (x > 2) or (y < 3)
True
```

We may use parentheses to improve readability.

```
>>> x > 2
True
>>> not x > 2
False
```

If the first one is True, then the opposite is not True (i.e., False).

6.10.1 Case Study: Music from Weighted Probabilities

The following program demonstrates how to develop more intricate algorithmic processes for setting up probabilities of musical events (e.g., pitches and durations) and mapping them into interesting musical artifacts.

First, we use parallel lists (here, pitches, durations, and chances) to create a probability table. For example,

```
pitches   = [C4, D4, E4, F4, G4, A4, B4, C5]
durations = [QN, EN, QN, EN, QN, EN, SN, QN]
chances   = [5,   1,  3,  2,  4,  3,  1,  5]
```

specify the chances associated with a particular note (pitch and duration combination). Pitches with a higher value in the chances list have a higher probability to appear in the resulting melody.

Definition: *Parallel lists* are lists of equal length, whose items are associated in terms of position on the list. In other words, items at the same position across all lists are conceptually grouped together—they refer to one thing.

For example, let's look at the items in the first position across lists pitches, durations, and chances above. These items, that is, C4, QN, and 5, signify that Note(C4, QN) has 5 chances to appear. Items at the second position, that is, D4, EN, and 1, signify that Note(D4, EN) has 1 chance to appear. Note(E4, QN) has 3 chances to appear, and so on.

These probabilities try to model what happens in human-composed music; that is, stable scale degrees such as I, V, and III appear more often than other scale degrees. In fact this frequency (of use) profile is what makes us perceive the sense of mode and key.

A nice characteristic of this approach is that the chances do not have to add up to 1.0 (as is usually the case with probabilities). This allows you to experiment with different probabilities easily. Just change a value in the chances list and the program will adjust the chances of the corresponding note (pitch-duration pair) accordingly.

First, let's see the code. Afterwards, we will discuss its interesting points.

```
# generativeMusic.py
#
# Demonstrates how to create music with weighted probabilities.

from music import *
from random import *


numNotes = 32   # how many random notes to play


# pitches and their chances to appear in output (the higher the
# chance, the more likely the pitch is to appear)
pitches   = [C4, D4, E4, F4, G4, A4, B4, C5]
durations = [QN, EN, QN, EN, QN, EN, SN, QN]
chances   = [5, 1, 3, 2, 4, 3, 1, 5]


####
# Create weighted lists of pitches and durations, where the number of
# times a pitch appears depends on the corresponding chance value.
# For example, if pitches[0] is C4, and chances[0] is 5, the weighted
# pitches list will get 5 instances of C4 added.
weightedPitches = []
weightedDurs = []
for i in range( len(chances) ):
   weightedPitches = weightedPitches + [ pitches[i] ] * chances[i]
   weightedDurs = weightedDurs + [ durations[i] ] * chances[i]
# now, len(weightedPitches) equals sum(chances)
# same applies to weightedDurs


# debug lines:
print "weightedPitches = ", weightedPitches
print "weightedDurations = ", weightedDurations


phrase = Phrase() # create an empty phrase


# now create all the notes
```

```
for i in range(numNotes):

    event = randint(0, len(weightedPitches)-1)
    note = Note(weightedPitches[event], weightedDurs[event])

    # the note has been found; now add this note
    phrase.addNote(note)
# now, all notes have been generated

# so, play them
Play.midi(phrase)
```

First, we create the parallel lists pitches, durations, and chances:

```
pitches   = [C4, D4, E4, F4, G4, A4, B4, C5]
durations = [QN, EN, QN, EN, QN, EN, SN, QN]
chances   = [5,  1,  3,  2,  4,  3,  1,  5]
```

Next, we create expanded versions of the pitches and durations lists, so that they contain enough copies of each original value (according to the corresponding chances value). For example, the first pitch value is C4 (i.e., index 0 in list pitches). Since the corresponding chance (index 0 in list chances) is 5, the expanded (or weighted) list of pitches will contain 5 instances of C4:

```
pitches   = [C4,...]
durations = [QN,...]
chances   = [5, ...]

weightedPitches = [C4, C4, C4, C4, C4,...]
```

In the next iteration (i.e., for index 1), the pitch value we are expanding is D4. The corresponding chance is 1, so the expanded list will become:

```
weightedPitches = [C4, C4, C4, C4, C4, D4,...]
```

Going once more (i.e., for index 2), the pitch value we are expanding is E4. The corresponding chance is 3, so the expanded list will become:

```
weightedPitches = [C4, C4, C4, C4, C4, D4, E4, E4, E4,...]
```

This is accomplished by the following code:

```
weightedPitches = []
weightedDurs = []
```

```
for i in range( len(chances) ):
    weightedPitches = weightedPitches + [ pitches[i] ] * chances[i]
    weightedDurs = weightedDurs + [ durations[i] ] * chances[i]
# now, len(weightedPitches) equals sum(chances)
# same applies to weightedDurs
```

At the end of this loop we are guaranteed that the length of weightedPitches is equal to the sum of all the chances. Why?

Notice how simple it becomes to create the expanded list weighted-Pitches starting with an empty list and adding incrementally each individual pitch (as a list) multiplied by the number of chances for this pitch. The same applies to weightedDurs.

Finally, the following code builds the musical phrase:

```
# now create all the notes
for i in range(numNotes):

    event = randint(0, len(weightedPitches)-1)
    note = Note(weightedPitches[event], weightedDurs[event])

    # the note has been found; now add this note
    phrase.addNote(note)
```

Notice how we get a random index from 0 to len(weightedPitches) − 1. (Why the − 1?) This index is used to get a random pitch from the list weightedPitches. Since this list was created to have as many repetitions of a particular pitch as the chance we wanted to get that pitch, we have achieved our goal, i.e., to sieve randomness using the result of randint() to different probabilities for different notes.

Notice how the program includes print statements to help you understand, at run time, how it operates. Consult the output of the program. Afterwards, feel free to comment out (or remove) these print statements.

This demonstrates how to use print statements to help debug a program. Print statements are your best friend to figure out what a piece of code is doing (or NOT doing). Use them frivolously during code implementation.

This is a useful accomplishment, as seen in the exercises below. Enjoy!

6.10.1.1 Exercises

1. Using the score for a favorite piece of music, count how often certain pitch-duration combinations occur. Update the lists of pitches, durations, and chances above using the data you collected. Use the updated code to create different melodies. Do you get anything interesting? Since the above program does not include any information

on higher-level musical structure, the music created will not sound perfect. However, it will contain portions that may inspire you to explore new melodies.

2. Repeat the above exercise with chord progressions. In other words, take a favorite piece of music and count how often certain chord-duration combinations occur. Update the lists of pitches, durations, and chances above using the data you collected. (Hint: Simply have the list of pitches contain chord lists (see section "Adding Chords with Lists" in Chapter 3).) Use the updated code to create different chord progressions. Do you get anything interesting?

3. Explore combining interesting results from exercises (1) and (2) to create ideas for new songs. And welcome to the world of computer-aided composition.

6.11 SUMMARY

This rather long chapter focused on the concept of randomness and creativity, i.e., how to harness the computer's capacity to generate pseudo-random events (e.g., sequences of numbers that appear unpredictable to humans) and turn them into creative artifacts. The trick is to learn ways to take something that appears truly random/chaotic and sieve it to produce balanced and aesthetically pleasing outcomes. This chapter offered some possibilities.

Additionally, we learned about if statements and how to use them inside for loops. We also learned about relational and logical operators, as well as the new (to us) Python boolean data type, which interestingly has only two values, true and false.

Finally, we learned about developing more intricate algorithmic processes for setting up probabilities of musical events (e.g., pitches, durations, chords—or, at a higher level, rhythmic patterns, phrases, and other musical parts) and mapping them into reasonable or plausible musical artifacts. These artifacts, although they are not perfect (some would say far from perfect), they can minimally serve as creative triggers. This is similar to other processes used by composers (such as throwing dice, observing birds on telephone wires, etc.) to develop seed musical material. However, given the richness of algorithmic expression provided by Python (and other programming languages), the computer may be a revolution waiting to happen for music composers (and other artists alike). This chapter suggests some ways in this direction. Many other possibilities exist.

## CHAPTER 7

Sonification and Big Data

Topics: Data sonification, mapValue() and mapScale(), Kepler, Python strings, music from text, Guido d'Arezzo, nested loops, file input/output, Python while loop, big data, biosignal sonification, defining functions, image sonification, Python images, visual soundscapes.

7.1 OVERVIEW

According to Scaletti (1993), "the idea of representing data in sound is an ancient one. For the ancient Greeks music was not an art-for-art's sake, practiced in a vacuum, but a manifestation of the same ratios and relationships as those found in geometry or in the positions and behaviors of the planets." Pythagoras, Plato, and Aristotle worked on quantitative expressions of proportion and beauty, such as the golden ratio. Pythagoreans, for instance, quantified harmonious musical intervals in terms of proportions (ratios) of the numbers 1, 2, 3, 4, and 5 (as we discussed earlier in Chapter 1). This became the basis for the scales, modes, and tuning systems used in Western music.

As Johannes Kepler (among others) observed, the universe is flowing and fluctuating at different time scales and size scales. We only hear a small range of these fluctuations as sound, namely, when they vibrate at frequencies between 20 and 20,000 times per second (Hertz or Hz). Similarly, we only see (a different) small range of these vibrations as visible light (red through purple). The limits of our senses make it difficult for us to fully appreciate how our cosmos is put together in harmonically (hence musically) meaningful ways. It is this harmony in the proportions and organizations of the parts that make our universe stay together, allowing us to be in it and experience it. It is this harmony that Kepler studied and tried to formalize and communicate through his three laws of planetary motion. Kepler also tried to map the harmony he observed in the astronomical data available at his time as sounds, that is, to sonify them (Kepler 1619).

Sonification allows us to capture and better experience phenomena that are outside our sensory range by mapping values into sound structures that we can perceive by listening to them. Data for sonification may come from any measurable vibration or fluctuation, such as planetary orbits, magnitudes of earthquakes, positions of branches on a tree, lengths of words in this chapter, and so on.

In addition to facilitating insight and analysis, sonification can also be used to make interesting music and sound, just like visualization of data can make interesting images. If music can involve the imitation of nature (mimesis), then sonification is a valuable tool for modern music making.

7.2 DATA SONIFICATION

When dealing with big data (i.e., large amounts of data), sometimes it is easier to hear patterns by turning the data into sound, as opposed to looking at, say, thousands or millions of numbers.

Definition: *Data sonification* involves mapping numbers to musical parameters, such as pitch, duration, volume, pan position, and timbre, in order to better perceive patterns in the data and appreciate their characteristics.

A clear and simple example of sonification is the Geiger counter, a device which measures radiation and turns this measurement to audible clicks. The number (frequency) of clicks is directly related to the radiation level in the vicinity of the device.

Another example of sonification is the biosignal monitor that we see at hospitals. These are the devices that monitor signals measured from the human body, such as heart rate. These monitors (similarly to the Geiger counter) provide constant sonic biofeedback, allowing the human operator to focus on other tasks.

7.2.1 The mapValue() Function

Mapping a value from one range to another is very common in sonification. For instance, you may want to map a list of numbers (say, from 20.0 to 110.0) to pitch values (say, from 30 to 90). This becomes more realistic if, say, you are interested in global warming and want to explore how temperatures change over time. By converting temperatures to pitch values, you can actually hear these changes.

The music library provides the `mapValue()` function precisely for this task. This function accepts the following arguments:

- value—the number to be mapped
- minValue—the lowest possible number to be mapped
- maxValue—the highest possible number to be mapped
- minResult—the lowest value of the destination range
- maxResult—the highest value of the destination range

For example,

```
>>> from music import * # mapValue() is defined in this library
>>> mapValue(24, 0, 100, 32, 212) # celsius to fahrenheit
75
```

This converts 24°C to 75°F. To verify that this conversion really works, let us try converting the extremes:

```
>>> mapValue(0, 0, 100, 32, 212) # lowest temperature
32
```

So, when we give the lowest temperature in the temperature range (i.e., 0), `mapValue()` returns the lowest pitch in the destination range (i.e., 32).

```
>>> mapValue(100, 0, 100, 32, 212) # highest temperature
212
```

Also, when we give the highest temperature in the temperature range (i.e., 100), `mapValue()` returns the highest pitch in the destination range (i.e., 212).

```
>>> mapValue(56.7, 0.0, 100.0, 32.0, 212.0) # arbitrary temperature
134.06
```

Using float numbers returns float numbers.

Having tested this, we can assume that the rest of the values are mapped linearly (as if we took two number lines, one from 0 to 100 and another from 32 to 212, and superimposed them).

A useful feature is that `mapValue()` will return a value using the data type of the destination range. So, for example, providing float (real) values in the destination range generates a float result:

```
>>> mapValue(56.7, 0.0, 100.0, 32.0, 212.0)
134.06
```

7.2.2 The mapScale() Function

The `mapScale()` function is similar to `mapValue()`, expect that it has an additional 6th argument:

• scale—(optional) the musical scale to be used in the destination range[*]

This function quantizes (sieves) the numeric results to fit pitches within the provided scale. The key of the scale is determined by `minResult`. For example, any C pitch (e.g., C4 (60), C5 (72), etc.) corresponds to the key of C any D pitch corresponds to the key of D and so on.

Accordingly, `mapScale()` returns integer values, since they are meant to be used as pitches.

The difference between `mapValue()` and `mapScale()` is demonstrated as follows. Note that the first examples use duration constants, demonstrating another useful musical application for `mapValue()`.

```
>>> from music import *
>>> from random import *

>>> mapValue(0.6, 0, 1, TN, WN)
2.4499999999999997

>>> mapValue(0.999, 0.0, 1.0, TN, WN)
3.996125

>>> mapScale(0.5, 0, 1, 0, 127, MAJOR_SCALE)
64

>>> mapScale(0.66, 0, 1, 40, 80, MINOR_SCALE)
66
```

Whereas `mapValue()` returns a float value with lots of accuracy, `mapScale()` returns an integer.

Here are some examples using a provided scale:

```
>>> for i in range(0, 13):
print i, "->", mapScale(i, 0, 12, 0, 12, MAJOR_SCALE)
0 -> 0
```

---

[*] The default is CHROMATIC_SCALE.

```
1 -> 0
2 -> 2
3 -> 2
4 -> 4
5 -> 4
6 -> 5
7 -> 7
8 -> 7
9 -> 9
10 -> 9
11 -> 11
12 -> 12
```

Here we can clearly see how the result is adjusted to fit the pitches in the C major scale. We know the scale is in C from the `minResult` (4th argument), which is 0. Pitch 0 is a C pitch, namely, C_1.[*]

Constraining pitches to a scale allows us to generate more aesthetically pleasing sonifications.

Finally, since scales are simply lists of pitch classes (between 0 and 11),[†] for example,

```
>>> MAJOR_SCALE
[0, 2, 4, 5, 7, 9, 11]
>>> MINOR_SCALE
[0, 2, 3, 5, 7, 8, 10]
>>>
```

it is possible to use your own list, such as `[0, 5, 8]`. The only constraint is that the numbers range from 0 to 11. So, for example,

```
>>> mapScale(97.5, 20, 110, 30, 90, [0, 5, 8])
80
```

Here the scale has pitches C (0), F (5), and G sharp (8). If we had used `mapValue()`, the result would have been 81.7, but rounding to the nearest C, F, or G sharp gives us GS5, which is 80.

To understand better how `mapScale()` works with scales, try different scales with the following loop:

```
for i in range(22):
    print mapScale(i, 0, 22, C3, C6, scale)
```

where `scale` is a particular scale or pitch class list.

---

[*] Actually, `mapScale()` has an extra, optional parameter, key, which could specify a key different from the one implied by `minResult`.

[†] In music theory, such a list is called a *pitch class set*. Pitch class sets have been used extensively by modern composers.

7.3 CASE STUDY: KEPLER—"HARMONIES OF THE WORLD" (1619)

In 1619 Johannes Kepler wrote his "Harmonices Mundi (Harmonies of the World)" book (Kepler 1619). While Pythagoreans only talked about the "music of the spheres," Kepler discovered physical harmonies in planetary motion. As a result, he became a key figure in the development of astronomy and modern physics.

Following Kepler's studies, in the late 1700s Johann Daniel Titius and Johann Elert Bode independently contributed to a model of the symmetries and proportions of our solar system. Their formula, known as the Titius–Bode law (or simply Bode's law), predicts the positions of the planets in our solar system. Actually, it also predicted the asteroid belt between Mars and Jupiter (long before it was discovered), but fail to account for the irregularly moving Neptune and the (now demoted nonplanet) Pluto.

The following program, `harmonicesMundi.py`, sonifies one aspect of the celestial organization of planets. In particular, it converts the orbital velocities of the planets to musical notes.

The planets' mean orbital velocities (in kilometers per second) are as follows:

Mercury	Venus	Earth	Mars	Ceres	Jupiter	Saturn	Uranus	Neptune	Pluto
47.89	35.03	29.79	24.13	17.882	13.06	9.64	6.81	5.43	4.74

We will map this range of velocities to a range of MIDI pitches, say, C1 to C6. To do this mapping, we can use the `mapScale()` function.

```
# harmonicesMundi.py
#
# Sonify mean planetary velocities in the solar system.
#

from music import *

# create a list of planet mean orbital velocities
# Mercury, Venus, Earth, Mars, Ceres, Jupiter,
# Saturn, Uranus, Neptune, Pluto
planetVelocities = [47.89, 35.03, 29.79, 24.13, 17.882, 13.06,
                    9.64, 6.81, 5.43, 4.74]

# get minimum and maximum velocities:
minVelocity = min(planetVelocities)
maxVelocity = max(planetVelocities)
```

```
# calculate pitches
planetPitches = []      # holds list of sonified velocities
planetDurations = []    # holds list of durations
for velocity in planetVelocities:
   # map a velocity to pitch and save it
   pitch = mapScale(velocity, minVelocity, maxVelocity, C1, C6,
                 CHROMATIC_SCALE)
   planetPitches.append( pitch )
   planetDurations.append( EN )   # for now, keep duration fixed

# create the planet melodies
melody1 = Phrase(0.0)      # starts at beginning
melody2 = Phrase(10.0)     # starts 10 beats into the piece
melody3 = Phrase(20.0)     # starts 20 beats into the piece

# create melody 1 (theme)
melody1.addNoteList(planetPitches, planetDurations)

# melody 2 starts 10 beats into the piece and
# is elongated by a factor of 2
melody2 = melody1.copy()
melody2.setStartTime(10.0)
Mod.elongate(melody2, 2.0)

# melody 3 starts 20 beats into the piece and
# is elongated by a factor of 4
melody3 = melody1.copy()
melody3.setStartTime(20.0)
Mod.elongate(melody3, 4.0)

# repeat melodies appropriate times, so they will end together
Mod.repeat(melody1, 8)
Mod.repeat(melody2, 3)

# create parts with different instruments and add melodies
part1 = Part("Eighth Notes", PIANO, 0)
part2 = Part("Quarter Notes", FLUTE, 1)
part3 = Part("Half Notes", TRUMPET, 3)
part1.addPhrase(melody1)
part2.addPhrase(melody2)
part3.addPhrase(melody3)

# finally, create, view, and write the score
score = Score("Celestial Canon")
score.addPart(part1)
score.addPart(part2)
score.addPart(part3)
View.sketch(score)
Play.midi(score)
Write.midi(score, "harmonicesMundi.mid")
```

Notice how we first create the list of planetary velocities. For this case study, we also include Ceres (i.e., the asteroid belt) and (the recently demoted non-planet) Pluto.

```
planetVelocities = [47.89, 35.03, 29.79, 24.13, 17.882, 13.06,
                    9.64, 6.81, 5.43, 4.74]
```

Notice how we use the list functions, `min()` and `max()`, to find the minimum and maximum velocity values.

```
minVelocity = min(planetVelocities)
maxVelocity = max(planetVelocities)
```

We could have done this by hand by scanning the list ourselves and hard-coded the min and max numbers in the program; however, this use of "magic numbers" (as they are called by seasoned software developers) can get us into trouble.

Good Style: Avoid using magic numbers in your programs.

Definition: A *magic number* is a number in our programs that seems to come out of nowhere, that is, it is not derived automatically (calculated) from the program's data.

Even if we document such magic numbers profusely (a minimum requirement if we have to use them), they become a place of future error, as the program evolves (and most programs do evolve). So as to avoid using magic numbers in the preceding code, we added code to calculate min and max from the list of velocities. Thus, if the list ever changes (or it is read from a data file, as we will see soon), `min()` and `max()` will continue to do the right thing. These functions, together with `mapValue()`, are very useful in sonification tasks.

Notice how `mapScale()` is utilized to sonify the orbital velocities. For example, assume that variable velocity contains the orbital velocity of a planet, and variables `minVelocity` and `maxVelocity` contain the smallest and largest of all planet velocities, respectively. Then this statement

```
pitch = mapScale(velocity, minVelocity, maxVelocity, C1, C6,
                 CHROMATIC_SCALE)
```

generates a pitch value positioned in the range C1 to C6 similarly to how velocity is positioned in the range minVelocity to maxVelocity.

Next we use a `for` loop to map each velocity in `planetVelocities` to the corresponding pitch (in the range 24 to 84). Notice how we create an empty list ahead of time (outside of the loop) and use the list operation `append()` to add each new pitch value, as it is created, to the end of the pitches list.

```
# calculate pitches
planetPitches = []         # holds list of sonified velocities
for velocity in planetVelocities:
    # map a velocity to pitch and save it
    pitch = mapValue(velocity, minVelocity, maxVelocity, C1, C6)
    planetPitches.append(pitch)
```

Now the list `planetPitches` is created and contains one pitch per planet. Notice how it is parallel to the list `planetVelocities`—for every planet velocity there is now a pitch, in the corresponding list positions.

Finally, we are ready to sonify. What follows is similar to the types of things we have seen in earlier chapters. We have many options for how we can sonify these data. One possibility is to just play the melody generated by the pitches. Another possibility would be to create a musical piece using the above melody as a component. For instance, we might use the original melody as a bass line, add chords, drums, and additional musical material.

Here, in the spirit of J.S. Bach and Arvo Pärt, we build a canon from the sonified orbital velocities. To do this, we treat the melody as the theme and use canonic devices (seen in Chapter 4) to create a celestial canon. We choose to play the melody concurrently, against itself, using different durations (see Figure 7.1). This is similar to Arvo Pärt's musical structure for `Cantus in Memoriam` (see case study in Chapter 4).

The preceding code implements this canon structure. Notice how we create three different melodies, `melody1`, `melody2`, and `melody3`, to contain the same pitch list (the one sonified earlier) but with different durations, adding up to 5 beats, 10 beats, and 20 beats, respectively. Then we adjust the start time of each melody (i.e., start at 0.0, 10.0, and 20.0), so that they are introduced incrementally, but end together.

7.3.1 Exercise

Modify the preceding sonification to use a different musical structure; there are many possibilities. Aim for something aesthetically pleasing. Consider using different scales with function `mapScale()`. Experiment with different scales, including your own pitch sets.

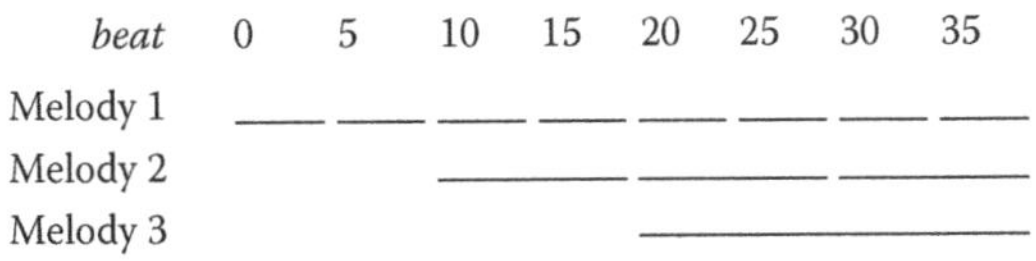

beat	0	5	10	15	20	25	30	35
Melody 1								
Melody 2								
Melody 3								

FIGURE 7.1 Diagram of canon structure. (*Note:* This canon idea, for this data, was proposed by Douglas McNellis and Ian Fricker.)

Search the Internet for interesting astronomical data, download them, and sonify them using a similar approach to the one explored here. For musical inspiration, listen to Olivier Messiaen's "Mode de valeurs et d'intensités" for piano (1949), in which he creates musical material in a systematic way focusing on durations and intensities.

7.4 PYTHON STRINGS

In Chapter 2, we discussed how Python uses strings to represent text data. String is another Python data type, such as int, float, list, and boolean. A string contains a sequence of characters. Strings are enclosed in quotes.

Definition: A Python string consists of zero or more characters, enclosed in quotes.

There are three types of quotes that may be used, single quotes (' '), double quotes (""), or three double quotes (""""""""). For example,

```
'This is a string - 1, 2, 3.'
```

Here is the same string using double quotes:

```
"This is a string - 1, 2, 3."
```

And again using three double quotes:

```
"""This is a string - 1, 2, 3."""
```

You can use any one of these three representations. Most programmers prefer the first two, since clearly they are most economical. The only reason to use the three double quotes is that such strings may contain newlines, that is, they can span several lines. This is not the case with the other quotes, for example,

```
"""This is a
string - 1,
2, 3."""
```

Strings are similar to lists in that they are sequences. As a result, everything we learned about lists, in terms of indexing, applies to strings as well. For example, normal indexing and slice operations apply:

```
>>> s = "This a string - 1, 2, 3."
>>> s[0]
```

```
'T'
>>> s[1]
'h'
>>> s[1:4]
'his'
>>> s[1:4] == s[10:13]
False
>>> s[2] == s[10]
True
>>> s[2]
'i'
```

Also, we can iterate through strings like we did with lists:

```
>>> s1 = "Hello!"
>>> for character in s1:
...     print character
...
H
e
l
l
o
!
```

In the last example, what would happen if we put a comma at the end of the print statement? Why? *

String characters are represented internally using numbers, according to the ASCII standard.† The complete listing of the ASCII characters (and the numbers used to represent them) is easily available on the Internet. Most of the time we can ignore this. However, sometimes knowing the internal representation can be useful.

In Python, we can use function ord() which, given a string with a single character as argument, returns the corresponding ASCII number. For example,

```
>>> ord('A')
65
>>> ord('a')
97
>>> ord(' ')
32
>>> ord('!')
33
>>> ord('0')
48
```

---

* The print statement was covered in Chapter 2.
† ASCII stands for American Standard for Information Interchange.

One thing to remember is that not all ASCII characters are printable, that is, some of them are used for formatting (e.g., newline character, tab, backspace, etc.). The printable ASCII characters range from space (ASCII value 32) to the tilde character, "~" (ASCII 126).

Having access to the ASCII numbers makes it straightforward to generate music from text. It is not that different from generating music from planetary orbital velocities, as we see in the following text.

7.4.1 Case Study: Music from Text

A simple way to algorithmically dictate musical structures is to follow some existing data patterns. One source of data patterns is astronomical data, as we saw earlier. Another source of patterns is natural language (e.g., English). Since languages have inherent structure—as described by Noam Chomsky (1957) and George K. Zipf (1949)—it is reasonable to expect that music based on text might maintain some of the expressiveness inherent in this structure. The sonification of text can, similar to all sonifications, use simple or complex mappings between the text and sound.

The following program, textMusic.py, demonstrates how to generate music from text. Using the ord() Python built-in function, this program converts the values of ASCII characters to MIDI pitches. For variety, note durations are randomized; other note properties (volume, etc.) are the same for all notes.

```
# textMusic.py
#
# It demonstrates how to generate music from text.
# It converts the values of ASCII characters to MIDI pitch.
# Note duration is picked randomly from a weighted-probability list.
# All other music parameters (volume, panoramic, instrument, etc.)
# are kept constant.
#

from music import *
from random import *

# Define text to sonify.
# Excerpt from Herman Melville's "Moby-Dick", Epilogue (1851)

text = """"The drama's done. Why then here does any one step forth?
- Because one did survive the wreck. """

##### define the data structure
textMusicScore  = Score("Moby-Dick melody", 130)
textMusicPart   = Part("Moby-Dick melody", GLOCK, 0)
textMusicPhrase = Phrase()
```

```
# create durations list (factors correspond to probability)
durations = [HN] + [QN]*4 + [EN]*4 + [SN]*2

##### create musical data
for character in text:  # loop enough times

   value = ord(character)  # convert character to ASCII number

   # map printable ASCII values to a pitch value
   pitch = mapScale(value, 32, 126, C3, C6, PENTATONIC_SCALE, C2)

   # map printable ASCII values to a duration value
   index = mapValue(value, 32, 126, 0, len(durations)-1)
   duration = durations[index]

   print "value", value, "becomes pitch", pitch,
   print "and duration", duration

   dynamic = randint(60, 120)   # get a random dynamic

   note = Note(pitch, duration, dynamic)    # create note
   textMusicPhrase.addNote(note)   # and add it to phrase

# now, all characters have been converted to notes

# add ending note (same as last one - only longer)
note = Note(pitch, WN)
textMusicPhrase.addNote(note)

##### combine musical material
textMusicPart.addPhrase(textMusicPhrase)
textMusicScore.addPart(textMusicPart)

##### view score and write it to a MIDI file
View.show(textMusicScore)
Play.midi(textMusicScore)
Write.midi(textMusicScore, "textMusic.mid")
```

Note that the string to sonify is at the top of the code. If you change this string, you will get different (yet similar music). Why? The music gen-erated depends on the relative probabilities of characters in the English language (and not on the actual words, or, even further, the meaning of those words). It would be interesting to explore how to somehow map the meaning of words (or actual words) to note pitch. This would involve more work—beyond the scope of this chapter, for sure—but definitely some-thing that could be explored using Python.

Also, notice how we use a for loop to iterate through every character in the string (stored in variable text). Actually, for loops work not only with lists, but also with any sequence, such as a string. (Remember that a string is a sequence of characters.) In the body of this loop, we get the ASCII number corresponding to each character and map it to a pitch.

ASCII characters are represented by integers from 0 to 127. However, in a regular text file, we could have mapped them directly to MIDI pitches (also from 0 to 127), that is,

```
pitch = ord(character)
```

But instead we opted for a more musical mapping utilizing the PENTATONIC_SCALE, that is,

```
pitch = mapScale(value, 0, 127, C2, C6, PENTATONIC_SCALE)
```

Notice the `print` statement in the loop, which is commented out. It can be used to output the original value and the mapped pitch and duration values.

Good Style: Use strategically placed print statements (as above) to figure out what the code is doing (especially if things do not work as expected). Delete them after the code is ready.

`Print` statements are a programmer's best friend. Use them extensively while you are developing your code. Then you can remove them (or comment them out, as above—they may come in handy again, after you have made a change in your code).

7.4.1.1 Exercise

Consider ways to make the earlier sonification more interesting.

- Perhaps by adding variety to parameters such as note lengths, dynamics, and panning.

- Consider how this variety might come from the text itself instead of simply harnessing randomness?

- What are some other text characteristics that we can use? (*Hint:* Some more ideas are introduced in the next case study.)

7.4.2 String Library Functions

Python provides a string library with useful string functions. Table 7.1 presents the most common ones (assume s is a string).

TABLE 7.1   Common string library functions (assume s is a string)

Function	Description
`capitalize(s)`	Turn first letter of `s` into upper case.
`upper(s)`	Turn every letter of `s` into upper case.
`lower(s)`	Turn every letter of `s` into lower case.
`count(s, substring)`	Count occurrences of `substring` in s.
`find(s, substring)`	Find leftmost occurrence of `substring` in s (returns the position in s, starting at `0`, or `-1` if not found).
`rfind(s, substring)`	Find rightmost occurrence of `substring` in s (returns the position in s, starting at `0`, or `-1` if not found).
`replace(s, old, new)`	Replace first occurrence of `old` with `new` in string s.
`strip(s)`	Remove whitespace from both ends of string s.
`lstrip(s)`	Remove whitespace from left side of string s.
`rstrip(s)`	Remove whitespace from right side of string s.
`split(s, char)`	Split string s into a list of substrings, using `char` as the delimiter (`char` is removed from the result). If `char` is omitted, it defaults to whitespace (space, tab, newline).
`join(listOfStrings, char)`	Join a list of strings into a single string, with `char` being the spacer placed in between the strings joined together (opposite of split).
`char in string`	Checks if `char` is in string.

Here are some examples:

```
>>> from string import *

>>> s = 'hello world!'

>>> capitalize(s)
'Hello world!'

>>> upper(s)
'HELLO WORLD!'

>>> s
'hello world!'
```

Note that string functions do *not* modify the string provided as an argument; they just return a new string. For example, after the above, s still has its original value.

To modify s, we need to do this:

```
>>> s = capitalize(s)
>>> s
'Hello world!'
```

Here is another example,

```
>>> s = 'hello world!'
>>> find(s, 'e')
1
```

The result, 1, indicates that the first occurrence of substring "e" appears in position 1 (where "h" is at position 0).*

Finally, the following demonstrates functions split() and join():

```
>>> s = 'hello world!'

>>> s1 = split(s, ' ')
>>> s1
['hello', 'world!']

>>> s2 = join(s1, ' ')
>>> s2
'hello world!'
```

These are a few of the string functions available in the Python string library. They should be sufficient for most tasks related to music making. Feel free to explore more online.†

7.4.3 Case Study: Guido d'Arezzo—"Word Music" (ca. 1000)

One of the oldest known algorithmic music processes is a rule-based algorithm that selects each note based on the letters in a text, credited to Guido d'Arezzo (991–1033). Originally, the intention was that the melody was a sung phrase and the text was the lyric to be sung. Each vowel in the text is associated with a pitch. The duration of notes comes from the word length.

Although d'Arezzo's original intention was simply to provide an approximate composition guide, here we formalize and automate these

---

* Python sequences (such as strings and lists) are always indexed starting at 0, that is, the first item is at position 0.
† Chances are you will find the functions you need in the string library. If not, you should be able to easily construct what you need building upon the tools in the library.

rules. This is an approximation to d'Arezzo's algorithm, adapted to text written in ASCII and to modern musical sensibilities.

```python
# guidoWordMusic.py
#
# Creates a melody from text using the following rules:
#
# 1) Vowels specify pentatonic pitch, 'a' is C4, 'e' is D4,
#    'i' is E4, 'o' is G4, and 'u' is A4.
#
# 2) Consonants extend the duration of the previous note (if any).
#

from music import *
from string import *

# this is the text to be sonified
text = """One of the oldest known algorithmic music processes is a
rule-based algorithm that selects each note based on the letters in a
text, credited to Guido d'Arezzo."""

text = lower(text)   # convert string to lowercase

# define vowels and corresponding pitches (parallel sequences),
# i.e., first vowel goes with first pitch, and so on.
vowels        = "aeiou"
vowelPitches  = [C4, D4, E4, G4, A4]

# define consonants
consonants = "bcdfghjklmnpqrstvwxyz"

# define parallel lists to hold pitches and durations
pitches   = []
durations = []

# factor used to scale durations
durationFactor = 0.1   # higher for longer durations

# separate text into words (using space as delimiter)
words = split(text)

# iterate through every word in the text
for word in words:

    # iterate through every character in this word
    for character in word:

        # is this character a vowel?
        if character in vowels:

            # yes, so find its position in the vowel list
            index = find(vowels, character)
            #print character, index

            # and use position to find the corresponding pitch
            pitch = vowelPitches[index]
```

```
             # finally, remember this pitch
             pitches.append( pitch )
             # create duration from the word length
             duration = len(word ) * durationFactor

             # and remember it
             durations.append( duration )

# now, pitches and durations have been created

# so, add them to a phrase
melody = Phrase()
melody.addNoteList(pitches, durations)

# view and play melody
View.notation(melody)
Play.midi(melody)
```

In this program, we import the string library to use functions `lower()`, `split()`, and `find()`.

We use function `lower()` to convert to lowercase the characters in variable text. We do this, because we have defined rules in terms of lowercase vowels. If we did not convert text to lowercase we would need to duplicate rules for uppercase letters.

Notice how we define two parallel sequences (a string and a list). The first one, `vowels`, holds the vowels we wish to associate with notes. The second one, `vowelPitches`, holds the pitches associated with those vowels. Since both are indexed starting at 0, we can use them to associate vowels with pitches.* For example,

```
>>> vowels = "aeiou"
>>> vowelPitches = [C4, D4, E4, G4, A4]
>>> vowels[0]
'a'
>>> vowelPitches[0]
60               # same as C4
```

Also, notice how we define two parallel lists, `pitches` and `durations`, to store musical data for the notes.

7.4.4 Python Nested Loops

Loops may be nested. This kind of hierarchical processing (loops within loops) is quite common as a programming pattern, and we will see another example of it later in this chapter. It is common in programming

---

* We have seen this idea before. For instance, in Chapter 3, we used parallel lists of pitches and durations to create phrases of notes.

because hierarchical structures are common in the world. As a geographic example of a hierarchy, consider how people live in suburbs, suburbs are in regions, regions are in states, states are in a country, and so on. This structure could be used as the basis for an algorithm to process details about all people in a particular geographic area.

Definition: Nested loops use two or more loops, one inside the other, to process data.

In the earlier program, the outer loop iterates as usual. For each iteration of the outer loop, the inner loop does a complete run (i.e., it does all its iterations).*

The outer loop repeats for the number of words in the text. For each word selected, the inner loop repeats for the number of characters in the word. The Python interpreter goes back and forth between the outer and inner loops, until all the text has been processed.

Let us now take a look at the nested loop in our recent musical case study in more detail:

```
# separate text into words (using space as delimiter)
words = split(text)

# iterate through every word in the text
for word in words:

    # iterate through every character in this word
    for character in word:

        # is this character a vowel?
        if character in vowels:

            # yes, so find its position in the vowel list
            index = find(vowels, character)
            #print character, index

            # and use position to find the corresponding pitch
            pitch = vowelPitches[index]

            # finally, remember this pitch
            pitches.append(pitch)

            # create duration from the word length
            duration = len(word) * durationFactor

            # and remember it
            durations.append(duration)
```

---

* For example, if the outer loop iterates 5 times and the inner loop iterates 10 times, then for each iteration of the outer loop, the inner loop will iterate 10 times, for a total of 50 times.

First, the `split()` function divides the text string using spaces as a delimiter (separator). This does mean that punctuation is considered part of "words", but we will live with that for now.

The variable `word` in the outer loop holds each word in the text in turn. The inner loop act on each word by iterating through each character looking for a vowel. When it finds a vowel, it creates a note by mapping the vowel to its pitch, and the current word length to duration. In particular,

```
if character in vowels:
```

if the current character is a vowel, we find its index in the list of vowels,

```
index = find(vowels, character)
```

Since `vowels` is parallel with `vowelPitches`, we use this index to find the corresponding pitch:

```
pitch = vowelPitches[index]
```

The note duration is calculated from the current word length. Because word lengths may typically range from 1 to 7 (or so), we multiply by a duration factor (0.1). This creates more appropriate note durations.

```
# create duration from the word length
duration = len(word) * durationFactor

# and remember it
durations.append(duration)
```

Once we have the pitch and duration, we add this to the `pitches` and `durations` lists:

```
pitches.append(pitch)
durations.append(duration)
```

The rest of the program is straightforward. After the loop, we use the two lists, pitches and durations, to construct a phrase. Finally, we view that phrase and play it.

7.4.5 Exercise

Explore adding musical rules for numbers, punctuation, and whitespace.

- Decide how such characters may affect the panning and/or dynamic values of, say, subsequent notes.

- Consider how to handle punctuation such as commas and full stops. For example, they may add rests. (*Hint:* The string library defines list punctuation to hold all punctuation characters. Notice how this is similar to the list vowels, in the above program.)

7.5 FILE INPUT AND OUTPUT

Python makes it very easy to input and output data through files. Files are an easy way to store and transfer information. This section explains how to have our programs open files in order to read in or write out information.

7.5.1 Reading Files

Opening a file is easy in Python—it is done with the `open()` function. For example, the following

```
data = open("someFile.txt", "r")
```

opens the file "someFile.txt" and stores in it the variable `data`. This file is expected to be in the same folder as your program.

Function `open()` expects two arguments, both strings. The first argument is the name of the file to be opened. The second argument determines if the file is opened for reading (i.e., "r"), for writing (i.e., "w"), or for appending (i.e., "a").

Once a file has been opened for reading, the following functions can be used to read information from it:

- `read()`—reads the complete file in as a single (long!) string

- `readline()`—reads the next line as a string

- `readlines()`—reads all lines together as a list of strings

For example, the following code:

```
book = open("book.txt", "r")
text = book.read()
book.close()
```

opens the file "book.txt," then reads the complete file as a string into the variable `text`. After reading the data, it is very important to close all files that have been opened. This is done via the `close()` function. Otherwise, open files are occupying memory for no reason.

Once you have read data from a file, you may use functions from the string library to extract whatever information you need. You must know the format an input file has if you are to successfully read it. Some possible input file formats include one word per line, two words separated by a semicolon, or, say, a sequence of numbers all on one line separated by whitespace. There are infinite possibilities; but as long as your program can anticipate (or better still knows) how data is structured inside a file (i.e., the file format), it can access what it needs through string functions.

The following is a more efficient way to read every line from a file. Actually, a file in Python is a sequence of lines. And since for loops operate on sequences (such as lists, strings, etc.), the following does not need to use readline() explicitly. In this example, we simply read every line and output it (via a print statement):

```
book = open("book.txt", "r")

for line in book:
    print line

book.close()
```

In your code, you should replace the print statement with whatever code is needed to process the input data (e.g., append it to a list, add it to a variable, etc.).

7.5.2 Writing Files

If a program needs to store data for later processing, this can be done by writing a file. Again, you open a file, but for writing this time. This is done using the open() function with the "w" argument (or "a" for appending). Then, we use the write() function to write information into the file.

* write(s)—writes string s to the file

Notice how this function requires a string. In other words, if we need to output other data (such as numbers), we need to first convert it to a string. This is done with the str() Python function.[*]

For example, the following

```
data = open("numbers.txt", "w")
```

---

[*] Python has similar functions to convert data to other Python data types as needed, such as int(), float(), bool(), etc.

```
for i in range(1000):
    data.write(str(i))
    data.write('\n')

data.close()
```

opens file `"numbers.txt"` for writing and stores a sequence of numbers into it. Notice how `write()` does not automatically output newlines. To get a newline, you need to explicitly output a newline character, `"\n"`.

Opening a file for writing overwrites (clears out) any earlier contents. To add data use "a" instead of "w," which appends new material to an existing file.

Notice how we close the file via the `close()` function, once done with writing data into it. Otherwise some data may be lost.

7.5.3 Exercises

1. Write a program that creates a file called "randomNumbers.txt," which contains a sequence of 20 random integers between 0 and 100 (inclusive). (*Hint:* See function `randint()` in Chapter 6.)

2. Go to Project Gutenberg on the internet and download Edgar Allan Poe's "The Gold-Bug." Modify the "Word Music" case study above to use words from a file. Decide what part of this book you would like to sonify and create an input text file accordingly from Poe's e-book.

7.6 PYTHON WHILE LOOP

In addition to the `for` loop, Python also provides the `while` loop. While loops have the following format:

```
while <condition>:
    <body>
```

where `condition` is a boolean expression (built with variables, relational and logical operators), similar to the ones we have used in `if` statements.

While the condition evaluates to `True`, the loop continues to iterate (hence its name—the `while` loop). As soon as the condition evaluates to `False`, the loop terminates.*

---

* Clearly, in order for a `while` loop to terminate, the variable (or variables) used in the loop condition have to be altered inside the loop body. Why?

For example, the following `while` loop iterates 100 times.

```
i = 0
while i < 100:
    print i
    i = i + 1
```

This is equivalent to this `for` loop:

```
for i in range(100):
    print i
```

Actually, `for` loops are useful for iterating over a sequence (i.e., a fixed number of times). On the other hand, `while` loops are useful when we do not know in advance how many times we need to iterate. For example, consider the following:

```
# get input from user (perform error checking)
pitch = input("Enter a MIDI pitch (0-127): ") # get first value

while pitch < 0 or pitch > 127: # if value is wrong, try again

    print "The value you entered,", pitch,
    print "is outside the specified range. Please try again."
    pitch = input("Enter a MIDI pitch (0-127): ")

# now, we have a proper pitch value
```

Notice how the number of times this will loop is unknown—it depends on the user's input. It may loop 0 times (if the user enters a valid input on the first attempt) or, say, 12 times, if the user makes 12 errors (highly unlikely, but possible).*

Incidentally, this example performs error checking for the last case study in Chapter 2. Modify that program and run it again. `while` loops are great for implementing error checking and providing the user with opportunities to correct input errors.

7.6.1 Exercise

Write a program that reads in a file called "randomNumbers.txt," which contains a sequence of random integers between 0 and 10 (inclusive).

---

* Actually, an even better test would include checking the data type of the input, pitch, to ensure that the user entered an integer. This would be accomplished using the function `type(pitch)` seen in Chapter 2.

- Have it generate notes using these input values as pitches. Make it terminate when the input value is 0. (*Hint:* This program should use the same `while` loop pattern as earlier, that is, read the first value outside the loop and then start looping. Inside the loop the program should create a note from the last integer (pitch) read in and then get the next integer from the input file. The loop should terminate when the input value is 0.)

- Verify that your program terminates when the first value in the file is 0. In this case, no notes should be generated.

7.7 BIG DATA

As a result of the proliferation of computers and technologies for data collection and storage over the last 60 years, society is encountering a deluge of data. This phenomenon of our times is called *Big Data*.

Definition: *Big Data* refers to modern data sets that have become too voluminous or complex to analyze and comprehend with traditional computing techniques.

Sources for such data sets include meteorology, astronomy and other physical systems, genomics, Internet social networks, and finance/business statistics.

Computers are perfect for doing repetitive tasks without getting tired. We can use them to create sonifications and/or visualizations of massive data sets, such as biosignals, weather data, etc. This can help to observe, analyze, and comprehend patterns in such data sets.

This section shows how to do this using Python and the music library. It also introduces the image library, which allows us to manipulate data in images.

7.7.1 Case Study: Biosignal Sonification

In this case study, we explore the processing and sonification of data from biological processes. Figure 7.2 displays heart data, captured by measuring blood pressure over time.

Additionally, Figure 7.3 displays skin conductance, captured by measuring electrical conductivity between two fingers over time (the more sweaty the fingers get, the higher the skin conductance).

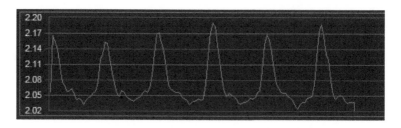

FIGURE 7.2   Sample raw heart data (x-axis is time, y-axis is pressure).

FIGURE 7.3   Sample skin-conductance data (x-axis is time, y-axis is skin conductance).

These images demonstrate the concept of visualization. Visualization is one way to experience big data.

Definition: *Visualization* involves mapping data to the visual characteristics of an image, such as lines, colors, size, shapes, etc., in order to perceive patterns and better appreciate their characteristics.[*]

Visualization is complementary to sonification, in that the human visual and auditory processing systems have evolved to perceive different types of patterns. Since the focus of this book is on music and programming, here we explore sonifying biological signals.[†]

The data presented in Figures 7.2 and 7.3 are actually stored in a data file. In order to sonify these data, we first need to understand their format (i.e., how they are stored in our data file).

In the following is a sample of the actual raw data displayed in Figures 7.2 and 7.3:

```
20:39:51.560          1.84          1.880
20:39:51.593          3.13          1.953
20:39:51.627          3.14          1.970
20:39:51.660          3.13          1.975
20:39:51.693          3.13          1.969
```

---

[*] Compare with the definition of data sonification at the beginning of the chapter.
[†] Nevertheless, the concepts presented here (e.g., data mapping) also apply to visualization.

```
20:39:51.727          3.14          1.978
20:39:51.760          3.13          2.027
20:39:51.793          3.13          2.315
20:39:51.827          3.14          2.489
20:39:51.860          3.14          2.466
```

The preceding data was captured at a rate of approximately 30 measurements per second. The complete data file is available online at http://jythonMusic.org.

The data format consists of three columns (fields). These are the time of measurement (e.g., 20:39:51.560), the skin conductance at that time (e.g., 1.84), and the particular blood pressure at the time (e.g., 1.880).

7.7.1.1 Sonification Design

To analyze data through sonification, we need to find a way to map these data to sound. We pose the following questions: How can we map characteristics of these data to musical parameters? Are there some characteristics that are more important than others? Are there certain musical parameters better suited to sonify these data characteristics? For a given data set, there may be many ways to answer these questions.

In the following are some possibilities for the preceding data set.

First of all, there is no correct way to map data to sound. Again, the trick is to decide what aspects of the data you would like to make easily perceivable by mapping them to sound parameters. Moreover, in the context of music making, you might also consider what aspects of the data might contribute to more interesting music.

Here are some possibilities:

- Map skin data to pitch (remember to scale to a preferred integer range, e.g., C3–C6).

- Map heart data also to pitch (e.g., add some variety to pitch).

- Map heart data to dynamic (remember to scale to 0–127).

The following program, sonifyBiosignals.py, demonstrates these rules:

```
# sonifyBiosignals.py
#
# Sonify skin conductance and heart data to pitch and dynamic.
#
# Sonification design:
#
```

```
# * Skin conductance is mapped to pitch (C3 - C6).
# * Heart value is mapped to a pitch variation (0 to 24).
# * Heart value is mapped to dynamic (0 - 127).
#
# NOTE: We quantize pitches to the C Major scale.
#

from music import *
from string import *

# first let's read in the data
data = open("biosignals.txt", "r")

# read and process every line
skinData = []   # holds skin data
heartData = []  # holds heart data
for line in data:

   time, skin, heart = split(line) # extract the three values
   skin = float(skin)              # convert from string to float
   heart = float(heart)            # convert from string to float
   skinData.append(skin)           # keep the skin data
   heartData.append(heart)         # keep the heart data

# now, heartData contains all the heart values

data.close()    # done, so let's close the file

##### define the data structure
biomusicScore   = Score("Biosignal sonification", 150)
biomusicPart    = Part(PIANO, 0)
biomusicPhrase  = Phrase()

# let's find the range extremes
heartMinValue   = min(heartData)
heartMaxValue   = max(heartData)
skinMinValue    = min(skinData)
skinMaxValue    = max(skinData)

# let's sonify the data
i = 0;   # point to first value in data
while i < len(heartData):   # while there are more values, loop

   # map skin-conductance to pitch
   pitch = mapScale(skinData[i], skinMinValue, skinMaxValue, C3, C6,
                    MAJOR_SCALE, C4)

   # map heart data to a variation of pitch
   pitchVariation = mapScale(heartData[i], heartMinValue,)
                             heartMaxValue, 0, 24, MAJOR_SCALE, C4)

   # also map heart data to dynamic
   dynamic = mapValue(heartData[i], heartMinValue, heartMaxValue,
                      0, 127)
```

```
# finally, combine pitch, pitch variation, and dynamic into note
note = Note(pitch + pitchVariation, TN, dynamic)

# add it to the melody so far
biomusicPhrase.addNote(note)

# point to next value in heart and skin data
i= i + 1

# now, biomusicPhrase contains all the sonified values

##### combine musical material
biomusicPart.addPhrase(biomusicPhrase)
biomusicScore.addPart(biomusicPart)

##### view score and write it to a MIDI file
View.sketch(biomusicScore)
Write.midi(biomusicScore, "sonifyBiosignals.mid")
Play.midi(biomusicScore)
```

First we import the necessary libraries and open the data file. Then we read every line from the file, using a loop, and split the data into its constituent parts. It so happens that the raw heart data is the third number on every line.*

```
# read and process every line
skinData = []   # holds skin data
heartData = []  # holds heart data
for line in data:

time, skin, heart = split(line)# extract the three values
skin = float(skin)                      # convert from string to float
heart = float(heart)                    # convert from string to float
skinData.append(skin)                   # keep the skin data
heartData.append(heart)                 # keep the heart data

# now, heartData contains all the heart values

data.close()    # done, so let's close the file
```

Notice how we create an empty list, heartData, to hold the heart rate data from the data file. Inside the loop, we get a line from the data file at a time, extract the value of interest (i.e., raw heart data), and append it to this list. This is a very common pattern in data manipulation, called the *accumulator pattern*.

---

* In most sonification (or other data processing) tasks, you may need to clean up the data file (e.g., remove a header line or some copyright text) before loading it into your program. Doing this manually (i.e., with a text editor) saves time and simplifies your code.

Definition: The *accumulator pattern* uses a loop to accumulate data, one item per iteration, in some container. For example, appending to a list, concatenating to a string, or adding to an integer variable.

Notice how we use the string function, `split(line)`, to separate the single line (a string) to be split into a list of three substrings. The statement

```
time, skin, heart = split(line)
```

automatically "unpacks" the list of three substrings returned by `split(line)` to three strings and assigns each to the corresponding variable.

One thing about big data is that the data files may need some "massaging" (preprocessing) to end up in a format we want. In the above example, we assume that the data file has gone through this step, and thus no error checking is required in the code.

7.7.1.2 Python Parallel Assignment

Python provides a shorthand operation to assign several variables, in parallel, to corresponding values.

For example,

```
>>> o, p, q = [4, 5, 6]
>>> o
4
>>> p
5
>>> q
6
```

Notice how the three variables, o, p, and q, automatically get assigned to each item in the list.

```
>>> a, b, c = "123"
>>> a
'1'
>>> b
'2'
>>> c
'3'
```

This also works with strings (since both strings and lists are sequences). Finally, parallel assignment allows us to easily swap values between two variables:

```
>>> x = 2
>>> y = 3
>>> x
```

```
2
>>> y
3
>>> x, y = y, x
>>> x
3
>>> y
2
```

Notice how x and y are originally 2 and 3, respectively. After the parallel assignment (i.e., x, y = y, x), their values have been reversed. Parallel assignment is a very convenient operation.

Let's get back to the program. The line

```
heart = float(heart)
```

converts the string representation of a number to a float number. This is necessary since data from a file is read in as a string.

Upon completion of the loop, all skin and heart data values are now accumulated in the skinData and heartData lists.

The rest of the program performs the sonification of these values. First we create the musical data structure,

```
##### define the data structure
biomusicScore       = Score("Biosignal sonification", 60)
biomusicPart        = Part(PIANO, 0)
biomusicPhrase      = Phrase()
```

Then we find the minimum and maximum data values (needed by mapValue() below).

```
# let's find the range extremes
minValue = min(heartData)
maxValue = max(heartData)
skinMinValue = min(skinData)
skinMaxValue = max(skinData)
```

We loop through the list of float numbers and sonify them. We have a few sonification rules:

- Each skin conductance value is mapped to a pitch in the range C3 to C6 (using the major scale).

- Each heart value is mapped to a pitch variation in the range 0 to 24. The idea is to combine both data values (skin conductance and heart) in a single melody. This makes sense since skin conductance tends to be very monotonous (without much fluctuation—see Figure 7.3).

The heart value is producing interesting heart-like fluctuations (see Figure 7.2). Alternatively, we could use two melodies (and observe the harmonies they generate).

- Since a note's dynamic level is an important musical parameter, we also map the heart values to it, in order to generate interesting dynamic variation.

- For simplicity, we leave the note duration constant (TN).*

This is just one of many possible mappings of the data. The following code implements this sonification design:

```
# let's sonify the data
i = 0;   # point to first value in data
while i < len(heartData):   # while there are more values, loop

    # map skin-conductance to pitch
    pitch = mapScale(skinData[i], skinMinValue, skinMaxValue, C3, C6,
                     MAJOR_SCALE, C4)

    # map heart data to a variation of pitch
    pitchVariation = mapScale(heartData[i], heartMinValue,
                              heartMaxValue, 0, 24, MAJOR_SCALE, C4)

    # also map heart data to dynamic
    dynamic = mapValue(heartData[i], heartMinValue, heartMaxValue,
                       0, 127)

    # finally, combine pitch, pitch variation, and dynamic into note
    note = Note(pitch + pitchVariation, TN, dynamic)

    # add it to the melody so far
    biomusicPhrase.addNote(note)

    # point to next value in heart and skin data
    i = i + 1
```

Notice the use of a `while` loop (instead of a `for` loop). This is equivalent to

```
for i in range(len(heartData)):
```

Notice how, with the `while` loop, we need to manually handle the initialization of i to 0 (right before the loop) and the increment of i (i.e., `i = i + 1`, at the very end of the loop).

Upon completion of the loop, `biomusicPhrase` contains all the notes resulting from sonifying the heart data. The rest of the program is

---

* Of course, note duration is another valuable musical parameter which could be used effectively to communicate patterns in data.

straightforward. Notice how the `view.sketch()` function creates a visualization (of the sonification) of the data. Actually, we could focus more on this visual representation, as opposed to the sonification itself. In other words, we could use the musical notes as an intermediate form to create visualizations of data. In the next section, we will see another way to possibly create visualizations, by manipulating image files. Still, the primary focus of this book remains making music with Python.

7.7.2 Exercises

1. Skin-conductance level could be sonified as a separate melody. Then both "melodies" could be played in parallel. Also, we could use different MIDI instruments for the two melodies by putting them in different parts on different MIDI channels. What other combinations are possible? See the Kepler case study, presented earlier, for inspiration.

2. What other ways of sonifying the heart data can you think of? For example, you could capture the difference between two values and use it as note duration. (*Hint*: Remember to take the absolute value of the difference, i.e., use function `abs()`, since the difference may be negative.) In such a sonification, the duration of the note signifies the change in heart contraction between two consecutive measurements. Since these measurements are equally spaced in time, the longer the note, the more work was exerted by the heart (during those two measurements). This is a useful pattern, which is currently lost in a sea of data. What other useful patterns can you think of?

3. The US National Climatic Data Center (http://www.ncdc.noaa.gov) has a vast archive of historical weather (and other) data. For example, use the provided average monthly temperatures for California from 1895 to 2004. Find ways to make apparent the trends of global warming through sonification (and visualization) of data.

4. Find other sources of data (e.g., stock market, population sizes, astronomical data) and try to answer important questions or present interesting aspects of these data through sonification.

7.8 PYTHON FUNCTIONS

So far, we have seen various useful Python functions, such as `abs()`, `round()`, `len()`, `mapValue()`, and `upper()`. These functions are already built into Python (or one of its libraries), so that you can easily perform a needed task. For example,

```
>>> x = -3
>>> x = abs(x)
>>> x
3

>>> y = 3.6
>>> y = round(y)
>>> y
4.0

>>> z = ["this", "is", "fun"]
>>> len(z)
3
```

A nice feature of Python is that it allows you to create your own functions.

This is useful when you have code that performs a common task, which you end up using many times in your program. Defining a function allows you to "package" that code, so that you can reuse it easily, as we saw earlier.

Definition: A Python *function* is a piece of code that performs a specific task. Once defined, it may be used many times, without having to write it again and again.

7.8.1 Defining Functions

Functions are very easy to define. Simply take the code you want to package as a function, indent it (by, say, 3 spaces), and add a header line:

```
def functionName():
   <your indented code goes here>
```

For example, let us define a function that plays a violin note:

```
# define function to play a C4 quarter note using violin sound
def playViolinNote():
   note = Note(C4, QN)
   phrase = Phrase()
   phrase.addNote(note)
   phrase.setInstrument(VIOLIN)
   Play.midi(phrase)
```

The first line, in bold, is called the *function header*. It specifies the name to use when calling this function. Any arguments the function may need to get its job done are listed between the parentheses. In the earlier example, function `playViolingNote()` requires no argument.

Now, we can call function `playViolinNote()` as follows*:

```
playViolinNote()
```

Every time you call this function, it plays a C4 quarter note using a violin timbre.

Functions are more useful (or general) when you can pass information to them, to customize what they do. For example, the preceding function definition can become more versatile by allowing us to specify the note's pitch and duration:

```
# define function to play a specified note using violin sound
def playViolinNote(pitch, duration):
   note = Note(pitch, duration)
   phrase = Phrase()
   phrase.addNote(note)
   phrase.setInstrument(VIOLIN)
   Play.midi(phrase)
```

Now, we can call this function as follows:

```
>>> playViolinNote(C4, WN)
>>> playViolinNote(E4, HN)
>>> playViolinNote(G4, QN)
```

to play different notes. Notice how much more versatile this function has become.

In summary, functions hide unnecessary details from the main code. Also, they allow us to reuse code efficiently, because we define a function once and use it many times. Without functions, our programs would be much larger and harder to understand and modify.

7.8.2 Exercise

Define a function called `playNote()` which accepts a pitch, duration, and instrument. (You may assume the music library has already been imported.) Try it out with different notes and instruments (see Appendix A).

---

* For simplicity, let us assume the music library has already been imported.

7.8.3 Returning Values

Python functions may return a value. This can be very useful when you are "packaging" code that produces a value (e.g., a number, a musical phrase, etc.) to be used by the caller.

For example, here is our own implementation of Python's `abs()` function. This function returns the absolute value of a number.

```
# returns the absolute value of a number
def absolute(number):
    if number < 0:   # is the argument negative?
        number = -1 * number   # yes, so make it positive

    # now, the number is positive (either way)
    return number
```

Notice the return statement at the bottom of the function body. When evaluated, this statement passes back the specified value to the caller of the function. For example,

```
>>> x = absolute(-2.3)
>>> x
2.3
```

In other words, when we call `absolute(-2.3)`, the argument, −2.3, is associated with the parameter number (see function definition) above. Then the body of the function executes with variable `number` having the value −2.3. The `if` statement's condition evaluates to True, and so the sign of number changes (multiplying a number by −1 changes its sign). Finally, the `return` statement takes the value of variable number and returns it to the caller (as seen in the preceding interpreter example).

A `return` statement is similar to a `print` statement. Both take a value (or expression) and send it somewhere else. The `print` statement sends it to the computer screen. The return statement sends the value back to the statement that called the function. The difference is that, unlike `print`, the `return` statement also stops the function from executing, so that a value may be returned to the caller, and so the caller can continue its work.

A function may have several `return` statements. As soon as one is encountered, the function stops executing, and control returns to the caller. However, it is considered bad style to have more than one `return` statement.

Good Style: A function should have at most one `return` statement.[*]

If a function does not have a `return` statement, it stops executing after the last statement in its body. Then control returns to the caller.

Let us see another example.

Python provides various functions for lists, such as `len()`, `sum()`, `min()`, and `max()`. However, Python does not provide a function for calculating the average of a list. Here we define such a function:

```
# returns the average of a numeric list
def avg(someList):

   total  = sum(someList)  # get the total of all items
   length = len(someList)  # get the number of items

   # next, convert to float (for accuracy) and calculate average
   result = float(total) / float(length)

   # finally, return result to caller
   return result
```

7.8.4 Exercises

1. Division by zero is always a problem. Update the preceding function to check if length is 0. If so, it should output an error message and return the special value None. Or you can use the statement:

   ```
   raise ZeroDivisionError(string)
   ```

   where string contains the error message. This will result in a typical Python error message similar to the ones we first saw in Chapter 2.

2. Write a function called `createMelody()`, which returns a phrase of notes with pitches and durations randomly selected from a scale and a list, respectively. The function should accept three parameters,

---

[*] This is a special case of the software engineering guideline "one-way in, one-way out" that suggests there should be only one way to get into a piece of code and only one way to get out of it. By following this guideline, your code becomes easier to understand and modify. In the case of multiple `return` statements, this guideline is violated. Most modern programming languages provide structures that enforce the one-way in, one-way out style of programming.

namely, `numNotes`, `scale`, and `durations` (a list of durations) to choose from. For example, the following call

```
phrase = createMelody(10, PENTATONIC_SCALE, [QN, SN, EN])
```

should generate a phrase of 10 notes from the pentatonic scale with durations randomly selected from the provided list. (*Hint:* See the "Pentatonic Melody Generator" case study in Chapter 6.)

3. Write a function named `createCantusVoice()`, which returns a phrase containing one of the voices in Arvo Part's "Cantus in Memoriam" seen in Chapter 4. The function should accept three parameters, namely, `pitches`, `durations`, and an `elongationAmount` (a float). For example, the following

```
pitches = [A5, G5, F5, E5, D5, C5, B4, A4]
durations = [HN, QN, HN, QN, HN, QN, HN, QN]
voice = createCantusVoice(pitches, durations, 1.0)
```

should generate a phrase with 8 notes starting with an A5 half note (HN), followed by a G5 quarter note (QN),... and ending with an A4 quarter note. Whereas the following

```
voice = createCantusVoice(pitches, durations, 2.0)
```

should generate a phrase with 8 notes starting with an A5 whole note (WN), followed by a G5 half note (HN),... and ending with an A4 half note. (*Hint:* You may use `Mod.elongate()` in the body of your function.)

4. Add two parameters to the `createCantusVoice()` function defined earlier, namely, transposition and repetitions. The first specifies the amount by which to transpose the pitches in the result, and the second the number of times to repeat the result. (*Hint:* You may use `Mod.transpose()` and `Mod.repeat()` in the function body.)

5. Using the `createCantusVoice()` function defined earlier, write a program that generates variations on Arvo Pärt's compositional idea, using different scales (e.g., C major scale) and different pitch movement (e.g., upward motion, or downward and upward motion in one scale, and so on). For example, the program may use the defined function as follows:

```
pitches = [A5, G5, F5, E5, D5, C5, B4, A4]
durations = [HN, QN, HN, QN, HN, QN, HN, QN]
```

```
voice1 = createCantusVoice(pitches, durations, 0, 1.0)
voice2 = createCantusVoice(pitches, durations, -12, 2.0)
voice3 = createCantusVoice(pitches, durations, -24, 4.0)
voice4 = createCantusVoice(pitches, durations, -36, 8.0)
```

to generate the different voices in the piece before adding them to a part.

7.8.5 Scope of Variables

In Python, variables that are declared outside a function are considered global; they have global scope. Global variables are visible anywhere within the program and can be accessed even from within functions. Variables that are defined within a function are only visible (accessible) within that function. We say that the scope of a variable is limited to the function.

Definition: The *scope* of a variable is the space within a program where that variable is visible (i.e., it can be accessed and possibly modified).

Ideally, a function should avoid using global variables.

Good Style: If a function needs any data, then that data should be passed to it through its parameter list.

Good Style: If a function changes any data outside its scope, it should return them via the return statement.

7.9 IMAGE SONIFICATION

One very interesting application of sonification is to make music from aesthetically pleasing or otherwise interesting images. Again, the idea is to find ways to map interesting patterns in one medium (in this case, images) to interesting patterns in another medium (in this case, sound). Image sonification is accomplished by reading in an image, which is a two-dimensional list of pixels (or pictures elements), and exploring ways to map these pixels to notes. Each pixel contains information about color in one small area of a picture.

7.9.1 Python Images

Your Python programs can read, manipulate, and write images through the `image` library. To access the `image` library, import it.

```
from image import *
```

An image is represented by Python as a two-dimensional table (i.e., matrix) of pixels.

Definition: An *image* is a table of x (width) * y (height) pixels.

For example, the following code creates a blank (black) image of 100 by 100 pixels.

```
img = Image (100, 100)
```

So an image has width * height number of pixels. The origin (0, 0) is at top left.

Definition: A *pixel* is a picture element or a point in an image.

Each pixel contains a list of three values; the red, green, and blue (or RGB) components of that point in the image. Each RGB value ranges from 0 to 255. For example,

- `[255, 255, 255]` is white

- `[0, 0, 0]` is black

- `[255, 0, 0]` is bright red

- `[150, 0, 0]` is a darker red

- `[0, 255, 0]` is bright green

- `[0, 0, 255]` is bright blue

- `[255, 255, 0]` is yellow

- `[150, 150, 150]` is a tone of gray

7.9.2 Image Library Functions

The Python `image` library provides various useful functions.[*] The most common ones are shown in Table 7.2.

The following case study applies some of these functions to generate music from images.

---

[*] The image library is not standard. Similarly to the music library, it is provided with this textbook.

TABLE 7.2 Image library functions (assume `img` is an image)

Function	Description
`Image(filename)`	Reads in a jpg or png file called `filename` (a string) and shows an image. It returns the image, so it should be stored in a variable, for example, `img = Image("sunset.jpg")`
`Image(width, height)`	Returns an empty (blank) image with provided `width` and `height`. It returns the image, so it should be stored in a variable, for example, `img = Image(200, 300)`
`img.getWidth()`	Returns the width of image `img`.
`img.getHeight()`	Returns the height of image `img`.
`img.getPixel(col, row)`	Returns this pixel's RGB values (a list, e.g., [255, 0, 0]), where `col` is the image column, and `row` is the image row. The image origin $(0, 0)$ is at top left.
`img.setPixel(col, row, RGBlist)`	Sets this pixel's RGB values, for example [255, 0, 0], where `col` is the image column, and `row` is the image row. The image origin $(0, 0)$ is at top left.
`img.show()`	Displays the image `img` in a window.
`img.hide()`	Hides the image window (if any).
`img.write(filename)`	Writes image `img` to the jpg or png `filename`.

7.9.3 Case Study: Visual Soundscape

A soundscape refers to a musical composition that incorporates sounds recorded from, and/or music that depicts the characteristics of, an environment (e.g., a city soundscape or a forest soundscape).

In this case study, we explore how, through sonification of image data using image library functions, we create interesting musical artifacts by mapping visual aspects of an image into corresponding musical aspects. Also, similarly to sonification of other media (e.g., biosignals, weather data, financial markets, etc.), we may capture and explore structural characteristics of images that may not be as evident visually.

As we mentioned earlier, images consist of pixels (or picture elements). A pixel is the elemental data that digitally represents a single point in the original scene (as captured by a camera). The number of pixels in an image depends on the quality (or resolution) of the digital camera. For example, Figure 7.4 (depicting a sunset at the Loutraki seaside resort in Greece) consists of 320×213 pixels.

Similarly to other sonification activities, the ways we can map pixels to sound are not prescribed. A rule of thumb is to find what inspires you

FIGURE 7.4   Loutraki sunset. (Available for download at jythonMusic.org.)

about a particular image and explore how you might convert that to sound. So image sonification involves imagination and artistic exploration.

The image in Figure 7.4 has a very nice gradient that gets brighter from left to right. The sun is not shown but can be imagined. There is a clear horizontal division between the sea and sky. The mountains, on the left, provide a contrast to the color of the sea and sky. Finally, the image gradient is interrupted by the (somewhat noisy) visual layers and the sea at the bottom half of the image. Clearly, there is enough structural variety in the visual domain to provide interesting analogies in the musical domain. All this can be exploited by selecting certain rows (or columns) of pixels, scanning the image left-to-right (or up-and-down), and converting individual pixels or areas of pixels to musical notes or passages.

7.9.3.1 Sonification Design
In this case study, we use the following sonification rules:

- Left-to-right pixel (column) position is mapped to time (actually, note start time);

- Brightness (or luminosity) of a pixel (i.e., average RGB value) is mapped to pitch (the brighter the pixel, the higher the pitch);

- Redness of a pixel (R value) is mapped to duration (the redder the pixel, the longer the note); and

- Blueness of a pixel (B value) is mapped to dynamic (the bluer the pixel, the louder the note).

These mappings were selected as a result of experimentation, simply because they sounded interesting, given this image. A little experimentation will go a long way.

Using the same sonification scheme with other images will likely generate interesting results. For instance, you could select a new image with this scheme in mind. Or you could create/modify an image (e.g., via Photoshop) with the particular sonification scheme in mind.

However, the most appropriate way is to pick an image and then design a set of sonification rules to use that matches its features. The image choice and sonification rules are intimately connected.

The following program implements the preceding rules to sonify the image in Figure 7.4.

```python
# sonifyImage.py
#
# Demonstrates how to create a soundscape from an image.
# It also demonstrates how to use functions.
# It loads a jpg image and scans it from left to right.
# Pixels are mapped to notes using these rules:
#
# + left-to-right column position is mapped to time,
# + luminosity (pixel brightness) is mapped to pitch within a scale,
# + redness (pixel R value) is mapped to duration, and
# + blueness (pixel B value) is mapped to volume.
#

from music import *
from image import *
from random import *


##### define data structure
soundscapeScore = Score("Loutraki Soundscape", 60)
soundscapePart  = Part(PIANO, 0)


##### define  musical parameters
scale = MIXOLYDIAN_SCALE

minPitch = 0           # MIDI pitch (0-127)
maxPitch = 127

minDuration = 0.8      # duration (1.0 is QN)
maxDuration = 6.0

minVolume = 0          # MIDI velocity (0-127)
maxVolume = 127

# start time is randomly displaced by one of these
# durations (for variety)
timeDisplacement = [DEN, EN, SN, TN]
```

```
##### read in image (origin (0, 0) is at top left)
image = Image("soundscapeLoutrakiSunset.jpg")

# specify image pixel rows to sonify - this depends on the image!
pixelRows = [0, 53, 106, 159, 212]
width = image.getWidth()            # get number of columns in image
height = image.getHeight()          # get number of rows in image

##### define function to sonify one pixel
# Returns a note wrapped in a phrase created from sonifying
# the RGB values of 'pixel' found on given column.
def sonifyPixel(pixel, col):

  red, green, blue = pixel  # get pixel RGB value

  luminosity = (red + green + blue) / 3          # calculate brightness

  # map luminosity to pitch (the brighter the pixel, the higher
  # the pitch) using specified scale
  pitch = mapScale(luminosity, 0, 255, minPitch, maxPitch, scale)

  # map red value to duration (the redder the pixel, the longer
  # the note)
  duration = mapValue(red, 0, 255, minDuration, maxDuration)

  # map blue value to dynamic (the bluer the pixel, the louder
  # the note)
  dynamic = mapValue(blue, 0, 255, minVolume, maxVolume)

  # create note and return it to caller
  note = Note(pitch, duration, dynamic)

  # use column value as note start time (e.g., 0.0, 1.0, and so on)
  startTime = float(col)        # time is a float

  # add some random displacement for variety
  startTime = startTime + choice( timeDisplacement )

  # wrap note in a phrase to give it a start time
  # (Phrases have start time, Notes do not)
  phrase = Phrase(startTime)    # create phrase with given start time
  phrase.addNote(note)          # and put note in it

  # done sonifying this pixel, so return result
  return phrase

##### create musical data

# sonify image pixels
for row in pixelRows:            # iterate through selected rows

  for col in range(width):    # iterate through all pixels on this row
```

```
# get pixel at current coordinates (col and row)
pixel = image.getPixel(col, row)

# sonify this pixel (we get a note wrapped in a phrase)
phrase = sonifyPixel(pixel, col)

# put result in part
soundscapePart.addPhrase(phrase)

# now, all pixels on this row have been sonified

# now, all pixelRows have been sonified, and soundscapePart
# contains all notes

##### combine musical material
soundscapeScore.addPart(soundscapePart)

##### view score and write it to an audio and MIDI files
View.sketch(soundscapeScore)
Write.midi(soundscapeScore, "soundscapeLoutrakiSunset.mid")
```

First, notice how the top-level comments describe what the program does and, in particular, the sonification rules employed by it. Keeping such comments at the beginning allows you to understand what a program does several months/years afterwards, without having to read the code again line for line. It takes great discipline to write quality comments. However, it saves time later on.* Actually, experienced programmers first write comments and then code. A common programming aphorism goes, "if you cannot comment it, you cannot code it." This is especially true for writing elegant code.

Notice how we have grouped the definitions of musical parameters at the top. These values are set once and are used throughout the program.

Next, we load the image and define a list of pixel rows to sonify. The program generates one melodic line from each pixel row. Figure 7.5 shows these pixel rows, where row 0 is at the top, row 53 is the second from the top, and so on.

```
pixelRows = [0, 53, 106, 159, 212]
```

Usually, it is a bad idea to hard-code such numbers in a program. For instance, what would happen if we changed the image we read in

---

* As mentioned earlier, most programs are written once, but modified over and over.

FIGURE 7.5   Loutraki sunset with highlighting of rows 0 (top), 53 (second from top), 106 (third), 159 (forth), and 212 (fifth).

(e.g., made it smaller or larger)? We make an exception here and hard code the numbers. Our rationale is that the choice of rows to sonify depends on the selected image. Hence, we can think of these as constants related to the selected image.

7.9.3.2 Defining a Function

Next comes the function definition. This function expects a pixel and its column coordinate. The pixel provides the RGB values for pitch, duration, and dynamic. The column is used to specify note start time (the piece evolves from left to right). By encapsulating the sonification scheme in a function, we help modularize our code. This makes it easier to change our code in the future, for example, to create different sonification schemes (and encapsulate them into different functions).

```
##### define function to sonify one pixel
# Returns a note wrapped in a phrase created from sonifying
# the RGB values of 'pixel' found on given column.
def sonifyPixel(pixel, col):

   red, green, blue = pixel  # get pixel RGB value

   luminosity = (red + green + blue) / 3   # calculate brightness

   # map luminosity to pitch (the brighter the pixel, the higher
   # the pitch) using specified scale
   pitch = mapScale(luminosity, 0, 255, minPitch, maxPitch, scale)
```

```
# map red value to duration (the redder the pixel, the longer
# the note)
duration = mapValue(red, 0, 255, minDuration, maxDuration)

# map blue value to dynamic (the bluer the pixel, the louder
# the note)
dynamic = mapValue(blue, 0, 255, minVolume, maxVolume)

# create note and return it to caller
note = Note(pitch, duration, dynamic)

# use column value as note start time (e.g., 0.0, 1.0, and so on)
startTime = float(col)        # time is a float

# add some random displacement for variety
startTime = startTime + choice( timeDisplacement )

# wrap note in a phrase to give it a start time
# (Phrases have start time, Notes do not)
phrase = Phrase(startTime)    # create phrase with given start time
phrase.addNote(note)          # and put note in it

# done sonifying this pixel, so return result
return phrase
```

Some general things to know about functions:

- Functions need to be defined before they are called.

- As with variable names, try to come up with meaningful function names (e.g., sonifyPixel). If the name of the function communicates what it does, this makes our code easier to read.

- Notice the two parameters, pixel and column. These variables will store the values provided when the function is called. We use these variables in the function body to make use of these values.

- The body of the function is indented. It contains all the code that implements what the function does. This is similar to the bodies of loops and if statements.

Now let us explore what this function does. The first statement,

```
red, green, blue = pixel
```

is parallel assignment. Variable pixel contains a list of three RGB values, for example, [243, 124, 0]. The parallel assignment provides a convenient way to "unpack" the three RGB values from pixel.

The rest of the function body uses red, green, and blue to calculate the note pitch, duration, and dynamic, according to the sonification scheme described earlier.

The following statements

```
startTime = float(col) # time is a float

# add some random displacement for variety
startTime = startTime + choice(timeDisplacement)
```

use the column coordinate of the pixel to specify start time for the note created for this pixel. Since notes do not have a start time, we wrap the note into a phrase and set the phrase's start time accordingly. Finally, since the columns are pretty rigid, that is, they take integral values (i.e., 0, 1, 2, 3, and so on), we add a random displacement (selected from the list timeDisplacement, i.e., [DEN, EN, SN, TN]). This makes the music less "square."

The last statement in the function

```
return phrase
```

returns the created phrase to the function caller.

7.9.4 Python Nested Loops (again)

Next we use two loops, one inside the other. Again, this is called a *nested loop*.

The outer loop iterates as usual. For each iteration of the outer loop, the inner loop does a complete run (i.e., it does all its iterations).* A helpful trick is to think of the inner loop as a simple statement. As all statements in the body of the outer loop, it needs to be executed completely.

```
##### create musical data

# sonify image pixels
for row in pixelRows:          # iterate through selected rows

  for col in range(width):     # iterate through all pixels on this row

    # get pixel at current coordinates (col and row)
    pixel = image.getPixel(col, row)

    # sonify this pixel (we get a note wrapped in a phrase)
    phrase = sonifyPixel(pixel, col)
```

---

* For example, if the outer loop iterates 5 times and the inner loop iterates 10 times, then for each iteration of the outer loop, the inner loop will iterate 10 times, for a total of 50 times.

```
  # put result in part
  soundscapePart.addPhrase(phrase)

 # now, all pixels on this row have been sonified

# now, all pixelRows have been sonified, and soundscapePart
# contains all notes
```

As the earlier comments indicate, every time the inner loop finishes, one more row of pixels has been sonified. The inner loop gets every pixel on this row, sonifies it (using the `sonifyPixel()` function), and adds the sonified result (a phrase) to `soundscapePart`.

The outer loop repeats this process for every row in the `pixelRows` list. Changing `pixelRows` (at the top of the program) will automatically adjust which areas of the image are sonified.

We end the program as usual, that is, by putting the part in a score, visualizing the score, and saving the score in a MIDI file.

7.9.5 Exercises

The earlier program is only the beginning. Many avenues for experimentation exist.

1. Try different musical scales, such as MAJOR_SCALE and PENTATONIC_SCALE.

2. Modify the above program to put each row into a different part (i.e., define a total of 5 parts). Experiment with different instruments per part. Also, experiment with transposing each part into a different register (octave).

3. Select a different image (e.g., try a symmetrical or fractal image). What sonification ideas come to mind? How can you map the symmetry to musical form most effectively?

4. Try other possibilities such as include scanning the image top-to-bottom (as opposed to left-to-right), averaging areas of the image to generate musical events, changing how we sonify each pixel, mapping pixels to chords (as opposed to single notes), opening several images at once, and so on.

7.10 SUMMARY

This chapter has explored some ideas and programming techniques for sonifying and visualizing data. Importantly, they may be used to perceive hidden, important patterns in big data. Once you understand the principles of sonification, the sky is the limit! Topics covered include data scaling using the `mapValue()` and `mapScale()` functions. A number of Python language features have been introduced. These include strings, reading and writing files, the `while` loop, and how to define your own functions. These have been applied to sonification case studies that demonstrate how these can be used for both artistic and scientific objectives. Explore and enjoy.

## CHAPTER 8

Interactive Musical Instruments

Topics: Computer musical instruments, graphical user interfaces, graphics objects and widgets, event-driven programming, callback functions, Play class, continuous pitch, audio samples, MIDI sequences, paper prototyping, iterative refinement, keyboard events, mouse events, virtual piano, parallel lists, scheduling future events.

8.1 OVERVIEW

This chapter explores graphical user interfaces and the development of interactive musical instruments. While it takes years to master playing a guitar or violin, it is much easier for beginners to play a computer-based musical instrument, especially if they already have experience with computer games and other graphical applications. The goal here is not to replace traditional instruments or to undermine the years of experience required to achieve expertise in performing with them, but to introduce and engage more people in musical performance and composition. Also, interactive computer-based musical instruments offer considerable versatility. They can be used by a single performer or by multiple performers in ensembles, like Laptop Orchestras or iPad Ensembles. It is also possible to have an ensemble that includes both traditional instruments and computer-based instruments. Computer-based musical instruments can be customized to particular compositions or musical styles. They may be used to perform music at impromptu musical happenings (such as at a party or a coffee shop) or be part of formally composed, avant garde orchestral pieces.

8.2 BUILDING MUSICAL INSTRUMENTS

To assist live interaction with our programs, we explore graphical user interfaces (GUIs) in this chapter. Through the various GUI primitives available, we can design unique interactive musical instruments for live performance. While such instruments use a GUI for interaction, under the hood they utilize the various Python building blocks we have seen so far. Additionally, developing interactive musical instruments through Python (i.e., using the GUI and music libraries) facilitates music making by other people, including your friends, who do not know how to program in Python. You can design a GUI for each specific piece or music-making activity. Then, through this GUI, the instrument players (i.e., the "end-users") can make music without having to know how to program in Python.

You (the programmer and GUI designer) become a "digital luthier" of sorts—a digital instrument maker, who can enable and shape the musical experience of others. Your ability to program musical interactions and design unique, innovative GUIs gives you immense power, because your design choices will affect the type of music your end-users can make and, ultimately, their creative expression.

The ability to shape (to both enhance and constrain) other people's expression through your programs results in software similar to that of any ready-made product, including ready-made music production software (such as GarageBand, Audacity, and Ableton Live). Such software tools impose conceptual and compositional constraints, which were put in place (either by design or inadvertently) by their developers. These constraints focus the types of musical expression possible through them, as discussed by Magnusson (2010). This is why it is so important to know how to develop your own tools—so you can explore new ways to channel your creative expression. This argument is made very clearly and forcefully by Douglas Rushkoff in *Program or Be Programmed* (Rushkoff 2010).

Of course, building effective, usable instruments (digital or physical) is hard work. In this chapter, we will lay the foundation by learning some basic human–computer interaction (HCI) ideas, including usability, design analysis concepts, and paper prototyping. We will also learn the basics of GUI development, using the provided GUI library (a simplified, yet powerful library for developing graphical user interfaces). Finally, we will further explore functions and event-driven programming.

8.3 GRAPHICAL USER INTERFACES

In this chapter we develop music applications that have a graphical user interface (GUI). For this we use the GUI library provided with the book. As with the music library, the GUI library follows Alan Kay's maxim that "Simple things should be simple, and complex things should be possible"

(Davidson 1993). Appendix C contains the complete list of graphical objects and GUI widgets available in this library.

To access the GUI library, you need to use the following statement:

```
from gui import *
```

From that point on, your program can access all the available GUI objects and functionality.

8.3.1 Creating Displays

A program's GUI exists inside a display (window). Displays contain other GUI components (graphics objects and widgets). A program may have several displays open. Displays may contain any number of GUI components, but they cannot contain another display.

The Display class is the most important component in the GUI library. Display objects have the following attributes:

- Title—a string, which appears at the top display border

- Width—a positive integer, which defines the number of horizontal pixels inside the display (x axis)

- Height—a positive integer, which defines the number of vertical pixels inside the display (y axis)

For example, the following:

```
d = Display("First Display", 400, 100)
```

creates a display with the given title, width, and height (as shown in Figures 8.1 and 8.2).

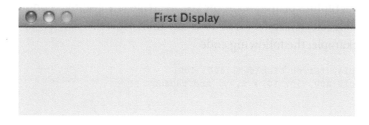

FIGURE 8.1   A sample display (Mac).

FIGURE 8.2 A sample display (Windows).

Notice how the GUI library uses the style of the operating system at hand. (This is because it is built on top of the Java Swing library, which has this desirable characteristic.)*

Once a display has been created, you can add GUI widgets and other graphical objects, using the following function:

```
d.add(object, x, y)
```

where object is a GUI widget or graphical object (presented below). The coordinates x, y specify where in the display to place the object.

Fact: The origin of a display (0, 0) is at the top-left corner.

8.3.2 Graphics Objects

Once you have created a display, you can add a wide variety of graphical objects. Adding a graphics object to a display actually draws the corresponding shape onto the display.

The available shapes are Line, Circle, Point, Oval, Rectangle, and Polygon. Each of these shapes can have a specific *color* and *thickness* (measured in pixels). Additionally, 2D shapes (i.e., circle, oval, rectangle, and polygon) can be drawn *filled* (i.e., solid) or not (i.e., outline).

These graphics primitives allow you to create any graphics composition imaginable (clearly, some easier to create than others).

For example, the following code

```
d = Display("First Display", 400, 100)
c = Circle(200, 50, 10) # x, y, and radius
d.add(c)
```

---

* From now on, for economy, we will show only the Mac version of displays (as the Windows version is similar).

```
r = Rectangle(180, 30, 220, 70) # left-top and right-bottom corners
d.add(r)
l = Line(160, 50, 240, 50) # x, y of two points
d.add(l)
ll = Line(200, 10, 200, 90)
d.add(ll)
```

creates the compound graphics shape in Figure 8.3.

8.3.2.1 Exercise

Create a function containing the code used to create the graphics shape in Figure 8.3. This function should have three arguments, namely, display, x, and y. The first argument is the display to draw the shape. The other arguments are the x and y coordinates of the shape's center on the display. *Hint:* Modify the above code to use relative coordinates from x and y. For example, the circle should be drawn at precisely x and y, whereas the rectangle's top-left corner should be at x-20 and y-20, etc.

Converting from absolute coordinates to relative coordinates (from a given location, e.g., center point) allow us to generalize and reuse code for drawing complex graphics shapes.

8.3.3 Showing Display Coordinates

For convenience, Display objects provide the function:

```
d.showMouseCoordinates()
```

which shows the coordinates of the cursor, using a tool tip.* This allows you to discover the coordinates of where to place various GUI components and thus simplifies GUI development.

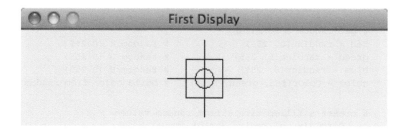

FIGURE 8.3   A sample shape drawn with graphics primitives.

---

* Since this uses a tool tip, you need to hover the mouse for a second or two before the mouse coordinates appear.

To turn off this functionality, use the function:

```
d.hideMouseCoordinates()
```

For an overview of available Display (and other graphics objects) functions, see Appendix C.

8.4 CASE STUDY: RANDOM CIRCLES

This case study demonstrates how to combine some of the programming building blocks we have learned so far (namely, randomness, loops, and GUI functions) to draw random circles on a display.

```
# randomCircles.py
#
# Demonstrates how to draw random circles on a GUI display.
#

from gui import *
from random import *

numberOfCircles = 1000        # how many circles to draw

# create display
d = Display("Random Circles", 600, 400)

# draw various filled circles with random position, radius, color
for i in range(numberOfCircles):

        # create a random circle, and place it on the display

        # get random position and radius
        x = randint(0, d.getWidth()-1)      # x may be anywhere on display
        y = randint(0, d.getHeight()-1)     # y may be anywhere on display
        radius = randint(1, 40)             # random radius (1-40 pixels)

        # get random color (RGB)
        red = randint(0, 255)               # random R (0-255)
        green = randint(0, 255)             # random G (0-255)
        blue = randint(0, 255)              # random B (0-255)
        color = Color(red, green, blue)     # build color from random RGB

        # create a filled circle from random values
        c = Circle(x, y, radius, color, True)

        # finally, add circle to the displayd.add(c)

# now, all circles have been added
```

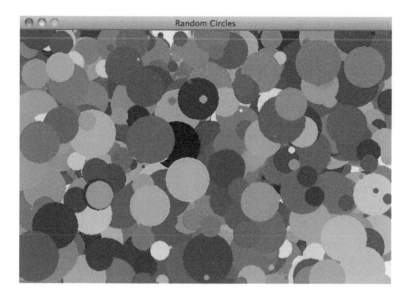

FIGURE 8.4 1000 random circles on a display.

Every time you run this program, it generates 1000 random circles and places them on the created display (see Figure 8.4). The display has a default width of 600 pixels and height of 400 pixels.

The center of each circle is placed at x, y coordinates which are selected randomly from 0 to d.getWidth() and 0 to d.getHeight(), respectively. As their names suggest, these display functions provide the actual dimensions of the display d.

Notice how the color of each circle is created using Color(red, green, blue), where red, green, and blue are the color's RGB color. Each is assigned a random value between 0 and 255 (which is the range of RGB values).

8.4.1 ■ Exercises

1. Write a program that draws a 1000 random shapes (include rectangles, lines, ovals, and polygons). For this exercise, utilize the compact way of drawing shapes on a display, that is, with the display functions drawCircle(), drawLine(), and so on (see Appendix C).

2. Write a program that draws a 1000 compound shapes similar to the one in Figure 8.3. *Hint:* Create a function that draws one such shape using relative coordinates (see earlier exercise). Then call this function 1000 times (in a loop), providing random x and y coordinates, ranging between 0 and the corresponding display border (width or height).

8.5 GUI WIDGETS

In addition to graphics objects, the GUI library supports a wide variety of widgets. Widgets are graphical components used for input and output. They receive user input and/or display information. Widgets are the interactive components out of which GUIs are built.

The available widgets include:

- Label—objects that present textual information.

- Button—objects that can be pressed by the user to perform an action.

- Checkbox—objects that can be selected (or deselected) by the user.

- Slider—objects that can be adjusted by the user to input a value.

- DropDownList—objects that contain items which can be selected by the user.

- TextBox—objects that allow the user to enter a single line of text.

- TextArea—objects that allow the user to enter multiple lines of text.

- Icon—objects that allow displaying of external images (.jpg or .png).

- Menu—objects that contain items which can be selected by the user. Menu objects are fixed at the menu bar (top), whereas DropDownList objects are placed anywhere on a display.

Below we will see various examples of how to use these widgets to build interactive musical instruments.

8.5.1 Event-Driven Programming

Interaction and building GUIs require a new way of thinking about programming. Up until now, our code was executed in the order specified by an algorithm (combining sequence, selection, and iteration). Every now and then, we introduced user interaction, through the input() function. This had the potential to change the order of execution, through if statements examining the value entered by the user (e.g., if the value is negative do *this*, otherwise do *that*). This interaction made our programs more dynamic and customizable, based on user input.

GUI widgets provide a new and more powerful way to get input from the user. Instead of the user typing text to a program prompt (via the input() function), GUI widgets allow the design of intricate user interfaces, which support diverse input scenarios and user interactions.

However, this requires a new way of thinking.

8.5.2 Callback Functions

With GUI widgets, we have no way of knowing *when* a user will interact with a GUI widget. Therefore, GUI widgets allow you to specify a function to be called when they are activated (i.e., the user interacts with them). In other words, the function gets called as a result of the user interaction. If the user does not interact with the widget, the function never gets called.

Definition: A *callback function* is a function associated with a GUI widget that gets called only when (and if) the user interacts with that widget.*

Callback functions are user-defined functions, which allow us to associate arbitrary functionality to a widget. It is through a combination of widgets and their callback functions that we implement the functionality of a GUI.

Definition: *Event-driven programming* refers to the style of programming introduced by GUIs, where code is not executed in the order specified by a fixed algorithm but, instead, code is executed in reaction to user-initiated events (if and when these may occur) from the user interface.

Event-driven programming and callback functions are not confined to GUIs. Other mechanisms for program interaction, for example, MIDI input/output and Open Sound Control (OSC), employ the same idea (as we will see in Chapter 9). In other words, you use components that receive user input (from wherever this input comes). Then you assign callback functions to be executed when (and if) user input arrives. These functions process the user input and make it available to other parts of the program

---

* Actually, callback functions may be associated with other types of objects, such as MIDI and OSC controls (see chapter 9). When the user interacts with such an object, the callback function is called to process the user event (input).

(if needed). The collective functionality of an event-driven program is put together through individual callback functions associated with interactive program elements (e.g., widgets).

8.6 CASE STUDY: A SIMPLE MUSICAL INSTRUMENT

The following program demonstrates event-driven programming. It creates the GUI shown in Figure 8.5. This GUI control surface allows the starting and the stopping of a single note. Soon we will see how to generalize this idea to create more versatile interactive musical instruments.

The GUI consists of two Button widgets. The first starts a note. The second stops the note. Under the hood, each button utilizes its own callback function, which performs the desired functionality when (and if) the button is pressed.

Fact: Callback functions have a fixed number of arguments, specified by the widget that calls them.

For example, callback functions for Button widgets must accept zero parameters.* If we use a function with the wrong number of arguments, this will cause a runtime error (i.e., the error will be reported when the button is pressed at runtime). So it is very important to follow documentation and test GUI code carefully. If you do not test a particular widget, you will not see an error (even if one is waiting to happen) until a user comes across it.

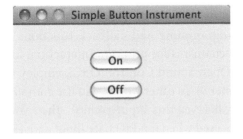

FIGURE 8.5   A single-note GUI control surface.

---

* See Appendix C for more information on required arguments for callback functions of different widgets.

The following program makes use of several Play class functions we have not seen before. These functions, namely, Play.noteOn() and Play.noteOff(), facilitate interactive performance by allowing the duration of a note to be determined in real time by the end user. This is in contrast to Play.midi(), which we have used extensively and is geared toward the playback of compositional, noninteractive, algorithmic music.

Play.noteOn() and Play.noteOff() are explained in more detail in the next section. For now, simply accept that they exist and that they do (pretty much) what their names suggest.

```
# simpleButtonInstrument.py
#
# Demonstrates how to create a instrument consisting of two buttons,
# one to start a note, and another to end it.
#

from gui import *
from music import *

# create display
d = Display("Simple Button Instrument", 270, 130)

pitch = A4        # pitch of note to be played

# define callback functions
def startNote():             # function to start the note

   global pitch              # we use this global variable

   Play.noteOn(pitch)     # start the note

def stopNote():              # function to stop the note

   global pitch              # we use this global variable

   Play.noteOff(pitch)    # stop the note

# next, create the button widgets and assign their callback functions
b1 = Button("On", startNote)
b2 = Button("Off", stopNote)

# finally, add buttons to the display
d.add(b1, 90, 30)
d.add(b2, 90, 65)
```

Notice the two callback functions, `startNote()` and `stopNote()` that trigger the sending of noteOn and noteOff messages.* The only thing that makes them callback functions is our intention to use them as such. They are regular user-defined functions, which happen to accept zero parameters. What makes them "callback functions" is that we assign them as callback functions to the two Button objects. Since Button objects require callback functions with zero arguments, we made sure that these functions do indeed require zero arguments.† Also, in order to assign them as callback functions, Python requires that they are defined before the Button objects are created.

Finally, notice the statement (used in both of the functions):

```
global pitch # we use this global variable
```

This states that functions `startNote()` and `stopNote()` use a variable defined outside them. Normally, such a variable would be passed through the parameter list. However, this is not the case here, as explained below.

8.6.1 Python Global Statement

As we mentioned above, callback functions have a fixed number of arguments, specified by the widget that calls them. If these functions require access to other variables (e.g., a label to update or a display to draw on), these variables need to be specified somehow. Normally, such variables are passed through the argument list. However, since Button objects (and other GUI widgets) control the parameter list of callback functions, we have to use the Python `global` statement.

For example, in the above functions, the statement

```
global pitch
```

informs the interpreter that the variable pitch, which is defined outside the function, is needed (i.e., will be used by) the function. This is equivalent to

---

* If you press the "On" button several times on some synthesizers, you may need to press the "Off" button the same number of times. On most synthesizers, however, a single press of the "Off" button will suffice.

† For more information on parameters expected by callback functions for different widgets, see Appendix C.

having this variable in the function argument list, except we do not have to physically put it there.[*]

Good Style: The `global` statement should be used minimally—mainly in callback functions. For normal functions, such variables should be passed in as arguments.

According to the Python manual, "variables that are only referenced inside a function are implicitly global. If a variable is assigned a new value anywhere within the function's body, it's assumed to be a local. If a variable is ever assigned a new value inside the function, the variable is implicitly local, and you need to explicitly declare it as global."

8.6.2 Exercise

Expand the above program to create a second set of buttons for a different instrument (your choice) playing another note (also your choice).

8.7 PLAY CLASS

The Play class, as its name suggests, contains functions related to playing music in real time.

In previous chapters we used the function `Play.midi()`. As you recall, this function allows the playing of any musical material (Note, Phrase, Part, or Score). Actually, you can have several calls to `Play.midi()` active at the same time—each will play its own musical material in parallel with the others.

Fact: `Play.midi()` makes extensive use of computational resources, so the number of parallel calls is limited by the computational load of your system.

As stated earlier, `Play.midi()` is geared toward compositional music making.

---

[*] Actually, we could forego argument lists and use the global statement instead. However, this would create functions that would be hard to reuse, since the name of the argument variable becomes fixed, and thus the function cannot be called in different contexts with different variable names. Therefore, the `global` statement should only be used with callback functions. To do otherwise makes functions harder to reuse and thus is bad programming style.

Additionally, the Play class provides several functions geared toward interactive (e.g., GUI) music making. These functions are intended for building interactive musical instruments:

Function	Description
`Play.noteOn(pitch, velocity, channel)`	Starts pitch sounding. Specifically, it sends a NOTE_ON message with `pitch` (0–127), at given `velocity` (0–127—default is 100), to be played on `channel` (0–15—default is 0) through the Java synthesizer.
`Play.noteOff(pitch, channel)`	Stops pitch from sounding. Specifically, it sends a NOTE_OFF message with `pitch` (0–127), on given `channel` (0–15—default is 0) through the Java synthesizer. If the pitch is not sounding on this channel, this has no effect.
`Play.allNotesOff()`	Stops all notes from sounding on all channels.
`Play.setInstrument(instrument, channel)`	Sets a MIDI `instrument` (0–127—default is 0) for the given `channel` (0–15, default is 0). Any notes played through `channel` will sound using `instrument`.*
`Play.getInstrument(channel)`	Returns the MIDI `instrument` (0–127) assigned to `channel` (0–15, default is 0).
`Play.setVolume(volume, channel)`	Sets the global (main) `volume` (0–127) for this channel (0–15). This is different from the velocity level of individual notes—see `Play.noteOn()`.
`Play.getVolume(channel)`	Returns the global (main) volume (0–127) for this channel (0–15).
`Play.setPanning(position, channel)`	Sets the global (main) panning `position` (0–127) for this `channel` (0–15). The default position is in the middle (64).
`Play.getPanning(channel)`	Returns the global (main) `position` (0–127) for this channel (0–15).
`Play.setPitchBend(bend, channel)`	Sets the pitch bend for this `channel` (0–15—default is 0) to the Java synthesizer object. Pitch bend ranges from −8192 (max downward bend) to 8191 (max upward bend). No pitch bend is 0 (which is the default). If you exceed these values the outcome is undefined (it may wrap around or it may cap, depending on the system.)

Continued

Function	Description
`Play.getPitchBend(channel)`	Returns the current pitch bend for this channel (0–15—default is 0).
`Play.frequencyOn(frequency, volume, channel)`	Starts a note sounding at the given frequency and volume (0–127—default is 100), on channel (0–15—default is 0). **Warning:** You should play only one frequency per channel. (Since this uses pitch bend indirectly, it will affect the pitch of all other notes sounding on this channel.)
`Play.frequencyOff(frequency, channel)`	Stops a note sounding at the given frequency on channel (0–15—default is 0). **Warning:** You should play only one frequency per channel. (Since the frequency gets translated to a pitch and a pitch bend, this will also affect notes with nearby frequencies on this channel.)
`Play.allFrequenciesOff()`	Same as `Play.allNotesOff()`. Stops all notes from sounding on all channels.

* You should avoid using both `Play.setInstrument()` and `Play.midi()`, since the latter also makes instrument assignments indirectly (as specified by its musical material, e.g., a Part).

Finally, the Play class provides an advanced function for scheduling notes to be played at a later time:

Function	Description
`Play.note(pitch, start, duration, volume, channel)`	Schedules a note with pitch (0–127) to be sounded after start milliseconds, lasting duration milliseconds, at given volume (0–127—default is 100), on channel (0–15—default is 0) through the Java synthesizer.

8.8 CASE STUDY: AN AUDIO INSTRUMENT FOR CONTINUOUS PITCH CONTROL

In this section, we explore how to incorporate audio samples (as opposed to MIDI) into our music-making activities. Being able to play arbitrary audio files through our programs opens new doors for composition, performance, and creative expression.

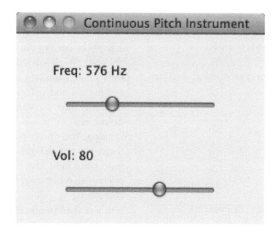

FIGURE 8.6   A continuous pitch GUI instrument.

This case study demonstrates how to use GUI functions to create a simple instrument for changing the volume and frequency of an audio loop in real time. The following program creates the GUI control surface shown in Figure 8.6.

This GUI surface consists of two sliders to control frequency and volume, respectively. It also uses two labels to provide feedback to the user about what the current sliders' values are (as they change, in real time).

```
# continuousPitchInstrumentAudio.py
#
# Demonstrates how to use sliders and labels to create an instrument
# for changing volume and frequency of an audio loop in real time.
#

from gui import *
from music import *

# load audio sample
a = AudioSample("moondog.Bird_sLament.wav")

# create display
d = Display("Continuous Pitch Instrument", 270, 200)

# set slider ranges (must be integers)
minFreq = 440   # frequency slider range
maxFreq = 880   # (440 Hz is A4, 880 Hz is A5)

minVol = 0      # volume slider range
maxVol = 127

# create labels
label1 = Label( "Freq: " + str(minFreq) + " Hz" )  # set initial text
```

```
label2 = Label( "Vol: " + str(maxVol) )

# define callback functions (called every time the slider changes)
def setFrequency(freq):  # function to change frequency

  global label1, a       # label to update, and audio to adjust

  a.setFrequency(freq)
  label1.setText("Freq: " + str(freq) + " Hz") # update label

def setVolume(volume):   # function to change volume

  global label2, a       # label to update, and audio to adjust

  a.setVolume(volume)
  label2.setText("Vol: " + str(volume))   # update label

# next, create two slider widgets and assign their callback functions
#Slider(orientation, lower, upper, start, eventHandler)
slider1 = Slider(HORIZONTAL, minFreq, maxFreq, minFreq, setFrequency)
slider2 = Slider(HORIZONTAL, minVol, maxVol, maxVol, setVolume)

# add labels and sliders to display
d.add(label1, 40, 30)
d.add(slider1, 40, 60)
d.add(label2, 40, 120)
d.add(slider2, 40, 150)

# start the sound
a.loop()
```

First, notice the class AudioSample is used to load an audio file into the program. AudioSample is very useful for building interactive musical instruments because it opens the door to recorded sound (as opposed to MIDI). This important class is presented in detail below. For now, observe how easy it is to load an audio file and play it in a loop. This is done with the following statements:

```
a = AudioSample("moondog.Bird_sLament.wav")
a.loop()
```

where "moondog.Bird_sLament.wav" is an audio sample (WAV audio file) containing the opening phrase of Moondog's piece "Bird's Lament" (1969). Available for download at http://jythonMusic.org this audio file should be stored in the same folder as the program. The second statement, a.loop(), starts the audio and repeats it indefinitely (see the next section for a complete list of AudioSample functions).

Notice the two functions, setFrequency() and setVolume(). Since they are used as slider callback functions, they are required to have only one argument, namely, the value of the slider. These functions are called every time the corresponding slider has been adjusted.

These functions, in addition to adjusting the sound, update the corresponding labels. Therefore, these variables, namely, label1 and label2, respectively, are identified via global statements (as discussed earlier).

Finally, notice the statement:

```
label1.setText("Freq: " + str(freq) + " Hz") # update label
```

This updates the text being displayed by label1. The label text is a string, so we use a combination of strings and the "+" (string concatenation) operator. Since variable freq is an integer (slider values are always integers), we use the str() conversion function to convert the value of freq to a string (so that it can be concatenated with the other strings). The result becomes the single string displayed in label1.

Good Style: Always provide feedback to the user via the GUI about the result of his/her actions. This makes users feel in control, which makes your software more usable.

8.9 AUDIOSAMPLE CLASS

The AudioSample class, as its name suggests, contains functions related to playing audio samples in real time. An audio sample is a sound object created from an external audio file, which can be played, looped, paused, resumed, and stopped.

8.9.1 Creating Audio Samples

The AudioSample class has the following attributes:

- filename – a string, the name of the audio file to be loaded (.wav or .aif)

- pitch – a MIDI pitch (0–127) which will be associated with the audio sample (default is A4)*

- volume – a positive integer (0–127), which sets the initial volume of playback (default is 127)

---

* For the purposes of a program.

The following function creates a new AudioSample, so you need to save it in a variable (in order to use it later).

Function	Description
AudioSample(filename, pitch, volume)	Creates an audio sample from the audio file specified in filename (supported formats are WAV and AIF— 16, 24, and 32 bit PCM, and 32-bit float). Parameter pitch (optional) specifies a MIDI note number to be used for playback (default is A4). Parameter volume (optional) specifies a MIDI note velocity to be used for playback (default is 127).

For example, an audio sample may be created as follows:

```
a = AudioSample("rainstorm.wav", C4, 100)
```

creates an audio sample, a, with pitch C4 and volume 100.

An audio sample can be played once, looped, stopped, paused, and resumed. Also, we can change its pitch or frequency in real time (through pitch shifting) by using setPitch() or setFrequency().

Once an audio sample, a, has been created, the following functions are available:

Function	Description
a.play() a.play(start, size)	Play the sample once. If start and size are provided, the sample is played from milliseconds start until milliseconds start+size (default is 0 and –1, respectively, meaning from beginning to end).
a.loop() a.loop(times, start, size)	Repeat the sample indefinitely. Optional parameters times specifies the number of times to repeat (default is –1, indefinitely). If start and size are provided, looping occurs between milliseconds start and milliseconds start+size (default is 0 and –1, respectively, meaning from beginning to end).
a.stop()	Stops sample playback immediately.
a.pause()	Pauses sample playback (remembers current position for resume).
a.resume()	Resumes sample playback (from the paused position).
a.isPlaying()	Returns True if the sample is still playing, False otherwise.

The following functions control audio sample playback parameters (pitch, frequency, and volume).

Function	Description
a.setPitch(pitch)	Sets the sample pitch (0–127) through pitch shifting from sample's base pitch.
a.getPitch()	Returns the sample's current pitch (it may be different from the default pitch).
a.setFrequency(freq)	Sets the sample pitch frequency (in Hz). This is equivalent to setPitch(), except it provides more granularity (accuracy). For instance, pitch A4 is the same as frequency 440 Hz.
a.getFrequency()	Returns the current playback frequency.
a.setVolume(volume)	Sets the volume (amplitude) of the sample (volume ranges from 0 to 127).
a.getVolume()	Returns the current volume (amplitude) of the sample (volume ranges from 0 to 127).
a.getFrameRate()	Returns the sample's recording rate (e.g., 44100.0 Hz).
a.setPanning(panning)	Sets the panning of the sample (panning ranges from 0–127).
a.getPanning()	Returns the current panning of the sample (panning ranges from 0–127).

8.9.2 Exercise

Using buttons, sliders, and other GUI elements, create a Music Production Controller (MPC).

- MPCs are electronic musical instruments (originally produced by Akai) that feature a grid of buttons that allow a user to play back (trigger) various samples.

- Use an external audio editor (such as Audacity) to capture and manipulate arbitrary audio files and live inputs.

- Then plan a music performance using the produced audio samples together with your custom MPC.

8.10 MIDISEQUENCE CLASS

The MidiSequence class provides playback functionality for MIDI sequences that is similar to the functionality described above for audio samples. This parallelism in the functionality of the two classes allows us to easily mix sounds from audio and MIDI material (files) in interactive musical instruments.

8.10.1 Creating MIDI Sequences

MidiSequence objects can be created from an external MIDI file (as well as Note, Phrase, Part, and Score objects).

The MidiSequence class has the following attributes:

- Material—a string, the name of a MIDI file to be loaded (.mid) or a Note, Phrase, Part, and Score object

- Pitch—a MIDI pitch (0–127) which will be associated with the material (default is A4)[*]

- Volume—a positive integer (0–127), which sets the initial volume of playback (default is 127)

The following function creates a new MidiSequence, so you need to save it in a variable (in order to use it later).

Function	Description
`MidiSequence(material, pitch, volume)`	Creates a MIDI sequence from the MIDI material specified in `material` (this may be a filename of an external MIDI file or a Note, Phrase, Part, and Score object. Parameter `pitch` (optional) specifies a MIDI note number to be used for playback (default is A4). Parameter `volume` (optional) specifies a MIDI note velocity to be used for playback (default is 127).

For example, a MIDI sequence may be created as follows:

```
m = MidiSequence("beat1.mid", C4, 100)
```

creates a MIDI sequence, m, with pitch c4 and volume 100.

A MIDI sequence can be played once, looped, stopped, paused, and resumed. Also, we may change its pitch, tempo, and volume. These changes happen immediately.

Once a MIDI sequence, m, has been created, the following functions are available:

Function	Description
`m.play()`	Play the MIDI sequence once.
`m.loop()`	Repeat the MIDI sequence indefinitely.
`m.stop()`	Stops MIDI sequence playback immediately.

Continued

---

[*] For the purposes of a program.

Function	Description
m.pause()	Pauses MIDI sequence playback (remembers current position for resume).
m.resume()	Resumes MIDI sequence playback (from the paused position).
m.isPlaying()	Returns True if the MIDI sequence is still playing, False otherwise.

The following functions control MIDI sequence playback parameters (pitch, frequency, and volume).

Function	Description
m.setPitch(pitch)	Set the MIDI sequence's playback pitch (0–127) by transposing the MIDI material.
m.getPitch()	Returns the MIDI sequence's playback pitch (0–127).
m.setTempo(tempo)	Set MIDI sequence's playback tempo in beats per minute (e.g., 60).
m.getTempo()	Return MIDI sequence's playback tempo (in beats per minute).
m.getDefaultTempo()	Return MIDI sequence's default tempo (in beats per minute).
m.setVolume(volume)	Returns the volume of the MIDI sequence (0–127).
m.getVolume()	Returns the current volume of the MIDI sequence (0–127).

8.10.2 Exercises

1. Extend the Music Production Controller (MPC) of the previous exercise to include MIDI sequences. Using algorithmic techniques explored in earlier chapters, develop a few interesting MIDI sequences. Plan a music performance using the combined audio samples and MIDI sequences through your custom MPC.

2. Given everything you have seen so far (in this chapter, and before), you can now create a wide variety of musical laptop instruments. Your imagination and musical creativity is the limit. Design an innovative interactive musical instrument. Interview a few musicians. Start with a paper prototype. Construct a functional prototype and refine it using input from your target audience. Plan a performance using this new instrument, possibly together with other traditional and software instruments.

8.11 PAPER PROTOTYPING

Paper prototyping is a technique from the field of human–computer interaction. It involves constructing a mock-up of the GUI on paper (usually via paper and pencil) (Moggridge 2007, Stone et al. 2005, Greenberg et al. 2011).

Definition: *Paper prototypes* are throwaway models of the GUI of a program, involving rough (usually hand-sketched) drawings.

Since drawing on paper is much faster than programming, paper prototyping allows you to explore and refine the GUI design with minimal effort. Once the design is finalized on paper, we can code it. Again, consider the programmer maxim "2 hours of design can save you 20 hours of coding." It also most definitely applies to GUI programming.

Fact: Although paper prototyping may appear simplistic, it is used widely in industry for GUI development.

Paper prototyping provides valuable user feedback early enough in the development process. This helps shape the GUI design with minimal effort. The earlier it is done, the better. The more often it is done, the better. It simply saves programming effort.

Consider, on the other hand, if you wait to seek user input on your design until after you have programmed your GUI. If you make the changes suggested by users, you throw away part of your program and start again. If you resist making those changes, the users will be unhappy with (and may avoid using) your program. Either way you end up wasting valuable effort.

Good Style: Use paper prototypes (even if rough sketches) whenever possible. They save considerable development time. Also they allow you to explore aesthetic choices before you implement a particular idea.

For example, Figure 8.7 shows the paper prototype for the GUI shown in Figure 8.6.

8.12 A SIMPLE METHODOLOGY FOR DEVELOPING GUIS

This section presents a simple methodology for developing GUIs through Python and the GUI library. It involves iterative refinement, through these three steps:

1. Develop a paper prototype. Get up to five representative users.* Show them the paper prototype. Ask them to imagine using it to

---

* Studies show that five users are enough to identify 75% of the problems that may exist in a GUI (Nielsen 2000).

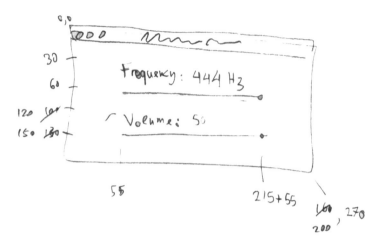

FIGURE 8.7   Paper prototype for the GUI in Figure 8.6.

perform the intended task (e.g., make music). Ask them for their thoughts and preferences. For example, do users like a horizontal slider for some functionality, or would they prefer something else? Where should this slider be placed? What should the labels contain? What should the colors be? And so on. When the users talk, listen. Do not explain (or even worse, defend) your initial design choices. Your goal is to learn from the users what they like (hence, the cheap implementation on paper—it's very easy to change). Iterate, as necessary, until the paper prototype captures the design that most users prefer. Once the paper prototype has evolved to capture all possible user ideas and preferences, you may move to the next step.

2. Create a blank Display object with the dimensions specified in the paper prototype. Use the showMouseCoordinates() function to identify the approximate position of the various GUI widgets and graphics objects. Notate those coordinates on the paper prototype. If necessary, make numeric adjustments on the paper prototype (see Figure 8.7).

3. Once you have the approximate coordinates for the various GUI components, write code that creates them and places them on the display. Running your program interactively, through the Python interpreter, use the display function move() to fine-tune the position

of the various GUI components relative to each other. If necessary, go back and adjust the size of the display. Notate those changes on the paper prototype. Make final adjustments in your code, and present to users. Be ready to make additional adjustments. If necessary, go back to step (1), especially if users suggest a radical change.

These steps may be repeated as more functionality and GUI components are added. By developing a system through this iterative methodology, you can focus on the most important design aspects first. Then you can add more details. By involving users early on in the design process, you are more likely to develop a system that is useable and appreciated by the intended user group. Your development effort will be well spent.

8.12.1 Listen, Listen, Listen

Listening is the most important skill of a GUI developer. Yet it is the hardest thing to learn to do.

When paper prototyping, beginner designers feel the need to explain why they designed something a certain way. At the moment you start explaining (or telling the user how to do something) during a paper prototyping session, you have biased the user toward your design and lost the opportunity to get new, unbiased ideas.

Expert GUI designers remain mostly silent during such sessions, and only ask questions to help probe the user's mind as to what the user expected to see. It is helpful to realize that, once you have collected all user preferences and desires, you can still decide *not* to use them. The choice is yours. The goal of paper prototyping is to collect these user preferences and desires, early on in the development effort, so that you can build the best, most usable system possible.

8.13 EVENT HANDLING

The GUI library provides a wide variety of functions to handle keyboard and mouse events. These functions are available for every GUI object (i.e., displays, graphics objects, and widgets). If desired, you can customize every component on your GUI to respond differently to user keyboard or mouse actions. This provides immense power for building various interactive behaviors and functionalities.

Every GUI object listens for user actions (e.g., mouse click, mouse drag, typing a key, etc.) and allows you to specify a callback function to be executed if and when a user action occurs. You decide which events to handle and on which objects.

8.13.1 Keyboard Events

Every widget in the GUI library supports the following functions: `onKey-Type()`, `onKeyDown()`, `onKeyUp()`. Each of these functions accepts one argument, namely, the callback function you have created (programmed) to handle this particular event.

For example, in the next case study, the function `clearOnSpacebar()` is given as an argument to the display's function `onKeyDown()`. In other words, `clearOnSpacebar()` will be called every time the user presses a key while the display is in focus.

Callback functions for keyboard events are passed the key that was typed, pressed, or released, accordingly. The callback function decides what to do with this information. As an example of this, see the function clearOnSpacebar() below.

8.13.2 Mouse Events

Every GUI object provides the following functions: `onMouseClick()`, `onMouseDown()`, `onMouseUp()`, `onMouseMove()`, `onMouseDrag()`, `onMouseEnter()`, `onMouseExit()`. Each of these functions accepts one argument, namely, the callback function you have created (programmed) to handle this particular event.

Callback functions for mouse events are given the x and y coordinates of the mouse. These are the coordinates of where the corresponding mouse event happened on the GUI. The callback function decides what to do with these coordinates. As an example of this, see the function `beginCircle()` below.

8.13.2.1 Example

To explore mouse events, let's start the Python interpreter and try the following:

```
>>> from gui import *
>>> d1 = Display("Playing with Events", 200, 355)
>>> d1.getWidth()
200
>>> d1.getHeight()
355
```

This creates a display with the given title, and with width 200 and height 355 (as demonstrated by the calls to display's getWidth() and getHeight() functions). Nothing new here.

Now let's draw a circle at position (50, 100) with radius 81 pixels, using the color orange. Leave the circle unfilled and make the outline 4 pixels thick:

```
>>> c1 = d1.drawCircle(50, 100, 81, Color.ORANGE, False, 4)
```

Let's move the circle to coordinates (80, 120):

```
>>> d1.move(c1, 80, 120)
```

Now let's create a function that moves the circle to those coordinates when called. Below, we define the function interactively. If we wish this function to be permanent, we need to define it inside a program. (Notice the ellipses, "...". These are automatically generated by the interpreter to indicate that it is expecting more input (i.e., the function body is not complete). To finalize the function body, we hit the <Enter> key twice.)

```
>>> def moveCircle():
... global c1, d1
... d1.move(c1, 80, 220)
```

Now let's create a button that calls this function when pressed. Also, let's add it to the display:

```
>>> b1 = Button("Move Circle", moveCircle)
>>> d1.add(b1, 25, 307)
```

Now let's press the button. This should move the circle to coordinates (80, 200). Then let's manually move the circle to coordinates (20, 20), through the code below.

```
>>> d1.move(c1, 20, 20)
```

Alternate back and forth between pressing the button and manually moving the circle.

Next, let's create another function. This function will be associated with a mouse event (below), so let's define it to accept two parameters, that is, the x and y coordinates of the mouse position (when the mouse event occurs).

```
>>> def moveCircle1(x, y):
... global d1, c1
... d1.move(c1, x, y)
```

Let's call this function manually a few times.

```
>>> moveCircle1(0, 0)
>>> moveCircle1(10, 30)
>>> moveCircle1(0, 0)
```

Finally, let's associate it with the display's mouse-click event:

```
>>> d1.onMouseClick(moveCircle1)
```

Now, clicking the mouse anywhere on the display moves the circle to that position. Pressing the button moves it back to coordinates (80, 220). Play with this interactive functionality a little to fully understand it.

8.13.2.2 Exercises

1. Make the circle follow the mouse movement (*Hint*: See display's `onMouseMove()` function.)

2. How could this be used to create a 2D slider? *Hint*: The only thing you need to do is to add a few more statements inside the `moveCircle1()` function, to adjust the parameters of, say, an AudioSample (e.g., connect x value (mouse coordinate) with frequency and y value with volume). There are numerous possibilities here to build innovative interactive instrument displays.

3. What other interactive possibilities can you think of for controlling a circle using GUI elements? Try some more things out. How about adding more graphical objects (with different colors, etc.)? What about using images (GUI icons) instead of graphical objects?

8.13.3 Case Study: Drawing Musical Circles

This program demonstrates how to use event handling to build a simple interactive musical instrument. The user plays notes by drawing circles (see Figure 8.8).

The diameter of the circle determines the pitch of the note (within the major scale). To draw a circle, the user presses the mouse on the display and starts moving the mouse. The point where the mouse is released determines the size of the circle (also the pitch of the note).

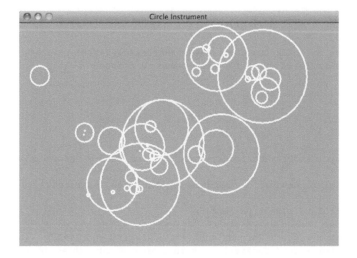

FIGURE 8.8 Screen snapshot of musical performance using a `simpleCircle`
`Instrument`.

The mathematics used to map the diameter of the circle is relatively
simple, but generates a musical effect that is interesting. Although the user
might expect that doubling of the diameter will double the pitch of the
note, that's not the case. It takes a little while to figure out how mouse
distance corresponds to pitch, and it is precisely this challenge that makes
this "game" interesting—not unlike regular video games.

```
# simpleCircleInstrument.py
#
# Demonstrates how to use mouse and keyboard events to build a simple
# drawing musical instrument.
#

from gui import *
from music import *
from math import sqrt

### initialize variables ######################
minPitch = C1 # instrument pitch range
maxPitch = C8

# create display
d = Display("Circle Instrument") # default dimensions (600 x 400)
d.setColor(Color(51, 204, 255))   # set background to turquoise

beginX = 0 # holds starting x coordinate for next circle
beginY = 0 # olds starting y coordinate
```

```
# maximum circle diameter - same as diagonal of display
maxDiameter = sqrt(d.getWidth()**2 + d.getHeight()**2) # calculate it

### define callback functions #####################
def beginCircle(x, y): # for when mouse is pressed

    global beginX, beginY

    beginX = x # remember new circle's coordinates
    beginY = y

def endCircleAndPlayNote(endX, endY):  # for when mouse is released

global beginX, beginY, d, maxDiameter, minPitch, maxPitch

    # calculate circle parameters
    # first, calculate distance between begin and end points
    diameter = sqrt( (beginX-endX)**2 + (beginY-endY)**2 )
    diameter = int(diameter)     # in pixels - make it an integer
    radius = diameter/2          # get radius
    centerX = (beginX + endX)/2 # circle center is halfway between...
    centerY = (beginY + endY)/2 # ...begin and end points

    # draw circle with yellow color, unfilled, 3 pixels thick
    d.drawCircle(centerX, centerY, radius, Color.YELLOW, False, 3)

    # create note
    pitch = mapScale(diameter, 0, maxDiameter, minPitch, maxPitch,
                    MAJOR_SCALE)

    # invert pitch (larger diameter, lower pitch)
    pitch = maxPitch - pitch

    # and play note
    Play.note(pitch, 0, 5000)    # start immediately, hold for 5 secs

def clearOnSpacebar(key):        # for when a key is pressed

    global d

    # if they pressed space, clear display and stop the music
    if key == VK_SPACE:
        d.removeAll()                 # remove all shapes
        Play.allNotesOff()            # stop all notes

### assign callback functions to display event handlers
#####################
d.onMouseDown(beginCircle)
d.onMouseUp(endCircleAndPlayNote)
d.onKeyDown(clearOnSpacebar)
```

First, notice the use of the math library.

```
from math import *
```

This library contains various mathematical constants (e.g., pi) and functions. Here we are using it for the square root function, that is, sqrt().* Notice the call to display's setColor() function:

```
d.setColor(Color(51, 204, 255)) # set background to turquoise
```

This sets the display background to the specified color using RGB values. However, if you call setColor() without parameters, it brings up a color chooser from which you can pick the desired color. This color picker also displays the RGB values, which allows you to hardcode them in your program to recreate a given color. This is a very useful feature.

8.13.3.1 Defining Callback Functions

As seen earlier, the callback functions are regular user-defined functions with specific argument lists. Functions intended to be mouse event-handlers need to accept two arguments, namely, the x and y coordinates of the mouse, when the event happens (e.g., mouse click). To be keyboard event-handlers, functions need to accept one argument, namely, the key pressed/released/typed (for more information see Appendix C).

In order for callback functions to do anything interesting, they usually require more information. But, since their argument lists are fixed and cannot be extended, this additional information is made available to them through global variables.

Good Style: Global variables should be avoided.[†]

Fact: In the case of callback functions (since their argument lists are specified by the event handler they are connected to), global variables are a necessity.

The above program's callback functions use a few global variables to provide information to each other and the main program (see the global statements in these functions). Of these, the variables beginX and beginY are of particular importance. They are set by the callback function

---

* The math library will be covered in detail in Chapter 10.
† Normally, if a function requires any information, this should be made available through its arguments; similarly, if a function generates useful information, this should be passed back through the return statement.

`beginCircle()`. This function is associated with the display's `onMouseDown` event, for example,

`d.onMouseDown(beginCircle)`

This indicates that the function `beginCircle()` will be called when the user presses down the (left) mouse button inside the display. Its arguments, x and y, will be automatically set to the mouse x and y coordinates at that moment. The function saves these x and y coordinates to the global variables `beginX` and `beginY` for further use by the program.

In particular, the callback function `endCircleAndPlayNote()` uses those variables to determine the diameter of the circle to draw, and from that to determine the pitch of the note to play. This function is associated with the display's `onMouseUp` event, as in

`d.onMouseUp(endCircleAndPlayNote)`

This means that function `endCircleAndPlayNote()` will be called when the user releases the mouse button. Its arguments, x and y, will be automatically set to the mouse x and y coordinates at that moment. The function uses these x and y coordinates to finalize and draw the circle, and to construct and play the corresponding note.

Finally, the callback function `clearOnSpacebar()` clears the display and stops all notes from sounding. This function is associated with the display's `keyDown` event, for example,

`d.onKeyDown(clearOnSpacebar)`

This means that the function `clearOnSpacebar()` will be called when the user presses any key on the keyboard. Its argument, key, will be automatically set to the virtual code of the key pressed. If the key is the spacebar (i.e., virtual key `VK_SPACE`), the function clears the display and stops all notes.

In summary, by associating callback functions with GUI events, we synthesize the functionality of the GUI control surface.

Figure 8.8 shows the image generated from one musical performance with this instrument. Part of the performance is lost, namely, the pitch and timing of the notes. What remains is still artistically interesting in its own right. This also demonstrates the creative possibilities of combining the GUI and music libraries to build interactive musical instruments.

8.13.3.2 Exercises

1. Explore different mappings between circle radius and note pitch. One possibility is to go up one octave when the radius is doubled. (Hint: For this, you might find function Note.freqToMidiPitch(freq) useful—see Appendix B.) Explore other possibilities. Finding a balance between challenge and predictability might make the instrument interesting in various ways.

2. Explore different ways to map interactions with sound for this instrument. Possibilities include connecting x position of circle with pitch, connecting y position of circle with volume, connecting radius of circle with volume, and/or connecting pitch with color of circle. Can you think of other possibilities?

3. What other musical instruments can you design, combining GUI controls, mouse/keyboard events, and music-making functions? There are many intriguing possibilities. Can you think of a few? Can you think of instruments that could be complementary to each other (for example, instruments you can use to create a laptop band)? What about instruments that involve game play as part of the interaction?

8.14 CASE STUDY: A VIRTUAL PIANO

This case study demonstrates how to create an interactive musical instrument that incorporates images. The following program combines GUI elements to create a realistic piano which can be played through the computer keyboard.

It uses an image of a complete piano octave, namely, iPianoOctave.png, to display a piano keyboard with 12 keys unpressed. Then, to generate the illusion of piano keys being pressed, it selectively adds the following images to the display (see Figure 8.9).

- iPianoWhiteLeftDown.png (used for "pressing" keys C and F),

- iPianoBlackDown.png (used for "pressing" any black key),

- iPianoWhiteCenterDown.png (used for "pressing" keys D, G, and A), and

- iPianoWhiteRightDown.png (used for "pressing" keys E and B).

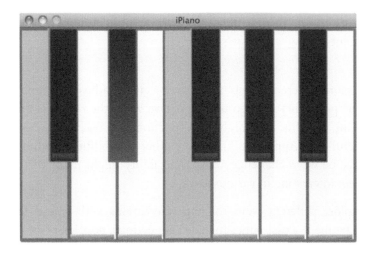

FIGURE 8.9   The iPiano GUI with keys C, D sharp, and F being pressed.

```
# iPianoSimple.py
#
# Demonstrates how to build a simple piano instrument playable
# through the computer keyboard.
#

from music import *
from gui import *

Play.setInstrument(PIANO)        # set desired MIDI instrument (0-127)

# load piano image and create display with appropriate size
pianoIcon = Icon("iPianoOctave.png")   # image for complete piano
display = Display("iPiano", pianoIcon.getWidth(),
                           pianoIcon.getHeight())
display.add(pianoIcon)           # place image at top-left corner

# load icons for pressed piano keys
cDownIcon      = Icon("iPianoWhiteLeftDown.png")   # C
cSharpDownIcon = Icon("iPianoBlackDown.png")       # C sharp
dDownIcon      = Icon("iPianoWhiteCenterDown.png") # D
# ...continue loading icons for additional piano keys

# remember which keys are currently pressed
keysPressed = []

###################################################################
# define callback functions
def beginNote(key):
    """This function will be called when a computer key is pressed.
       It starts the corresponding note, if the key is pressed for
       the first time (i.e., counteracts the key-repeat function of
       computer keyboards).
    """
```

```
      global display        # display surface to add icons
      global keysPressed    # list to remember which keys are pressed

      print "Key pressed is " + str(key)    # show which key was pressed

      if key == VK_Z and key not in keysPressed:
         display.add(cDownIcon, 0, 1)    # "press" this piano key
         Play.noteOn(C4)                 # play corresponding note
         keysPressed.append(VK_Z)        # avoid key-repeat
      elif key == VK_S and key not in keysPressed:
         display.add(cSharpDownIcon, 45, 1) # "press" this piano key
         Play.noteOn(CS4)                # play corresponding note
         keysPressed.append(VK_S)        # avoid key-repeat

      elif key == VK_X and key not in keysPressed:
         display.add(dDownIcon, 76, 1)   # "press" this piano key
         Play.noteOn(D4)                 # play corresponding note
         keysPressed.append(VK_X)        # avoid key-repeat

      #...continue adding elif's for additional piano keys


def endNote(key):
    """This function will be called when a computer key is released.
       It stops the corresponding note.
    """

      global display        # display surface to add icons
      global keysPressed    # list to remember which keys are pressed

      if key = = VK_Z:
         display.remove(cDownIcon)       # "release" this piano key
         Play.noteOff(C4)                # stop corresponding note
         keysPressed.remove(VK_Z)        # and forget key

      elif key = = VK_S:
         display.remove(cSharpDownIcon)  # "release" this piano key
         Play.noteOff(CS4)               # stop corresponding note
         keysPressed.remove(VK_S)        # and forget key

      elif key = = VK_X:
         display.remove(dDownIcon)       # "release" this piano key
         Play.noteOff(D4)                # stop corresponding note
         keysPressed.remove(VK_X)        # and forget key

      #...continue adding elif's for additional piano keys


#################################################################
# associate callback functions with GUI events
display.onKeyDown( beginNote )
display.onKeyUp( endNote )
```

Let's explore this code. First, the following statement

```
Play.setInstrument(PIANO) # set desired MIDI instrument (0-127)
```

sets the timbre generated by this instrument.

Then the following statements

```
# load piano image and create display with appropriate size
pianoIcon = Icon("iPianoOctave.png") # image for complete piano
display = Display("iPiano", pianoIcon.getWidth(),
                            pianoIcon.getHeight())
```

load the full piano image (actually one octave), create a display with the exact size (given the image dimensions), and place the image on it.*

Now that we have the full (one-octave) keyboard shown the display, we can create the illusion of different keys getting pressed by superimposing (i.e., adding) various images of pressed keys onto the display. So, next we load those images, one image for each key we want pressed. The following code demonstrates how to load such images. It is left as an exercise for you to complete it.

```
# load icons for pressed piano keys
cDownIcon = Icon("iPianoWhiteLeftDown.png")     # C
cSharpDownIcon = Icon("iPianoBlackDown.png")    # C sharp
dDownIcon = Icon("iPianoWhiteCenterDown.png")   # D
#...continue loading icons for additional piano keys
```

Next we create a list to remember which keys are pressed. This is needed because when we keep a key pressed, most computer keyboards initiate the key-repeat function, i.e., they type multiple instances of the same character. (While this is convenient when we type, say, an email, it is quite disturbing when trying to play a piano; we do not want to get several note repeats from a single key press. For this reason, when a key is pressed, we add it to the keysPressed list. We remove it when the key is released. This way we can differentiate between a new key press and one that was generated by the computer keyboard's key-repeat functionality.

```
# remember which keys are currently pressed
keysPressed = []
```

---

* If we wanted to create a longer piano (e.g., two octaves) by loading the image twice, we could create a display with double the width (i.e., pianoIcon.getWidth() * 2), and place the two images on it.

Next we define the two callback functions to be used when a key is pressed and when a key is released, respectively.

The first callback function, `beginNote()`, is assigned to the display's `onKeyDown` event handler. As mentioned above, since computer keyboards usually have a key-repeat action (i.e., if you keep pressing a key, it repeats), this function will be called repeatedly when a key is held pressed. This would have the effect of restarting the corresponding note, which is undesirable. For this reason, we use the `keysPressed` list to remember which keys are currently being pressed, so that when the key-repeat action causes repeated calls of `beginNote()` for the same key press, we can ignore them.

This function associates a given computer key with a piano key, using the following code:

```
if key == VK_Z and key not in keysPressed:
    display.add( cDownIcon, 0, 1 )   # "press" this piano key
    Play.noteOn( C4 )                # play corresponding note
    keysPressed.append( VK_Z )       # avoid key-repeat
```

In the above case, if the key pressed is "Z" (i.e., virtual key VK_Z)*:

- We superimpose the proper key-pressed image (i.e., image cDownIcon) at the proper place on the display (i.e., x, y coordinates 0, 1).

- We start playing the proper note (i.e., C4).

- Finally, we add this virtual key to the `keysPressed` list. This way, when the key-repeat action starts and function `beginNote()` is called again with the same virtual key, the if condition (i.e., key not in `keysPressed`) will prevent the code from executing again.

Similarly, when the key is released, the `endNote()` callback function is called. This function associates a given computer key with a piano key, using the following code:

```
if key == VK_Z:
    display.remove( cDownIcon )    # "release" this piano key
    Play.noteOff( C4 )             # stop corresponding note
    keysPressed.remove( VK_Z )     # and forget key
```

In the above case, if the key pressed is "Z" (i.e., virtual key VK_Z):

- We remove the pressed-down image (so that it creates the illusion that the piano key has been released).

---

* For a list of available virtual keys, see the keyboard events section earlier in this chapter.

- We stop the corresponding note from sounding.

- Finally, we remove the released key from the `keysPressed` list, so that if it is pressed again, we can act on it.

The last part of the code associates the two callback functions with the corresponding keyboard events on the display:

```
# associate callback functions with GUI events
display.onKeyDown(beginNote)
display.onKeyUp(endNote)
```

8.14.1 Exercise

Complete the functionality of the above program. In other words, make the remaining piano keys (F sharp, G, G sharp, A, A sharp, and B) functional. Keep in mind that, if we wish to use the same key-pressed image for several keys, we need to load it several times (i.e., once for each key we want to use it with). This allows us to have several keys appear to be pressed simultaneously.

Note: To improve typing accuracy, some computer keyboards do not allow certain keys to be typed together, such as Z, S, and X. Therefore, you may notice that your virtual piano has the same limitation.

8.14.2 A Variation, Using Parallel Lists

As you work through the above exercise, you will soon realize the repetitive nature of the task. For each of the computer keys you wish to associate with a virtual piano key, you have to add a special `elif` case in each of the two callback functions (in addition to loading the corresponding key-pressed icon). This can take some time to implement.

The following variation takes advantage of the repetitive nature of this task. It introduces parallel lists to hold related information:

- `downKeyIcons` holds the icons corresponding to "pressed" piano keys,

- `virtualKeys` holds the virtual keys (e.g., VK_Z, etc.) corresponding to the above piano keys (pressing these keys on the computer keyboard "presses" the corresponding piano keys),

- `pitches` holds the note pitches corresponding to the above (piano) keys, and

- iconWidths holds the X coordinate of each "pressed" piano key icon, so it perfectly aligns with the underlying "unpressed" piano octave icon.

These lists can be extended to support more keys (currently, only 6 keys are functional).

```
# iPianoParallel.py
#
# Demonstrates how to build a simple piano instrument playable
# through the computer keyboard.
#

from music import *
from gui import *

Play.setInstrument(PIANO)       # set desired MIDI instrument (0-127)

# load piano image and create display with appropriate size
pianoIcon = Icon("iPianoOctave.png")  # image for complete piano
d = Display("iPiano", pianoIcon.getWidth(), pianoIcon.getHeight())
d.add(pianoIcon)                        # place image at top-left corner

# NOTE: The following loads a partial list of icons for pressed piano
#       keys, and associates them (via parallel lists) with the
# virtual keys corresponding to those piano keys and the corresponding
# pitches. These lists should be expanded to cover the whole octave
# (or more).

# load icons for pressed piano keys
# (continue loading icons for additional piano keys)
downKeyIcons = [] # holds all down piano-key icons
downKeyIcons.append(Icon("iPianoWhiteLeftDown.png"))    # C
downKeyIcons.append(Icon("iPianoBlackDown.png"))        # C sharp
downKeyIcons.append(Icon("iPianoWhiteCenterDown.png"))  # D
downKeyIcons.append(Icon("iPianoBlackDown.png"))        # D sharp
downKeyIcons.append(Icon("iPianoWhiteRightDown.png"))   # E
downKeyIcons.append(Icon("iPianoWhiteLeftDown.png"))    # F

# lists of virtual keys and pitches corresponding to above piano keys
virtualKeys = [VK_Z, VK_S, VK_X, VK_D, VK_C, VK_V]
pitches     = [C4, CS4, D4, DS4, E4, F4]

# create list of display positions for downKey icons
#
# NOTE: This as hardcoded - they depend on the used images!
#
iconLeftXCoordinates = [0, 45, 76, 138, 150, 223]

keysPressed = [] # holds which keys are currently pressed
```

```python
########################################################################
# define callback functions
def beginNote( key ):
    """Called when a computer key is pressed.  Implements the
       corresponding piano key press (i.e., adds key-down icon on
       display, and starts note).  Also, counteracts the key-repeat
       function of computer keyboards.
    """

    # loop through all known virtual keys
    for i in range(len(virtualKeys)):

        # if this is a known key (and NOT already pressed)
        if key == virtualKeys[i] and key not in keysPressed:

            # "press" this piano key (by adding pressed key icon)
            d.add(downKeyIcons[i], iconLeftXCoordinates[i], 0)
            Play.noteOn(pitches[i])   # play corresponding note
            keysPressed.append(key)   # avoid key-repeat

def endNote( key ):
    """Called when a computer key is released. Implements the
       corresponding piano key release (i.e., removes key-down icon,
       and stops note).
    """

    # loop through known virtual keys
    for i in range(len(virtualKeys)):

        # if this is a known key (we can assume it is already pressed)
        if key == virtualKeys[i]:

            # "release" this piano key (by removing pressed key icon)
            d.remove( downKeyIcons[i] )
            Play.noteOff( pitches[i] )     # stop corresponding note
            keysPressed.remove( key )      # and forget key

########################################################################
# associate callback functions with GUI events
d.onKeyDown( beginNote )
d.onKeyUp( endNote )
```

This revision demonstrates that for most programming tasks there may be different algorithms that accomplish the same task. In this case, the first variation is simpler to understand, but more lengthy, repetitious, and thus error-prone. The second variation is more elegant and easier to extend—to add more functioning keys to the virtual piano you simply add more items to the parallel lists at the top of the program.

Notice the use of `for` loops in the two callback functions to iterate through the known virtual keys (`virtualKeys` list) to see if we need to act on the key generating the keyboard event. For example,

```
# loop through all known virtual keys
for i in range(len(virtualKeys)):

   # if this is a known key (and NOT already pressed)
   if key == virtualKeys[i] and key not in keysPressed:

      # "press" this piano key (by adding pressed key icon)
      d.add( downKeyIcons[i], iconLeftXCoordinates[i], 0 )
      Play.noteOn( pitches[i])    # play corresponding note
      keysPressed.append( key )   # avoid key-repeat
```

Notice how we use the index of the virtual key, `i`, to access the corresponding information in the parallel lists. This replaces the repetitive sequence of `if/elif` statements for each of the known virtual keys seen in the original program.

8.14.2.1 Exercises

1. Complete the functionality of the above program. In other words, make the remaining piano keys (F sharp, G, G sharp, A, A sharp, and B) functional. Keep in mind that, if we wish to use the same key-pressed image for several keys, we need to load it several times (i.e., once for each key we want to use it with). This allows us to have several keys appearing to be pressed simultaneously.

2. Extend the above program to display a two octave piano. (*Hint*: Load the "iPianoOctave.png" image twice and place the second instance at the end of the first instance. Use the icon `getWidth()` function to make this more general.)

3. Create a second display with an additional piano. Figure out which computer keyboard keys to assign to which piano display and notes. Assign a harpsichord timbre and pick registers accordingly to create a four octave, "two-decker" harpsichord instrument.

4. Create your own GUI version of audio mixer slider (e.g., mixer master volume). It should consist of two images, that is, the base of the slider and the knob. The knob should be moved by dragging the mouse. (*Hint*: Use the icon's `onMouseDrag()` function.) Constrain the knob's movement to be only vertical (i.e., ignore the mouse x coordinate);

also constrain the vertical movement to begin and end using the base image's position and height.

5. Once your GUI version of the audio mixer slider is ready (see above), use it to control the volume of an AudioSample. (*Hint*: Similarly to the example in the Mouse Events section above, include all necessary code inside the callback function associated with slider's `onMouseDrag` event.)

6. Design a collection of GUI control knobs, ribbons, and related components to be used in building more advanced surfaces for interactive musical instruments. This collection can be easily implemented using Python classes (presented in the next section).

8.15 SCHEDULING FUTURE EVENTS

So far we have seen how to capture user input through graphical user interfaces to drive our program's behavior, that is, event-driven programming.

Sometimes it is also useful to schedule computer-initiated (as opposed to user-initiated) events to happen sometime in the future. This allows us, among other things, to build buttons that temporarily change color (or shape) when pressed, etc. To do so, we simply change the color (or shape) of a graphical object when it is clicked on and then schedule another change in color (or shape) to occur a fraction of a second later.

This is accomplished using Timer objects. Timer objects are given a function to call, and a delay specifying after how much time to call it.* The next case study demonstrates how to do this.

8.15.1 Case Study: Random Circles with Timer

This case study presents a generative music application consisting of circles drawn randomly onto a display.

- Every circle is connected to a note— as the circle is drawn, a pitch is sounded.

- The color of the circle determines the pitch of the note—the redder the color, the lower the note.

- The circle's size (radius) determines the volume of the note—the bigger the radius, the louder the note.

---

* Among other things, Timer objects can be used to do animation. Animation is covered in Chapter 10.

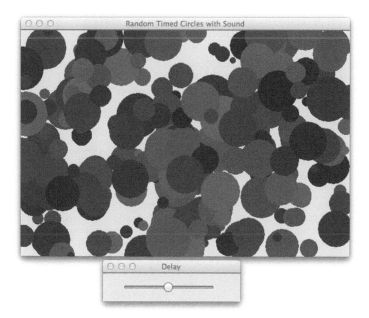

FIGURE 8.10   Graphical user interface of "Random Circles with Timer" program.

To make the music aesthetically pleasing, pitches are selected from the major scale. This concept is inspired by Brian Eno's "Bloom" musical app for smartphones.

Circles are initially drawn once per 0.5 seconds (500 milliseconds), or at a rate of 2 circles per second. This rate works quite well with adding circles to a display, possibly reminding the viewer of drops of rain. Additionally, the program presents a secondary display (see Figure 8.10), which allows the user to control the delay between successive drops (i.e., circle-notes).

Here is the code:

```
# randomCirclesTimed.py
#
# Demonstrates how to generate a musical animation by drawing random
# circles on a GUI display using a timer.  Each circle generates
# a note - the redder the color, the lower the pitch; also,
# the larger the radius, the louder the note.  Note pitches come
# from the major scale.
#

from gui import *
from random import *
from music import *

delay = 500 # initial delay between successive circle/notes
```

```
##### create display on which to draw circles #####
d = Display("Random Timed Circles with Sound")

# define callback function for timer
def drawCircle():
   """Draws one random circle and plays the corresponding note."""

   global d    # we will access the display

   x = randint(0, d.getWidth())       # x may be anywhere on display
   y = randint(0, d.getHeight())      # y may be anywhere on display
   radius = randint(5, 40)            # random radius (5-40 pixels)

   # create a red-to-brown-to-blue gradient (RGB)
   red = randint(100, 255)                # random R component (100-255)
   blue = randint(0, 100)                 # random B component (0-100)
   color = Color(red, 0, blue)            # create color (green is 0)
   c = Circle(x, y, radius, color, True)  # create filled circle
   d.add(c)                               # add it to the display

   # now, let's create note based on this circle

   # the redder the color, the lower the pitch (using major scale)
   pitch = mapScale(255-red+blue, 0, 255, C4, C6, MAJOR_SCALE)

   # the larger the circle, the louder the note
   dynamic = mapValue(radius, 5, 40, 20, 127)

   # and play note (start immediately, hold for 5 secs)
   Play.note(pitch, 0, 5000, dynamic)

# create timer for animation
t = Timer(delay, drawCircle)    # one circle per 'delay' milliseconds

##### create display with slider for user input #####
title = "Delay"
xPosition = d.getWidth()/3 # set initial position of display
yPosition = d.getHeight() + 45
d1 = Display(title, 250, 50, xPosition, yPosition)

# define callback function for slider
def timerSet(value):
   global t, d1, title    # we will access these variables
   t.setDelay(value)
   d1.setTitle(title + "(" + str(value) + "msec)")

# create slider
s1 = Slider(HORIZONTAL, 10, delay*2, delay, timerSet)
d1.add(s1, 25, 10)

# everything is ready, so start animation (i.e., start timer)
t.start()
```

First, notice the delay variable used to initialize the initial delay between successive circles.

```
delay = 500 # initial delay between successive circle/notes
```

The next section of the code creates the main display, d, and a callback function, called drawCircle(). This function will be called by the Timer object repeatedly, every delay (e.g., 500) milliseconds, to draw a new circle and play a note.

The Timer object is created as follows:

```
# create timer for animation
t = Timer(delay, drawCircle) # one circle per 'delay' milliseconds
```

As seen in the next section, the Timer function expects a delay and a call-back function. It creates a Timer object which automatically calls this function every delay milliseconds.

A program may have several timers going on at once. Timers may be started (so that they do their intended function) or stopped. Also, we may change their delay time. Actually, the rest of the program does precisely that, that is, it creates a secondary display consisting of a slider, which allows the end-user to change the timer's delay.

```
##### create control surface #####
title = "Delay"
xPosition = d.getWidth()/3 # set initial position of display
yPosition = d.getHeight() + 45
d1 = Display(title, 250, 50, xPosition, yPosition)
```

Notice the positioning of this secondary display. It is placed below the main display, relative to the main display's width (i.e., xPosition = d.getWidth()/3) and height (i.e., yPosition = d.getHeight() + 45).

The slider is created with a callback function, timerSet(), which expects the value of the slider and uses this value to set the Timer object's delay, as well as the title of the secondary display.

```
# define callback function for slider
def timerSet(value):
global t, d1, title
t.setDelay(value)
d1.setTitle(title + "(" + str(value) + " msec)")
# create slider

s1 = Slider(HORIZONTAL, 10, delay*2, delay, timerSet)
d1.add(s1, 25, 10)
```

Notice how the slider is created to range from 10 milliseconds to twice the initial delay that may be set at the beginning of the program, and with an initial value equal to `delay`. Remember that, as mentioned in the previous chapter, every time the user moves the slider button the system calls the provided (callback) function, `timerSet()`, passing to it the slider's new value.

Finally, in order for the application to come alive, we need to start the timer:

```
# everything is ready, so start animation (i.e., start timer)
t.start()
```

The following section provides more information on creating and managing Timer objects.

8.15.2 The Timer Class

The GUI library supports scheduling future tasks through the Timer class. Timer objects are used to schedule when and how often to perform a certain task (i.e., to call a given function). These tasks are not confined to graphical events; as demonstrated in the previous case study, they can be any computer-generated task, including musical tasks.

8.15.2.1 Creating Timers

Timer objects are used to schedule functions to be executed after a given time interval, repeatedly or once. The following function creates a new Timer, so you need to save it in a variable (so you can use it later).

Function	Description
`Timer(delay, function, parameters, repeat)`	Creates a new Timer to execute function after delay time interval (in milliseconds). The optional parameter parameters is a list of parameters to pass to the function (when called). The optional parameter repeat (boolean—default is True) determines if the timer will go on indefinitely.

For example, the following:

```
t = Timer(500, Play.noteOn, [A4], True)
```

creates a Timer `t`, which will call function `Play.noteOn(A4)` repeatedly every 500 milliseconds (i.e., half second). In order for a timer to operate, it needs to get started:

```
t.start()
```

Once a Timer t has been created, the following functions are available:

Function	Description
t.start()	Starts timer t.
t.stop()	Stops timer t.
t.getDelay()	Returns the delay time interval of timer t (in milliseconds).
t.setDelay(delay)	Sets a new delay time interval for timer t (in milliseconds). This allows us to change the speed of the animation, after some event occurs.
t.isRunning(delay)	Returns True if timer t is running (has been started), False otherwise.
t.setFunction(function, parameters)	Sets the function to execute. The optional parameter parameters is a list of parameters to pass to the function (when called).
t.getRepeat()	Returns True if timer t is set to repeat, False otherwise.
t.setRepeat(flag)	If flag is True, timer t is set to repeat (this also starts the timer, if stopped). Otherwise, if flag is False, timer t is set to not repeat (this stops the timer, if running).

8.16 SUMMARY

This chapter focused on the design and development of graphical user interfaces (GUIs) and interactive software instruments. We discussed a number of new real-time musical functions, including Play.noteOn() and Play.noteOff(), as well as how to load in audio samples and MIDI sequences and make them part of our interactive software. We also saw the GUI library which opens new interaction possibilities. We talked about paper prototypes and how industry experts use them extensively (along with similar GUI development methodologies) while developing interactive software. Although starting with code might feel like the right thing to do, in the case of GUIs, a little paper prototyping goes a long way. We also saw how the various GUI widgets make the instrument interfaces come alive with visual animations synchronized to user input and sounds. Finally, we presented the Timer class, which enables the scheduling of functions to be executed in the future. This sets the stage for some power possibilities that will be explored further in the remaining chapters.

## CHAPTER 9

Making Connections

Topics: Connecting to MIDI devices (pianos, guitars, etc.), the Python MIDI library, MIDI programming, Open Sound Control (OSC) protocol, connecting to OSC devices (smartphones, tablets, etc.), client-server programming via OSC messages, the Python OSC library, creating hybrid (traditional + computer) musical instruments.

9.1 OVERVIEW

In the previous chapter, we began designing unique interactive musical instruments for live performance, employing the GUI library in the process. In this chapter, we extend this thread by exploring connections between a computer (e.g., a laptop) and external devices, such as MIDI instruments, synthesizers, various control surfaces, and smartphones. This is accomplished through two additional Python libraries (provided with this textbook), which allow your programs to connect to any device that communicates using MIDI or OSC (Open Sound Control) protocols.

This capability is quite enabling because, in addition to developing GUIs that turn your computer into a musical instrument, now we can design and implement *hybrid* musical instruments, incorporating traditional musical instrument interfaces (e.g., piano or guitar). These traditional instruments need to be MIDI or OSC enabled (as discussed below), so they can be connected to a Python program, with the latter "deciding" what sounds to generate.* Additionally, you may create futuristic (at the time of this writing) instruments, which may consist of smartphones and or tablets that somehow drive, guide, or contribute to the making of sound. Again you, the programmer, become a digital instrument maker, who can shape the musical experience of others. Your ability to design and program unique, innovative hybrid instruments

---

* The choice about sounds to generate rests with you, the programmer. As should be obvious by now, there are endless possibilities. You may transform an existing musical instrument to a digital instrument that has never existed before. Also, instead of buying expensive pedals and sound processors, you can develop your own.

gives you immense power, as your design choices will affect the type of music (style, timbres, etc.) that your end-users can make and, ultimately, will channel their creative expression. This allows you to make connections not only with arbitrary external instruments and devices, but also to design social events and community musical experiences that bring people together in various contexts of shared performance and music making.[*]

9.2 MIDI DEVICES—CONNECTING TO PIANOS, GUITARS, ETC.

This section looks at how to connect Python programs to external MIDI devices. This is accomplished by utilizing the MIDI library (provided with this textbook). There are two types of possible connections, *input* and *output*:

* Using MIDI input (one or more), we allow our program to receive incoming MIDI data from external devices, such as a MIDI-enabled instrument (guitars, pianos, violins, saxophones).

* Using MIDI output, we enable our programs to drive external synthesizers.

In other words, we are not limited anymore to receiving musical (or other) input only through the computer keyboard and mouse, and possibly interacting with a graphical user interface (as described in the previous chapter). Now we can extend our input framework to include elaborate, multimodal instruments (i.e., instruments with various input modalities), connecting virtual (GUI) displays with MIDI pedals, and, most importantly, traditional (yet MIDI-enabled) instruments, as shown in Figure 9.1.

Additionally, we are no longer limited to the MIDI synthesizers provided with our computer (i.e., the synthesizer of our computer's operating system or the Java synthesizer). We now can connect to any available MIDI synthesizer. This is very empowering as it allows us to make music of higher timbral quality and to construct more elaborate musical instruments (limited by the connectivity of our computer and our imagination), as well as more creative, avant garde music performance scenarios and projects.

---

[*] This is similar to how a DJ helps create musical contexts for dancing and shared musical listening.

FIGURE 9.1 A MIDI-enabled guitar and a MIDI pedal board connected to a Python program running on a laptop, along with various other devices and instruments (photo courtesy of Dave Brown).

9.2.1 Case Study: Make Music with a MIDI Instrument

This case study is a variation of the last case study in Chapter 8 ("Random Circles with Timer"). Here it modifies the program to receive input from an external MIDI instrument, such as a MIDI guitar, piano, or saxophone (among others). It eliminates the Timer object, since timing information will now be provided by the user through entering notes on a real musical instrument.

Here is the (much shorter) program:

```
# randomCirclesThroughMidiInput.py
#
# Demonstrates how to generate a musical animation by drawing random
# circles on a GUI display using input from a MIDI instrument.
# Each input note generates a circle - the lower the note, the lower
# the red+blue components of the circle color.  The louder the note,
# the larger the circle. The position of the circle on the display
# is random.  Note pitches come directly from the input instrument.
#
```

```
from gui import *
from random import *
from music import *
from midi import *

##### create main display #####
d = Display("Random Circles with Sound")

# define callback function for MidiIn object
def drawCircle(eventType, channel, data1, data2):
    """Draws a circle based on incoming MIDI event, and plays
       corresponding note.
    """

    global d                        # we will access the display

    # circle position is random
    x = randint(0, d.getWidth())    # x may be anywhere on display
    y = randint(0, d.getHeight())   # y may be anywhere on display

    # circle radius depends on incoming note volume (data2)
    radius = mapValue(data2, 0, 127, 5, 40)     # ranges 5-40 pixels

    # color depends on on incoming note pitch (data1)
    red = mapValue(data1, 0, 127, 100, 255) # R component (100-255)
    blue = mapValue(data1, 0, 127, 0, 100)  # B component (0-100)
    color = Color(red, 0, blue)             # create color (green is 0)

    # create filled circle from parameters
    c = Circle(x, y, radius, color, True)

    # and add it to the display
    d.add(c)

    # now, let's play the note (data1 is pitch, data2 is volume)
    Play.noteOn(data1, data2)

# establish a connection to an input MIDI device
midiIn = MidiIn()

# register a callback function to process incoming MIDI events
midiIn.onNoteOn( drawCircle )
```

First, we need to import the MIDI library

```
from midi import *
```

This library (as discussed below) provides objects that allow our programs to connect to external input and output MIDI devices.

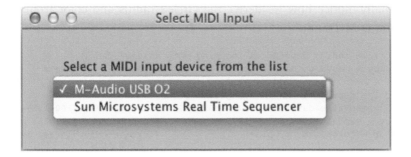

FIGURE 9.2 Display opened when creating a MidiIn object. It contains all available input MIDI devices.

To establish a connection to a MIDI object, we do the following:

```
midiIn = MidiIn()
```

This creates a MidiIn object, which allows us to establish communication with a MIDI instrument connected to our computer. When the above statement is executed, it opens the GUI shown in Figure 9.2. This GUI contains a drop-down list populated with all available input MIDI devices. When we select one of these devices, the MidiIn object is connected to that device (and the GUI disappears).

A program may contain several MidiIn objects, one for each desired input MIDI device. This opens endless possibilities for communication and processing of incoming MIDI data.

Now, to receive and process input, we register a callback function with the MidiIn object, as follows:

```
midiIn.onNoteOn(drawCircle)
```

This callback function needs to accept four parameters, namely, `eventType`, `channel`, `data1`, `data2`. These parameters are defined by the MIDI standard, and are explained in the next section. For NOTE-ON events, `eventType` is always 144, channel represents which channel is being used by the MIDI instrument, `data1` holds the pitch of the note, and `data2` holds the velocity of volume of the note.

The MidiIn object provides support for several other events. As seen in the next section, the information stored in the parameters (i.e., eventType, channel, data1, data2) can vary depending on the type of the particular event.

Finally, notice how we derive the circle's radius and color from the incoming MIDI note:

```
# circle radius depends on incoming note volume (data2)
radius = mapValue(data2, 0, 127, 5, 40)  # ranges 5-40 pixels

# color depends on on incoming note pitch (data1)
red = mapValue(data1, 0, 127, 100, 255)  # R component (100-255)
blue = mapValue(data1, 0, 127, 0, 100)   # B component (0-100)
color = Color(red, 0, blue)              # create color (green is 0)
```

In particular, the circle's radius depends on the MIDI note's velocity (which is data2 for note events), whereas both the red and blue components of the circle's color depend on the MIDI note's pitch (data1 for note events). Other mappings, of course, are possible.

9.2.1.1 Exercise
Currently, notes generated by the above program continue to sound even after the user has stopped playing them (on the input MIDI instrument, e.g., a digital piano). This is because the callback function, drawCircle(), is executed when a note starts and this contains a Play. noteOn() statement. However, the program does not take into account the ending of notes in order to execute a corresponding Play.noteOff() statement.

Using the MidiIn object's capability to capture NOTE-OFF events, add a second callback function, stopNote(), to terminate those notes. *Hints:* This function should have the same four parameters as drawCircle(), namely, eventType, channel, data1, data2. This function should contain a Play.noteOff() statement. The pitch to be turned off is provided in data1. To register this function with the MidiIn object, so that it is called when a NOTE-OFF event occurs, include the following:

```
midiIn.onNoteOff(stopNote)
```

9.2.2 The MIDI Library
The MIDI library enables the connection to various input and output MIDI devices connected to your computer.

9.2.2.1 The MidiIn Class
As we have seen, MidiIn objects are used to act as an interface between our Python code and MIDI devices that generate input events (e.g., a MIDI

guitar, keyboard, or control surface). The following function creates a new MidiIn, so you need to save it in a variable (so you can use it later).

Function	Description
MidiIn()	Creates a new MidiIn object to connect to an input MIDI device. When called, it presents the user with a GUI to select one from the available MIDI devices (see Figure 9.2).

For example, the following:

```
mInput = MidiIn()
```

creates a MidiIn object mInput. Once such an object has been created, the following functions are available:

Function	Description
mInput.onNoteOn(function)	When a NOTE_ON event happens on the mInput device (i.e., the user starts a note), the system calls the provided function. This function should have four parameters, eventType, channel, data1, data2. For NOTE_ON events, the eventType is always 144, the channel ranges from 0 to 15, data1 is the note pitch (0–127), and data2 is the volume of the note (0–127).
mInput.onNoteOff(function)	When a NOTE_OFF event happens on the mInput device (i.e., the user ends a note), the system calls the provided function. This function should have four parameters, eventType, channel, data1, data2. For NOTE_OFF events, the eventType is always 128, the channel ranges from 0 to 15, data1 is the note pitch (0–127), and data2 is ignored.
mInput.onSetInstrument(function)	When a SET_INSTRUMENT (also known as CHANGE_PROGRAM) event happens on the mInput device (i.e., the user selects a different timbre), the system calls the provided function. This function should have four parameters, eventType, channel, data1, data2. For SET_INSTRUMENT events, the eventType is always 192, the channel ranges from 0 to 15, data1 is the MIDI instrument (0–127), and data2 is ignored.

Continued

Function	Description
`mInput.onInput(eventType, function)`	Associates an incoming `eventType` with a callback `function`. When the specified eventType event happens on the `mInput` device, the system calls the provided function. This function should have four parameters, `eventType`, `channel`, `data1`, `data2`.
	Can be used repeatedly to associate different event types (128–224) with different callback functions (one function per event type).
	If eventType is `ALL_EVENTS`, then `function` will be called for all incoming events that have not yet been assigned callback functions.

Notice how the function `onInput()` makes it easy to specify callback functions for specific MIDI event types. This is done through the `eventType` parameter (i.e., the number associated with a particular message type). Moreover, `onInput()` can be used repeatedly to associate different event types with different callback functions (one function per event type).

Again, if `eventType` is ALL_EVENTS, the associated callback function is called for all events not handled already. This function should figure out which `eventType` it is dealing with (i.e., it will probably have a sequence of `if/elif` statements) and then specify what to do for different event types.

Finally, if `onInput()` is used more than once for the same `eventType`, only the latest callback function is retained.

9.2.2.2 The MidiOut Class

MidiOut objects are used to act as an interface between our Python code and MIDI devices that accept output events (e.g., an external MIDI synthesizer). The following function creates a new MidiOut, so you need to save it in a variable (so you can use it later).

Function	Description
`MidiOut()`	Creates a new `MidiOut` object to connect to an output MIDI device. When called, it presents the user with a GUI to select one from the available MIDI devices (see Figure 9.3).

FIGURE 9.3   Display opened when creating a MidiOut object. It contains all available output MIDI devices.

For example, the following:

```
mOutput = MidiOut()
```

creates a MidiOut object mOutput. Once such an object has been created, the following functions are available. (These functions are similar to the Play functions seen in Chapter 8, except that class Play is connected to the Java synthesizer.)

Function	Description
`mOutput.noteOn(pitch, velocity, channel)`	Sends a NOTE_ON message with `pitch` (0–127), at a given `velocity` (0–127 — default is 100), to channel (0–15 — default is 0) on the `mOutput` device.
`mOutput.noteOff(pitch, channel)`	Sends a NOTE_OFF message with `pitch` (0–127), on given channel (0–15 — default is 0) on the `mOutput` device. If the pitch is not sounding on this channel, this has no effect.
`mOutput.setInstrument(instrument, channel)`	Sets a MIDI instrument (0–127 — default is 0) for the given channel (0–15, default is 0) on the `mOutput` device. Any notes played through the channel will sound using the instrument.
`mOutput.playNote(pitch, start, duration, velocity, channel)`	Schedules playing of a note with `pitch` at the given `start` time (in milliseconds from now), with `duration` (in milliseconds from start time), `velocity` (0–127 — default is 100), to channel (0–15 — default is 0) on the `mOutput` device.
`mOutput.play(material)`	Play music library material (`Score`, `Part`, `Phrase`, `Note`) on the `mOutput` device.

Notice how function `playNote()` makes it easy to schedule notes to be played at a future time.

Finally, notice how function `play()` allows playback through any synthesizer connected to your computer for musical material developed with the algorithmic techniques presented in the earlier chapters.

9.3 OSC DEVICES—CONNECTING TO SMARTPHONES, TABLETS, ETC.

This section describes how to connect various everyday devices, such as smartphones, to your music-making programs, through the Open Sound Control (OSC) protocol.* This can be done through existing OSC applications and/or the OSC library provided with this book.

OSC is similar to MIDI in that it allows different devices to connect to and communicate with each other. It is different from MIDI in that it does not require the devices to be connected via special cables. Actually, OSC devices do not even need to be in the same room, or the same country or continent for that matter. This is because OSC uses the Internet to send and receive messages, and as long as the devices you wish to connect to are on the network they can "talk" to each other.

To turn any device (such as a smartphone, a tablet, or a laptop) into an OSC device, you need to connect them to the same network and run an OSC application (or program) that either sends or receives OSC messages.

Definition: An application/device that sends OSC messages is called an *OSC client*.

Definition: An application/device that receives OSC messages and acts on them is called an *OSC server*.

It is possible for a single device to be both an OSC client and server, in that it may send OSC messages to other devices as well as receive OSC messages from those (or other) devices. The connectivity options are quite flexible.

---

* Open Sound Control is a freely available protocol widely used in the computer music community. For more information see the main OSC web site: http://opensoundcontrol.org/

9.3.1 OSC Messages

OSC messages look similar to the URLs we type in browsers. They usually consist of an address pattern and optional arguments:

- OSC address patterns form a hierarchical name space, such as "/hello/world/". It is up to the OSC client and server to decide what addresses will be used to communicate.[*]

- Arguments mainly are integers, floats, strings, and booleans. OSC messages may include zero or more arguments, as agreed between OSC clients and servers.

The advantages of OSC over MIDI are primarily connectivity over the Internet (no dedicated wires!), flexibility of setting up message types (OSC addresses), and use of standard data types (e.g., integer, float, string, boolean).

We will now explore how to write simple OSC programs in Python. These programs can be combined with existing OSC applications (clients or servers) running on other devices (or the same computer).

9.3.2 Case Study: Hello (OSC) World!

This case study shows a simple example to get two devices (an OSC client and an OSC server) to talk to each other. The OSC client device will control the OSC server device.

It is preferred that you use two different computers. However, for simplicity, you may use a single computer (just run the programs on two separate command windows).[†]

9.3.2.1 Program for OSC Server Device

First, here is the program to receive the OSC messages. (Even if you use a single computer to execute both programs, the messages will be delivered via the Internet, as if you were using two different machines.)

```
# oscServer.py
#
# Demonstrates how to create an OSC server program.
#

from osc import *
from music import *
```

---

[*] This is more flexible than the MIDI standard, which restricts the types of messages to a few predefined possibilities.

[†] Of course, it is much more convincing to see two different computers "talk" to each other via OSC.

```
###### create an OSC input object ######
oscIn = OscIn( 57110 )      # receive incoming messages on port 57110

# define two message handlers (functions) for incoming OSC messages
def simple(message):
  print "Hello world!"

def complete(message):
  OSCaddress = message.getAddress()
  args = message.getArguments()

  # print OSC message time and address
  print "\nOSC Event:"
  print "OSC In - Address:", OSCaddress,

  # also, print message arguments (if any), all on a single line
  for i in range( len(args) ):
  print ", Argument " + str(i) + ": " + str(args[i]),
  print

###### now, associate above functions with OSC addresses ######

# callback function for incoming OSC address "/helloWorld"
oscIn.onInput("/helloWorld", simple)

# callback function for all incoming OSC addresses
# (specify that as "/.*")
oscIn.onInput("/.*", complete)
```

The above program creates an OscIn object, which sets up the computer for incoming OSC traffic. This specifies which port is being used. For more information, see the next section.

Then it defines two callback functions, each to be called for different OSC message addresses.

Finally, it associates each of the callback functions with the corresponding addresses.

When you run this program, it will output the following:

```
OSC Server started:
Accepting OSC input on IP address xxx.xxx.xxx.xxx at port 57110
(use this info to configure OSC clients)
```

where "xxx.xxx.xxx.xxx" is the IP address of the receiving computer (e.g., "192.168.1.223"). You should share this IP address and the port number with the people running the OSC clients from which you wish to receive messages. Of course, if you are behind a firewall, you also need to ensure the firewall allows incoming traffic at the given port (e.g., "57110").

9.3.2.2 Program for OSC Client Device

Now here is the program to run on the OSC client (i.e., the computer sending the OSC messages).

```
# oscClient.py
#
# Demonstrates how to create an OSC client program.
#

from osc import *

###### create an OSC output object ######

# where to send outgoing OSC messages - you may replace "localhost"
# with the IP address of a receiving computer (e.g., "192.168.1.223")
oscOut = OscOut( "localhost", 57110 )   # use port 57110

###### that's it - now, send some test OSC messages ######

# send a "/helloWorld" message without arguments
oscOut.sendMessage("/helloWorld")

# send a "/itsFullOfStars" message with an integer, a float,
# a string, and a boolean argument
oscOut.sendMessage("/itsFullOfStars", 1, 2.35, "wow!", True)
```

where "xxx.xxx.xxx.xxx" is the IP address of the receiving computer (e.g., "192.168.1.223"). You may use this IP address to set up your OSC clients. (The IP address is the Internet address you would use if you wanted to connect to your computer via the Internet.) Of course, if you are behind a firewall, you would also need to ensure that the firewall allows incoming traffic at the above port (e.g., "57110").

9.3.2.3 Exercises

1. Keep the OSC server running. Run the OSC client program a few times to see how the incoming messages arrive on the server.

2. Modify the OSC client program to send a different message via the `oscOut.sendMessage()` function. Since the OSC server program has a generic function, that is, `complete()` that handles all incoming messages, see the output generated by your message. Run a few times.

3. Modify the OSC server program to have a specific function for your new message. First, you will need to define that function (use `complete()` as a model). Then you need to associate this function with your new OSC message address; for that you use the `oscIn.onInput()` function. (Use the OSC server program as a model. Also see the next section.)

9.3.3 The OSC Library

The Open Sound Control (OSC) library provides functionality for OSC communication between programs running on your computer (servers) and OSC devices (clients).

9.3.3.1 The OscIn Class

OscIn objects are used to receive messages from OSC devices. The following function creates a new OscIn object, so you need to save it in a variable (so you can use it later).

Function	Description
OscIn(port)	Creates a new OscIn object to receive incoming messages from an OSC device (such as a smartphone or tablet) on the given port. The port number is an integer from 1024 to 65535 that is not being used by another program.[*]

For example, the following:

```
oscIn = OscIn(57110)
```

creates an OscIn object oscIn. You may create different OSC input objects (servers) to receive and handle OSC messages. As mentioned above, you just need to provide unique port numbers (of your choice). When created, the OscIn object prints out its host IP number and its port:

```
OSC Server started:
Accepting OSC input on IP address 192.168.1.223 at port 57110
(use this info to configure OSC clients)
```

This done is for convenience. Use the above information to set up OSC clients to send messages to this object.

To send messages to oscIn you may use objects of the OscOut class below, or another OSC client, such as TouchOSC or Control. The latter is most enabling, as it allows programs to be driven by external devices, such as smartphones, tablets, and other computers. This way you may build sophisticated musical instruments and artistic installations.

---

[*] Each OscIn object requires its own port. So pick port numbers not used by other applications. For example, TouchOSC (a mobile app for Android and iOS devices), defaults to 8000 for sending OSC messages and 9000 for receiving messages. In general, any port from 1024 to 65535 may be used, as long as no other application is using it. If you have trouble, try changing port numbers. The best bet is a port in the range 49152 to 65535 (which is reserved for custom purposes).

Once such an object has been created, the following function is available:

Function	Description
oscIn. onInput(address, function)	When an OSC message with address arrives, call function. OSC addresses look like a URL, e.g., "/first/second/ third". The function should expect one parameter, the incoming OSC message.

Notice how the function onInput() makes it easy to specify callback functions for specific OSC addresses. For example,

```
oscIn = OscIn( 57110 )  # receive incoming messages on port 57110

def simple(message):    # define a simple message handler (function)
   print "Hello world!"

# callback function for incoming OSC address "/helloWorld"
oscIn.onInput("/helloWorld", simple)
```

If an incoming OSC address is "/helloWorld", the oscIn object will automatically call the function simple(). This function simply outputs "Hello world".

Similarly, the following code associates the function complete() with *all* incoming OSC address:

```
def complete(message):
   OSCaddress = message.getAddress()
   args = message.getArguments()

   # print OSC message time and address
   print "\nOSC Event:"
   print "OSC In - Address:", OSCaddress,

   # also, print message arguments (if any), all on a single line
   for i in range( len(args) ):
     print ", Argument " + str(i) + ": " + str(args[i]),
   print

# callback function for all incoming OSC addresses
# (specify that as "/.*")
oscIn.onInput("/.*", complete)
```

Notice the special OSC address "/.*" — this matches for all incoming addresses. Actually, the OSC library allows the use of *regular expressions* to specify OSC addresses used with function onInput(). Regular expressions allow matching of one or more different OSC addresses in one statement. Regular expressions are beyond the scope of this book.* You should be able to handle most of your needs with simple OSC addresses, similar to "/gyro" and "/accelerometer", as shown in the previous case study.

The function `complete()` in the code snippet above (as well as the functions `gyro()` and `accel()` in the previous case study) demonstrates how to extract arguments from incoming OSC messages. Initially in your code development, using a function like `complete()` allows you to see all incoming OSC messages (including their addresses and arguments). This allows you to decide which OSC addresses and arguments to use (i.e., to write functions specifically tailored to extract/capture only the needed information and ignore everything else). Again, see functions `gyro()` and `accel()` above as examples.

9.3.3.2 The OscOut Class

OscOut objects are used to send messages to other OSC devices. The following function creates a new OscOut object, so you need to save it in a variable (so you can use it later).

Function	Description
`OscOut(IPaddress, port)`	Creates a new `OscOut` object to send outgoing messages to an OSC device (such as a smartphone or tablet) to the given `IP address` (a string, e.g., "192.168.1.223") and `port` (an integer in the range 1024 to 65535).

For example, the following:

```
# where to send outgoing OSC messages
# ("192.168.1.223" is the IP address of a receiving computer)
oscOut = OscOut( "192.168.1.223", 57110 )
```

creates an OscOut object `oscOut`.

You may create different OSC output objects (clients) to send OSC messages to different OSC servers (as defined by the provided IP address and port number).†

---

* Several tutorials are available online if you are interested in learning more about writing regular expressions.

† If you are planning to connect to OscIn objects (as mentioned in the previous section) when created, OscIn objects output (print out) their host IP number and port (for convenience). Use this info to set up your OscIn objects (OSC clients).

Once such an OscOut object has been created, the following function is available:

Function	Description

OscOut.
  sendMessage(address,
  arg1, arg2,...)
	Sends an OSC message with address and 0 or more arguments to the OSC device associated with OscOut.

For example, the following code:

```
oscOut.sendMessage("/helloWorld")
oscOut.sendMessage("/itsFullOfStars", 1, 2.3, "wow!", True)
```

sends two messages to the OSC input shown as an example in the previous section. Notice how the first message has no arguments. On the other hand, the second message demonstrates the different Python data types you may send via an OSC message (i.e., integer, float, string, and boolean).

9.3.4 Case Study: Make Music with your Smartphone

In this case study, we modify the "Random Circles Through Midi Input" program (see earlier in this chapter) to receive input from a smartphone, using the OSC protocol.

For this example, we used an iPhone running the free OSC app *Control* (see Figure 9.4). We selected this app because it is free and also available on Android devices. Of course, there are many other possibilities. Below we explain how to adapt this code for use with other devices.

9.3.4.1 Performance Instructions

This program creates a musical instrument out of your smartphone. It has been specifically designed to allow the following performance gestures:

- **Ready Position:** Hold your smartphone in the palm of your hand, flat and facing up, as if you are reading the screen. Make sure it is parallel with the floor. Think of an airplane resting on top of your device's screen, its nose pointing away from you, and its wings flat across the screen (left wing superimposed with the left side of your screen, and right wing with the right side of your screen).

- **Controlling Pitch:** The pitch of the airplane (the angle of its nose - pointing up or down) corresponds to musical pitch. The higher the nose, the higher the pitch.

- **Controlling Rate:** The roll of the airplane (the banking of its wings to the left or to the right) triggers note generation. You could

FIGURE 9.4   Using a smartphone to create music on a laptop (via OSC messages). The smartphone may be in a different room or a different continent (as long as it is connected to the Internet).

visualize notes falling off the device's screen, so when you roll/bank the device, notes escape (roll off).

- **Controlling Volume:** Device shake corresponds with loudness of notes. The more intensely you shake or vibrate the device as notes are generated, the louder the notes are.

That's it. To summarize, the smartphone's orientation (pointing from zenith to nadir) corresponds to pitch (high to low). Shaking the device plays a note—the stronger, the louder. Tilting the device clockwise produces more notes.

On the server side, that is, the program you are controlling with your smartphone:

- Note pitch is mapped to the color of circles (lower means darker/browner, higher means brighter/redder/bluer).

- Shake strength is mapped to circle size (radius).

- Finally, the position of the circle on the display is random.

All these settings could easily be changed. We leave that as an exercise.

9.3.4.2 Setting up Your Smartphone (OSC Client)

This section describes how to setup your smartphone so that you may use it send OSC data to the program below.*

First, it is up to the OSC client (e.g., *Control*) to capture the necessary data from your device (e.g., smartphone or tablet computer), package it as OSC message, and send it to the right computer at the right port.

When setting up your OSC client running on your smartphone (or any other device, e.g., a tablet or laptop), you need to tell the OSC client where to send its messages. This consists of

- the receiving computer's IP address (e.g., 192.168.1.223) and

- the appropriate port (e.g., 57110).

How you set up this information depends on the OSC client software you are using.†

The program below helps you by printing out the necessary information (its IP address and port) when it starts.

9.3.4.3 Setting up Your Computer (OSC Server)

Below is the Python program you need to run on your computer. It creates the OSC server and starts listening for incoming messages from your smartphone. It also prints out the IP address and port, so that you can use these values to set up the OSC client (see previous section).

Here is the code:

```
# randomCirclesThroughOSCInput.py
#
# Demonstrates how to create a musical instrument using an OSC device.
# It receives OSC messages from device accelerometer and gyroscope.
#
```

---

* Actually, if you do not have a smartphone, it is very easy to write a (small) Python program to send OSC data. It takes about three lines of code (see section "The OscOut Class").

† If you choose to use the *Control* app, then there are setup instructions at the following URL and you can use the *Gyro + Accelerometer* interface preset. http://charlie-roberts.com/Control/?page_id=51

```
# This instrument generates individual notes in succession based on
# its orientation in 3D space.  Each note is visually accompanied by
# a color circle drawn on a display.
#
# NOTE: For this example we used an iPhone running the free OSC app
# "Control OSC". (There are many other possibilities.)
#

from gui import *
from random import *
from music import *
from osc import *

# parameters
scale = MAJOR_SCALE        # scale used by instrument
normalShake  = 63          # shake value at rest (using xAccel for now)
shakeTrigger = 7           # deviation from rest value to trigger notes
                           # (higher, less sensitive)
shakeAmount  = 0           # amount of shake
devicePitch  = 0           # device pitch (set via incoming OSC messages)

##### create main display #####
d = Display("Smartphone Circles", 1000, 800)

# define function for generating a circle/note
def drawCircle():
  """Draws one circle and plays the corresponding note."""

  global devicePitch, shakeAmount, shakeTrigger, d, scale

  # map device pitch to note pitch, and shake amount to volume
  pitch = mapScale(devicePitch, 0, 127, 0, 127, scale) # use scale
  volume = mapValue(shakeAmount, shakeTrigger, 60, 50, 127)
  x = randint(0, d.getWidth())              # random circle x position
  y = randint(0, d.getHeight())             # random circle y position
  radius = mapValue(volume, 50, 127, 5, 80) # map volume to radius

  # create a red-to-brown gradient
  red = mapValue(pitch, 0, 127, 100, 255)   # map pitch to red
  blue = mapValue(pitch, 0, 127, 0, 100)    # map pitch to blue
  color = Color(red, 0, blue)               # make color (green is 0)
  c = Circle(x, y, radius, color, True)     # create filled circle
  d.add(c)                                  # add it to display

  # now, let's play note (lasting 3 secs)
  Play.note(pitch, 0, 3000, volume)

##### define OSC callback functions #####
# callback function for incoming OSC gyroscope data
def gyro(message):
  """Sets global variable 'devicePitch' from gyro OSC message."""

  global devicePitch       # holds pitch of device
```

```
    args = message.getArguments()    # get OSC message's arguments


    # output message info (for exploration/fine-tuning)
    #print message.getAddress(),     # output OSC address
    #print list(args)                # and the arguments


    # the 4th argument (i.e., index 3) is device pitch
    devicePitch = args[3]


# callback function for OSC accelerometer data
def accel(message):
    """
    Sets global variable 'shakeAmount'.  If 'shakeAmount' is higher
    than 'shakeTrigger', we call function drawCircle().
    """

    global normalShake, shakeTrigger, shakeAmount

    args = message.getArguments()  # get the message's arguments

    # output message info (for exploration/fine-tuning)
    #print message.getAddress(),     # output the OSC address
    #print list(args)                # and the arguments

    # get sideways shake from the accelerometer
    shake = args[0]     # using xAccel value (for now)

    # is shake strong enough to generate a note?
    shakeAmount = abs(shake - normalShake)     # get deviation from rest
    if shakeAmount > shakeTrigger:
        drawCircle()  # yes, so create a circle/note

##### establish connection to input OSC device (an OSC client) #####
oscIn = OscIn( 57110 )    # get input from OSC devices on port 57110

# associate callback functions with OSC message addresses
oscIn.onInput("/gyro", gyro)
oscIn.onInput("/accelerometer", accel)
```

Notice the various parameters defined at the beginning:

```
# parameters
scale = MAJOR_SCALE       # scale used by instrument
normalShake  = 63         # shake value at rest (using xAccel for now)
shakeTrigger = 7          # deviation from rest value to trigger notes
                          # (higher, less sensitive)
shakeAmount  = 0          # amount of shake
devicePitch  = 0          # device pitch (set via incoming OSC messages)
```

As the comments indicate, normalShake and shakeTrigger parameters allow
fine-tuning of the OSC instrument. The last two parameters, shakeAmount

and `devicePitch`, are used as global variables. These are set by the OSC callback functions (see below) when the corresponding OSC messages arrive; they are mapped to circle and note attributes.

As we mentioned earlier, the use of global variables is bad style. The next section, on Python classes, will provide a neat way to avoid using global variables and improve reuse of our code.*

Function `drawCircle()`, when called, draws a circle and plays the corresponding note. Among other things, it maps the smartphone's tilt to note pitch and the smartphone's shake intensity to note volume, as follows:

```
# map device pitch to note pitch, and shake amount to volume
pitch = mapScale(devicePitch, 0, 127, 0, 127, scale)   # use scale
volume = mapValue(shakeAmount, shakeTrigger, 60, 50, 127)
```

As it turns out, the device and OSC app used (i.e., iPhone and Control) return `devicePitch` values that roughly range between 0 and 127, so no real mapping is needed. The mapping code above was included for convenience, in case you wish to use this code with another device (which returns a different range of values).

Interestingly, the values returned for `shake` also range between 0 and 127. However, since we ignore `shake` values that are too weak, in this case, the second mapping statement is useful, as it stretches the incoming `shake` values across the complete MIDI volume range (0 to 127). Since the shake value returned at rest on the iPhone is 63, the following statements capture both leftward shake (which generates shake values smaller than the rest value) and rightward shake (which generates shake values larger than the rest value):

```
# map device pitch to note pitch, and shake amount to volume
pitch = mapScale(devicePitch, 0, 127, 0, 127, scale)   # use scale
volume = mapValue(shakeAmount, shakeTrigger, 60, 50, 127)
```

If the absolute difference of the shake value (either leftward or rightward) is greater than the specified threshold (`shakeTrigger`), we start generating notes.

An important step in this case study is how to establish the connection with the OSC device and how to extract the data of interest.

---

* Code reuse refers to how easy it is to reuse existing code in different applications.

The OSC connection is accomplished with the last few statements in this program:

```
##### establish connection to input OSC device (an OSC client) #####
oscIn = OscIn( 57110 )    # get input from OSC devices on port 57110

# associate callback functions with OSC message addresses
oscIn.onInput("/gyro", gyro)
oscIn.onInput("/accelerometer", accel)
```

The first line creates an OSC input object listening to port 57110. When this statement is executed it opens an OSC connection on the specified port. Any other device on the Internet can now send OSC messages to that particular port (for more information on OSC messages and selecting ports, see the next section).

OSC messages look similar to the URLs we type in browsers. They are automatically created by the software we use to capture and send data (e.g., the Control app mentioned above). In this case, Control sends messages with data from the smartphone's gyroscope using the "/gyro" address pattern. It also sends messages with data from the smartphone's accelerometer using the "/accelerometer" address. This could be changed, but we wrote the above code using Control's defaults.

The last two lines above simply associate each of the two incoming OSC message addresses ("/gyro" and "/accelerometer") with a corresponding callback function. In other words, when "/gyro" messages come in, Python will automatically call function gyro() to process them. For "/accelerometer" messages, it will call function accel().

Each of these functions accepts one parameter, namely, the OSC message, which will automatically be passed to them by Python. They use the following code to extract the desired information from the message:

```
args = message.getArguments()  # get the message's arguments

# output message info (for exploration/fine-tuning)
#print message.getAddress(),    # output OSC address
#print list(args)               # and the arguments
```

The first line extracts the list of arguments sent with the OSC message. The other two lines (currently commented out) can be used (as the comment suggests) to see (explore) what the incoming arguments look like for this OSC message. Then you can decide which arguments to use. In our case, the `gyro()` function uses the 4th argument of the "/gyro" message, that is,

```
# the 4th argument (i.e., index 3) is devicePitch
devicePitch = args[3]
```

whereas the `accel()` function uses the first argument of the "/accelerometer" message, i.e.,

```
# get sideways shake from the accelerometer
shake = args[0] # sideways shake
```

Finally, as discussed earlier, it is up to the OSC client (e.g., Control) to capture the necessary data from the OSC device (e.g., smartphone), package it as OSC message, and send it to the right computer at the right port. Now that you understand the above program, it should be straightforward to update it, so that you can use it with any OSC client device. This is left as an exercise. Enjoy!

9.3.4.4 Exercises

1. Modify this code to work with another device (e.g., an Android smartphone or tablet).

2. Modify this code to work with a different OSC app—there are several.

3. Create a different mapping between gyroscope/accelerometer data and music or drawing outcomes. Again, there are many possibilities.

4. Different OSC devices have different sensors. For instance, your device may not have a gyroscope, but it may have another (possibly unique) sensor. Use the above program to observe and discover the types of OSC messages generated by your device. Write them down. What types of musical (or other) actions can you envision associating with these incoming messages?

5. Design a new music-making application using incoming OSC data (e.g., using a different OSC app/device combination). There are endless possibilities. Be innovative.

9.3.5 Hybrid Musical Instrument Projects

Here are some more advanced exercises and design/implementation activities for creating hybrid (traditional + computer musical instruments):

1. Design, on paper, a hybrid instrument consisting of a MIDI guitar (piano or control surface) and a Python program. Describe what type of musical experience you are trying to create. Also what type of musical background is required by your performer (e.g., how to play the guitar, or none).

2. Design on paper a hybrid instrument consisting of a collection of smartphones communicating over OSC with a Python program. Describe what type of musical experience you are trying to create. Also what type of musical background is required by your performers.

3. Design on paper a hybrid instrument consisting of a collection of a MIDI instruments and smartphones communicating over OSC with a Python program. Describe what type of musical experience you are trying to create. Also what type of musical background is required by your performers.

4. Implement a prototype of one of the above hybrid instruments using Python and the libraries provided in this chapter.

5. What other possibilities can you think of for creating immersive, shared musical performance spaces utilizing some of the tools presented so far?

9.4 SUMMARY

This chapter discussed how to connect your musical programs to external devices via the MIDI and OSC protocols. These features will enable you to write programs that interact with other devices and instruments in a way that will open up a whole world of music performance and interaction opportunities.

Using MIDI connections allows you to interact with electronic musical instruments and various control surfaces. MIDI is a well-established protocol and many digital music devices support it. The chapter explored how to interface between MIDI devices and your programs and how to send and receive performance data (e.g., note messages) to either play back your music on a MIDI synthesizer or to capture performance information from a MIDI instrument.

Using OSC allows you to use Internet-enabled mobile computing devices (like smartphones and tablets) to connect and control your programs. The final exercises open the door to a whole new way of thinking about computer programming and instrument design. For instance, you may design innovative applications and hybrid musical instruments, which combine traditional instrument interfaces (e.g., piano) with algorithmic processes. Since you can control what functions are called in your programs, through arbitrary events on a musical instrument (or a smartphone), the sky is the limit. Also, you may design innovative performance projects, where you might allow many OSC clients (e.g., smartphones in the audience) control aspects of your performance on stage. This gives a whole new meaning to connecting with your audience!

## CHAPTER 10

Music, Number, and Nature

Topics: Connecting nature, music and number, Pythagorean theorem, music from math curves, sin() and cos() functions, the Python math library, visualizing oscillations, the harmonograph, sonifying oscillations, Kepler's harmony of the world revisited.

10.1 OVERVIEW

In the previous chapters, we studied essential building blocks of music and computer science. We now know enough about music and programming to return to the themes introduced in Chapter 1. In this chapter we will deepen our exploration into the connections between music, number, and nature, and introduce you to ideas that will hopefully inspire and guide you in your own personal journey into music and programming.

The Pythagoreans discovered that music harmony can be modeled by numbers. As Aristotle mentions, they thought that the principles of mathematics "were the principles of all things. Since, of these principles, numbers are by nature the first, and in numbers there seemed to see many resemblances to the things that exist and come into being" (Aristotle 1992, pp. 70–71).

The Pythagoreans observed that strings exhibit *harmonic* proportions, that is, they resonate at integer ratios of their length (i.e., 1/1, 1/2, 1/3/, 1/4, 1/5, etc.). They also observed that these proportions are aesthetically pleasing to the human ear. Accordingly, they developed musical modes based on these ratios, which formed the basis of our modern-era musical scales.

Among other things, the Pythagoreans quantified "harmonious" musical intervals in terms of proportions (ratios) of the first few whole numbers: a unison is 1:1, octave is 2:1, perfect fifth is 3:2, perfect fourth is 4:3, and so on (Miranda 2001, p. 6). The Pythagorean scale was refined over centuries to produce well-tempered and equal-tempered scales (Livio 2002, pp. 29, 186).

Aristotle supported the Pythagorean view that "[the interplay] between opposites is the beginning of all beings" (Aristotle 1992, pp. 72–73). Plato, Euclid, and others provided a more precise description of this interplay in the form of proportional analogies (e.g., "A is to B as C is to D"). The apogee of this exploration may have been the discovery of the golden mean. This is a special proportion (i.e., 1.61803398875 ...) that humans find aesthetically very pleasing. It is found in natural or human-made artifacts (Beer 2008; Calter 2008, pp. 46–57; Hemenway 2005, pp. 91–132; Livio 2002; May 1996; Pickover 1991, pp. 203–205). It is also found in the human body (e.g., the bones of our hands, the cochlea in our ears, etc.). The golden ratio reflects a place of balance in the structural interplay of opposites.

How did the Pythagoreans accomplish this much? And why did they care? What drove them to this?

Here is a possible answer, called "The Great Theme":

"Picture to yourself, if you can, a universe in which everything makes sense. A serene order presides over the earth around you, and the heavens above revolve in sublime harmony. Everything you can see and hear and know is an aspect of the ultimate truth: the noble simplicity of a geometric theorem, the predictability of the movements of heavenly bodies, the harmonious beauty of a well-proportioned fugue—all are reflections of the essential perfection of the universe" (James 1995, p. 3)

10.2 ORIGINS AND REPRESENTATIONS

Although Pythagoreans coined the term *mathematics,*[*] they were not the first to study it. The origins of exploring patterns in nature (e.g., seasons, astronomical patterns, etc.) and trying to comprehend the principles behind them goes back to Chinese emperors, Babylonian priests, prehistoric cave people in France, and the Mayans, among others (Pickover 2002, pp. 11–35).

We know that these different peoples were studying patterns in nature and were developing mathematical knowledge and skills, because they left behind various representations (inscriptions and objects) of that knowledge. Naturally, as we examine representations from these diverse societies, we see that they and the knowledge they depict changed over time.

Which brings us to an important realization.

---

[*] Derived from μάθημα = lesson, μαθαίνω = learn. The Pythagoreans divided people who attended their school to three groups, the akoustikoi (from άκουσμα = listening, i.e., those who listen), to mathematikoi (from μάθημα = lesson, i.e., those who learn), and the physikoi (from φύση = nature, those who study nature, the natural philosophers) (Cooke 2005, p. 271).

Fact: The *representation* of knowledge is independent from the actual knowledge.

For example, let's explore some the history of one of those expressions of "the ultimate truth" captured in "the noble simplicity of a geometric theorem," the Pythagorean Theorem.

10.2.1 Pythagorean Theorem

The Pythagorean Theorem revolves around a balance emerging from right triangles. Why did the Pythagoreans care about triangles? Probably because triangles emerge from first principles (i.e., ab initio, or "from the beginning"), as we will see in the next section. Why *right* triangles? Probably because their orthogonality* may have been a way to model the "[interplay] between opposites is the beginning of all beings" (Aristotle, pp. 72–73).

Fact: In right triangles, the (area of the) square of the hypotenuse always equals the (areas of the) squares of the two other sides (see Figure 10.1).

Interestingly, the Pythagorean Theorem did not originate with the Pythagoreans. As we know now, sometime around 1800 BCE, a Babylonian clay tablet was inscribed using cuneiform script with instances of Pythagorean triples (see Figure 10.2). Pythagorean triples are numbers that demonstrate the property of right triangles captured by the Pythagorean Theorem (e.g., 3, 4, 5).

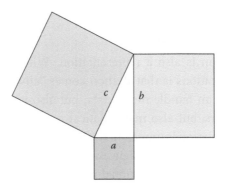

FIGURE 10.1   A Pythagorean triangle (source Wikipedia).

---

* Orthogonal, in addition to perpendicular, refers to two independent forces or artifacts.

FIGURE 10.2 A Babylonian clay tablet with instances of Pythagorean triples (no 322, Plimpton Collection, Columbia University) (source Wikipedia—public domain).

As the centuries passed, people developed different languages to represent this significant piece of knowledge. In our culture, we use the following representation:

$$c^2 = a^2 + b^2$$

One of the advantages of this notation is that it is succinct. One of its disadvantages is that it succinct, and thus "compresses" a lot of knowledge in few symbols (which, linguistically, makes it hard to interpret). It is easy to produce, but it is hard to deconstruct, in order to understand what's being said.

Similarly to Western notation for music, this modern notation for mathematics is only one of several possible representations.

10.2.2 Python as a Representation

Interestingly, Python is also a representation. What sets it apart from the above representations is that Python comes "alive" when you run it on a computer. It can handle music data, but also musical processes. It can handle numbers, but also mathematical processes. Such power in a knowledge representation is unseen in the history of civilization (however, the Antikythera mechanism design and implementation comes close). In summary, Python's representational power enables us to explore connections between music, number, and nature.[*]

---

[*] This holds for other programming languages as well.

Since in previous chapters we have seen how to model music data and musical processes, this chapter will focus on modeling mathematical (and physical) processes. Since mathematical processes have been used for centuries to model nature (e.g., physics and astronomy), we can tap into this knowledge, translate it to Python, and then write programs that are driven by these processes to create music and sound. The possibilities are endless.

Now we will return to the Pythagorean Theorem and explore its possible connection to the Pythagorean fascination with musical (and other) harmonies.

Let's begin.

10.3 CASE STUDY: MUSIC FROM MATH CURVES

A mathematical function describes a line or curve. These curves could be thought of as describing a musical gesture, for example, a melodic contour. A function can be mapped to any musical parameter, that is, pitch in the case of a melodic curve, or volume in the case of an amplitude envelope. The mapping between the function values and the musical parameters is entirely up to the composer; for instance, a function may influence more than one parameter at a time.

A simple mathematical function is the sine. It describes a smooth repetitive oscillation (i.e., a wave). The sine function is a trigonometric function.[*] It is used in measuring right triangles.

The sine of angle α (see Figure 10.3a) is equal to the ratio of two triangle sides:

- the opposite side to α, compared to

- the hypotenuse.

Alternatively (see Figure 10.3b), if point B was traveling the circumference of a unit circle (a circle with radius 1), the sine of angle α corresponds to how high B is from the horizontal centerline (its vertical distance).

Notice that as point B moves around the circle, the sine traverses an oscillation.

- sin(0) is 0 (point B is on the centerline),

- sin(90) is 1 (point B is on the vertical centerline), and so on.[†]

---

[*] From τρίγωνο = triangle + μέτρηση = measurement.

[†] Recall that 90 degrees equals $\pi/2$ radians, 180 degrees equals π radians, and so on.

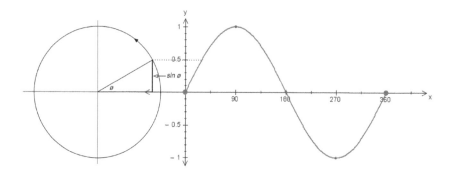

FIGURE 10.3   Two complementary views of the sine function: (a) as a measurement of a right triangle, and (b) as the rise of a point traversing a unit circle. (source Wikipedia).

Python provides this and other useful math functions. These are defined in the math library (see next section). One thing to remember about the sin() function is that, in Python, it works with radians. If your angle is in degrees, you can convert it with function radians(). For example:

```
>>> from math import *

>>> sin(0)
0.0
>>> sin(pi/2)
1.0
>>> radians(90) == pi/2
True
>>> sin(radians(90))
1.0
>>> radians(180) == pi
True
```

and so on. Similarly, if the angle is in radians, you can convert with function degrees().

```
>>> degrees(pi/2)
90.0
>>> degrees(pi)
180.0
```

10.3.1 Hearing the Music

The following program demonstrates how to create a simple melodic contour using the sin() function. This program creates the piano roll

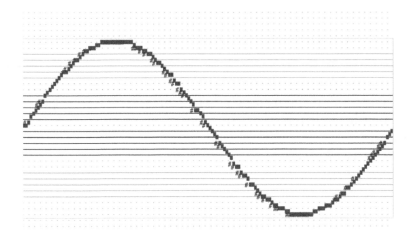

FIGURE 10.4   Melodic contour from a sine wave.

shown in Figure 10.4, which traces the sine wave oscillation discussed above.

The "sineMelody.py" program creates notes from one cycle of a sine wave function, that is, 0 radians (or 0 degrees) to 2*pi radians (or 360 degrees). It creates many notes, whose pitches correspond to the sine value, inside the for loop. The density variable adjusts how many notes are spread across each cycle of the sine wave. The program uses the music library mapValue() function to map sine wave values to the appropriate musical parameters.

Here is the code:

```
# sineMelody.py
#
# This program demonstrates how to create a melody from a sine wave.
# It maps the sine function to a melodic (i.e., pitch) contour.
#

from music import *
from math import *

phr = Phrase()
density = 25.0                   # higher for more notes in sine curve
cycle = int(2 * pi * density)    # steps to traverse a complete cycle

# create one cycle of the sine curve at given density
for i in range(cycle):
   value = sin(i / density)       # calculate the next sine value
   pitch = mapValue(value, -1.0, 1.0, C2, C8)   # map to range C2-C8
   note = Note(pitch, TN)
   phr.addNote(note)
```

```
# now, all the notes have been created

View.pianoRoll(phr)     # so view them
Play.midi(phr)          # and play them
```

The constant `pi` is approximately 3.141592653589793. It corresponds to half a circle (or 180 degrees). Accordingly, `2*pi` is a full circle (or 360 degrees).

Notice the following:

```
density = 25.0                   # higher for more notes in sine curve
cycle = int(2 * pi * density)    # steps to traverse a complete cycle
```

It is used to adjust how many data points (i.e., notes) to generate while traversing the circle. The higher the density, the more data points (notes) will be generated. The more data points, the slower the music will be (since every note has duration TN).

Next we connect additional musical parameters to the sine function, namely, note duration, dynamic, and panning. The piano roll generated by the updated program is shown in Figure 10.5. Notice the distortion in the sine wave graph. Why does that happen? The sine wave graph assumes steady movement on the x-axis (time), that is, constant note durations. Since we connected the sine function to note

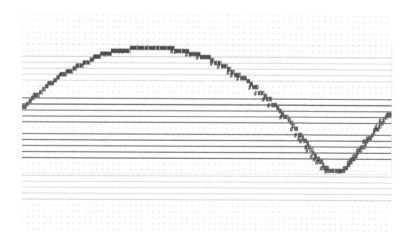

FIGURE 10.5 Melodic contour from sine wave also mapped to duration of notes.

duration, the resultant note durations fluctuate, hence the distortion in the graph.*

Here is the code:

```
# sineMelodyPlus.py
#
# This program demonstrates how to create a melody from a sine wave.
# It maps the sine function to several musical parameters, i.e.,
# pitch contour, duration, dynamics (volume), and panning.
#

from music import *
from math import *

sineMelodyPhrase = Phrase()
density = 25.0                       # higher for more notes in sine curve
cycle = int(2 * pi * density)        # steps to traverse a complete cycle

# create one cycle of the sine curve at given density
for i in range(cycle):
   value = sin(i / density)       # calculate the next sine value
   pitch = mapValue(value, -1.0, 1.0, C2, C8)    # map to range C2-C8
   #duration = TN
   duration = mapValue(value, -1.0, 1.0, TN, SN)        # map to TN-SN
   dynamic = mapValue(value, -1.0, 1.0, PIANISSIMO, FORTISSIMO)
   panning = mapValue(value, -1.0, 1.0, PAN_LEFT, PAN_RIGHT)

   note = Note(pitch, duration, dynamic, panning)
   sineMelodyPhrase.addNote(note)

View.pianoRoll(sineMelodyPhrase)
Play.midi(sineMelodyPhrase)
```

To best appreciate the results (e.g., the smooth oscillation in panning), use a set of good headphones.

10.3.2 Exercises

1. Currently, panning fluctuates from center to right, to left, and back to center. How would this change if you used the cosine function, cos(), instead? Why?

2. Experiment with additional math functions to create interesting melodic and other contours. See the next section for some possibilities. Also see Watkins, 2001.

---

* This sine wave inside a sine wave is the basis for FM synthesis (or frequency modulation) — a technique used for creating realistic, rich sounds for synthesizers.

10.4 MATH LIBRARY

The Python math library provides various math constants and operations. Similarly to the other libraries we have seen, you have to import it in your programs:

```
from math import *
```

Among others, this gives you access to the following (in the examples below, assume a is 2, and b is 3):

Operator	Description	Example
pi	The mathematical constant π	3.141592653589793
E	The mathematical constant e	2.718281828459045
sqrt(x)	The square root of x	sqrt(25) evaluates to 5
log(x, base)	The logarithm of x (the power by which to raise base in order to get x)	log(9, 3) evaluates to 2.0
degrees(x)	Converts x radians to degrees	degrees(pi) evaluates to 180.0
radians(x)	Converts x degrees to radians	radians(180) evaluates to 3.14159...
sin(x)	The sine of x in radians	sin(0) evaluates to 0.0
cos(x)	The cosine of x in radians	cos(0) evaluates to 1.0
hypot(x, y)	The hypotenuse of the Pythagorean triangle whose other sides have lengths x and y (or Euclidean distance)	hypot(3, 4) evaluates to 5

10.5 CASE STUDY: THE HARMONOGRAPH

Although the Pythagorean Theorem might appear unrelated to our musical narrative, it is not. It is encompassed in a very interesting device called the harmonograph.

The harmonograph is a device used to study what happens when you combine harmonic oscillations. It is closely related to the Pythagorean discoveries about musical intervals and their relationship to harmonic ratios to various other musical tunings and scales (Ashton, 2003).

Definition: A harmonograph is a device with a pen attached to two pendulums moving in *orthogonal* directions to each other (see Figure 10.6). As the pendulums move, the pen draws geometric shapes on a paper.

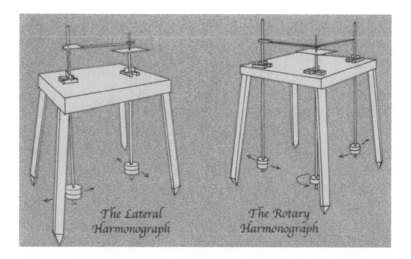

FIGURE 10.6 Two types of harmonographs. Left is a lateral, two-pendulum har-monograph (note the orthogonal direction of the pendulums). Right is a rotary, three-pendulum harmonograph (third pendulum has a rotary bearing, which allows circular motion). (From Ashton, A. (2003). *Harmonograph: A Visual Guide to the Mathematics of Music*. Glastonbury, UK Wooden Books., p. 19.)

In the lateral harmonograph (Figure 10.6, left), each pendulum swings in a plane; the two planes are perpendicular. The right pendulum has a square paper atop its shaft; the paper oscillates back and forth in one direction, forced by the pendulum's movement. The left pendulum has an attached arm that moves a pen back and forth in a direction perpendicular to the direction the paper's movement.

In the rotary harmonograph (Figure 10.6, right), the paper is attached to a third shaft that traces a circular path. The two pendulums push the pen in two perpendicular directions.

Some harmonographs have additional components (e.g., an additional pendulum) to help model more complicated harmonic relationships and movements.

Fact: Harmonographs are used to explore and easily visualize the relation-ships between different harmonic ratios.

For example, see Figure 10.7. It demonstrates the shapes generated from common harmonic ratios (i.e., 1:1, 2:1, and 3:2). On the left are the shapes

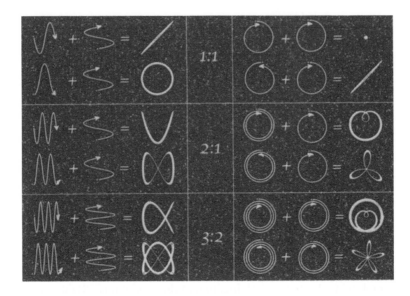

FIGURE 10.7 Using two harmonographs, lateral (left) and rotational (right) to generate shapes for different ratios (e.g., 1:1, 2:1, etc.). (From Ashton, A. (2003). *Harmonograph: A Visual Guide to the Mathematics of Music.* Glastonbury, UK Wooden Books., p. 19.)

generated from a lateral harmonograph with pendulums moving in same (or opposing) phase. On the right are the shapes generated from a rotary harmonograph with pendulums moving in same (or opposing) circular motion. Both versions will be shown below in Python.

Notice how simple geometric shapes naturally emerge from harmonic ratios (e.g., the octave, or 2:1 ratio) of *orthogonal* movement. Harmonograph drawings are related to Chladni patterns (seen in Chapter 1) and clearly help elucidate the principles of harmonics, overtones, and tunings of musical instruments.

Harmonographs are similar to the popular spirograph children's toy (which uses plastic gears to accomplish a similar result).*

The Antikythera mechanism (seen in Chapter 1) demonstrates that the ancients understood how to model different ratios using gears. It also demonstrates they knew how to create same and opposite circular motion (Vallianatos, 2012).

---

* Both harmonographs and spirographs combine rotary movements at different ratios to trace orbits generated from this combination. The relationship of such devices with models of astronomical circular movement should be obvious.

The Pythagoreans explored ways to *sonify* existing knowledge about harmonious ratios. Apparently, this knowledge came from ancient Egyptian and Mesopotamian scholars, who had been studying the periodicities of the night sky. These observations originated with early humans being curious about the bright lights moving in the night sky, at the beginning of civilization. Eventually, these observations coalesced into theories and (eventually) models, possibly over thousands of generations and across cultures (each inheriting some, but not necessarily all of the knowledge of its predecessors).

Next we will see that harmonographs can be easily described with simple Python trigonometric functions, i.e., `sin()` and `cos()`. The function `cos()` is similar to `sin()`.

Going back to Figure 10.3b, sin(α) corresponds to how high B is from the horizontal centerline (its y distance). Accordingly, cos(α) corresponds to how far B is from the vertical centerline (its x distance).

These two functions all that we need. They provide a concise representation for the movement of the two pendulums (as shown below). It is all that we need to see this "[interplay] between opposites" which, according to the Pythagoreans, was "the beginning of all beings" (Aristotle, pp. 72–73).

10.5.1 Lateral Harmonograph

A lateral harmonograph is a device with a pen attached to two pendulums moving in orthogonal directions to each other. As the pendulums move, the pen draws geometric shapes on a paper. Our controls include:

- The length of the pendulums—this affects the frequency of their oscillation. By combining different frequency ratios (e.g., 2:3), we get the shapes shown in Figure 10.7 (and more).

- The phase of the pendulums, relative to one another. In Python this is modeled by either using two `sin()` functions (same phase) or one `sin()` and one `cos()` function (opposite, or orthogonal) phase.

The program below creates a display onto which it draws points that trace the movement of the virtual pen connected to the pendulums. It also outputs the current ratio setting (see Figure 10.8). If you are wondering why the use of sin() and cos() accurately describe the motion of the pen on the paper, the answer lies in Figures 10.3 and 10.7 (top left).

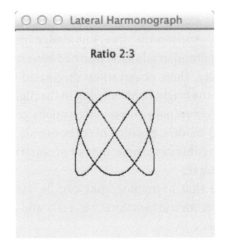

FIGURE 10.8 Output generated by lateral harmonograph program (2:3 ratio, same phase).

Here is the code:

```
# harmonographLateral.py
#
# Demonstrates how to create a lateral (2-pendulum) harmonograph
# in Python.
#
# See Ashton, A. (2003), Harmonograph: A Visual Guide to the
# Mathematics of Music, Wooden Books, p.19.
#

from gui import *
from math import *

d = Display("Lateral Harmonograph", 250, 250)
centerX = d.getWidth() / 2         # find center of display
centerY = d.getHeight() / 2

# harmonograph parameters
freq1 = 2        # holds frequency of first pendulum
freq2 = 3        # holds frequency of second pendulum
ampl  = 50       # the distance each pendulum swings

density = 250                      # higher for more detail
cycle = int(2 * pi * density)      # steps to traverse a complete cycle
times = 6                          # how many cycles to run

# display harmonograph ratio setting
d.drawText("Ratio " + str(freq1) + ":" + str(freq2), 95, 20)
```

```
# go around the unit circle as many times requested
for i in range(cycle * times):

    # get angular position on unit circle (divide by a float
    # for more accuracy)
    rotation = i / float(density)

    # get x and y coordinates (run and rise)
    x = sin( rotation * freq1 ) * ampl        # get run (same phase)
    #x = cos( rotation * freq1 ) * ampl       # get run (opposite phase)
    y = sin( rotation * freq2 ) * ampl        # get rise

    # convert to display coordinates (move display origin to center,
    # from top-left)
    x = x + centerX
    y = y + centerY

    # draw this point (pixel coordinates are int)
    d.drawPoint( int(x), int(y) )
```

Notice how we calculate the center of the display:

```
centerX = d.getWidth()/2 # find center of display
centerY = d.getHeight()/2
```

This is used later in the program to move the origin of the virtual pen from top-left to the center, so that the shapes are drawn centered in the display.

Notice the various harmonograph parameters (defined as constants at the program's beginning). Again, this makes it easy to adjust the harmonograph to new settings.

```
# harmonograph parameters
freq1 = 2        # holds frequency of first pendulum
freq2 = 3        # holds frequency of second pendulum
ampl = 50        # the distance each pendulum swings
```

The comments should be self-explanatory.

Parameter `times` is used to determine how many cycles to let the simulated pendulums oscillate (i.e., how many times to go back and forth).

Fact: When using integer ratios (e.g., 3:2), the generated shapes are complete after one cycle. Additional cycles simply retrace the original orbit (or shape).

Fact: When using noninteger ratios (e.g., 5.1:4), the generated shapes take more cycles to complete.*

Notice how we can change from same to opposite phase by switching the comments in the following:

```
# get x and y coordinates (run and rise)
x = sin(rotation * freq1) * ampl        # get run (same phase)
#x = cos(rotation * freq1) * ampl       # get run (opposite phase)
```

Finally, we shift the origin of the shape from (0, 0), the top-left of the display, to the display's center. And we draw the point.

```
# convert to display coordinates (move display origin to center,
# from top-left)
x = x + centerX
y = y + centerY

# draw this point (pixel coordinates are int)
d.drawPoint( int(x), int(y) )
```

10.5.2 Rotary Harmonograph

A rotary harmonograph is a device with a pen attached to two pendulums moving in orthogonal directions to each other. As the pendulums move, the pen draws geometric shapes on a paper. Additionally, the paper is placed on another pendulum mounted on a rotary bearing (i.e., gimbals), which adds a third circle (oscillation) to the drawing.

The next program creates a display onto which it draws points that trace the movement of the virtual pen connected to the pendulums. It also outputs the current ratio setting (see Figure 10.9).

Let's see the code:

```
# harmonographRotary.py
#
# Demonstrates how to create a rotary (3-pendulum) harmonograph
# in Python.
#
# Here, the position of the pen is determined by two pendula,
# and is modeled by either (sin, sin) or (cos, sin).
# The third pendulum has its own sin() and cos() to model the second
# circle.
#
# See Ashton, A. (2003), Harmonograph: A Visual Guide to the
# Mathematics of Music, Wooden Books, p.19.
#
```

---

* Some ratios, actually, never terminate (i.e., produce chaotic behavior). We will return to this soon.

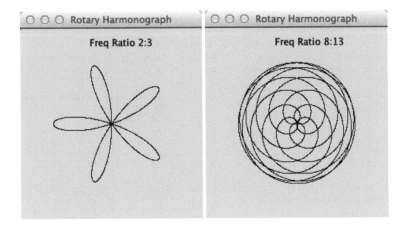

FIGURE 10.9 Output generated by rotary harmonograph program (left) 2:3 ratio, opposite rotation—(right) 8:13 ratio, same rotation.

```
from gui import *
from math import *

d = Display("Rotary Harmonograph", 250, 250)
centerX = d.getWidth() / 2      # find center of display
centerY = d.getHeight() / 2

# harmonograph parameters
freq1 = 2       # holds frequency of first pendulum
freq2 = 3       # holds frequency of second pendulum

ampl1 = 40      # holds swing of movement for pair of pendulums
                # (radius of first circle)

ampl2 = ampl1   # holds swing of movement for third pendulum
                # (radius of second circle)

#friction = 0.0003      # how much energy is lost per iteration

density = 250                   # higher for more detail
cycle = int(2 * pi * density)   # steps to traverse a complete cycle
times = 4                       # how many cycles to run

# display harmonograph ratio setting
d.drawText("Freq Ratio " + str(freq1) + ":" + str(freq2), 80, 10)

# go around the unit circle as many times requested
for i in range(cycle * times):

    # get angular position on unit circle (divide by a float
    # for more accuracy)
    rotation = i / float(density)
```

```
# get x and y coordinates (run and rise)
#x1 = sin( rotation * freq1 ) * ampl1      # get run (same phase)
#y1 = cos( rotation * freq1 ) * ampl1      # get rise
x1 = cos( rotation * freq1 ) * ampl1       # get run (opposite phase)
y1 = sin( rotation * freq1 ) * ampl1       # get rise

x2 = sin( rotation * freq2) * ampl2        # get run (second pendulum)
y2 = cos( rotation * freq2) * ampl2        # get rise

# combine the two oscillations
x = (x1 - x2)
y = (y1 - y2)

# convert to display coordinates (move display origin to center,
# from top-left)
x = x + centerX
y = y + centerY

# draw this point (pixel coordinates are int)
d.drawPoint( int(x), int(y) )

# loss some energy due to friction
#    ampl1 = ampl1 * (1 - friction)
#    ampl2 = ampl2 * (1 - friction)
```

This program is very similar to the previous one. Notice how we can change from same to opposite rotation by switching the comments in the following:

```
# get x and y coordinates (run and rise)
#x1 = sin( rotation * freq1 ) * ampl1      # get run (same phase)
#y1 = cos( rotation * freq1 ) * ampl1      # get rise
x1 = cos( rotation * freq1 ) * ampl1       # get run (opposite phase)
y1 = sin( rotation * freq1 ) * ampl1       # get rise
```

Notice the second pair of coordinates, x2 and y2, calculated with a second pair of sin() and cos() functions:

```
x2 = sin( rotation * freq2) * ampl2        # get run (second pendulum)
y2 = cos( rotation * freq2) * ampl2        # get rise
```

These contribute the second circle (caused by the rotation of the third pendulum). These are coordinates of the paper plane, as it rotates, being attached to the third pendulum. Also notice the addition of a second amplitude variable; again, this controls the amount of swing of the third pendulum.

Finally, notice how we have added a way to dampen both oscillations, if we wish (using variable friction). Simply uncomment the last three statements. Damping the oscillations introduces another element of control, which produces more interesting shapes.

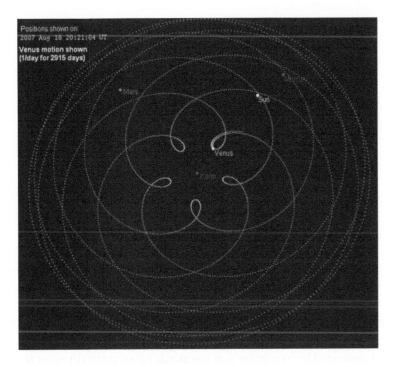

FIGURE 10.10 Successive inferior conjunctions of Venus as seen from Earth (from Wikipedia—public domain).

10.5.3 Exercises

1. Compare Figure 10.9 (right) with Figure 10.10. The former is created by the harmonograph from a 8:13 ratio, using same (concurrent) rotation. The second is the trajectory of Venus on the Earth night sky (the Earth orbits 8 times for every 13 orbits of Venus). This creates a pentagram on the night sky.

2. Research the importance of the pentagram to the Pythagoreans.

3. Research the connection between the pentagram and the golden ratio. As you will soon realize, the pentagram contains the golden ratio in many ways. What does this mean?

10.5.4 Noninteger Ratios

Noninteger ratios correspond to musical intervals that are not harmonious, that is, not pleasing to the ear.

Interestingly, such ratios generate chaotic behavior in the path traced by the harmonograph pen. When exploring, increase the value of variable

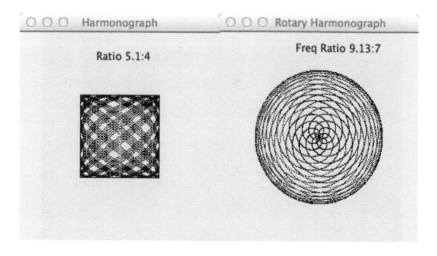

FIGURE 10.11 (Left) Lateral harmonograph (5.01:4 ratio, same phase, 6 times/cycles). (Right) Rotational harmonograph (5.01:4 ratio, same phase, 6 times/cycles).

times to allow the pen to trace orbits over several cycles, so that you can begin to see the behavior that emerges. For example, Figure 10.11 shows the shapes generated by particular non harmonic ratios (with times set to 6).

Certain ratios will never converge, that is, no matter how many cycles you run, they will always add more detail (as allowed by the resolution of your computer screen).

10.6 CASE STUDY: KEPLER'S HARMONY OF THE WORLD, NO. 2

The following program explores a more advanced sonification of the planetary velocities we explored in Chapter 7. Here we apply some of the same processes used in the lateral harmonograph to convert planetary velocities to music (or to sonify planetary velocities).

Here is the code:

```
# harmonicesMundiRevisisted.py
#
# Sonify mean planetary velocities in the solar system.
#

from music import *
from math import *
from random import *


# Create a list of planet mean orbital velocities
# Mercury, Venus, Earth, Mars, Ceres, Jupiter, Saturn, Uranus,
```

```
# Neptune. (Ceres is included in place of the 5th missing planet
# as per Bode's law).
planetVelocities = [47.89, 35.03, 29.79, 24.13, 17.882, 13.06, 9.64,
                    6.81, 5.43]
numNotes        = 100        # number of notes generated per planet
durations       = [SN, QN]   # a choice of durations
instrument      = EPIANO     # instruent to use
speedFactor     = 0.01       # decrease for slower sound oscillations


score = Score(60.0)          # holds planetary sonification


# get minimum and maximum velocities:
minVelocity = min(planetVelocities)
maxVelocity = max(planetVelocities)


# define a function to create one planet's notes - returns a Part
def sonifyPlanet(numNotes, planetIndex, durations, planetVelocities):
  """Returns a part with a sonification of a planet's velocity."""


  part = Part(EPIANO, planetIndex)     # use planet index for channel
  phr  = Phrase(0.0)


  planetVelocity = planetVelocities[planetIndex]     # get velocity


  # create all the notes by tracing the oscillation generated using
  # the planetary velocities
  for i in range(numNotes):


    # pitch is constant
    pitch = mapScale(planetVelocity, minVelocity, maxVelocity,
            C3, C6, MIXOLYDIAN_SCALE, C4)


    # panning and dynamic oscillate based on planetary velocity
    pan = mapValue(sin(i * planetVelocity * speedFactor * 2),
                   -1.0, 1.0, PAN_LEFT, PAN_RIGHT)
    dyn = mapValue(cos(i * planetVelocity * speedFactor * 3),
                   -1.0, 1.0, 40, 127)


    # create the note and add it the the phrase
    n = Note(pitch, choice(durations), dyn, pan)
    phr.addNote(n)


  # now, all notes have been created


  part.addPhrase(phr)    # add phrase to part
  return part            # and return it


# iterate over all plants
for i in range( len(planetVelocities) ):
  part = sonifyPlanet(numNotes, i, durations, planetVelocities)
  score.addPart(part)
```

```
View.sketch(score)
Write.midi(score, "harmonicesMundiRevisisted.mid")
Play.midi(score)
```

After importing the required libraries, the program starts by declaring
some variables to hold constant values that will be used through the work.
The most important of these is a list of the planetary velocities.

```
planetVelocities = [47.89, 35.03, 29.79, 24.13, 17.882, 13.06, 9.64,
                    6.81, 5.43]
numNotes        = 100       # number of notes generated per planet
durations       = [SN, QN]  # a choice of durations
instrument      = EPIANO    # instruent to use
speedFactor     = 0.01      # decrease for slower sound oscillations
```

The core musical mapping happens in this code fragment:

```
# create all the notes by tracing the oscillation generated using
# the planetary velocities
for i in range(numNotes):

  # pitch is constant
  pitch = mapScale(planetVelocity, minVelocity, maxVelocity,
              C3, C6, MIXOLYDIAN_SCALE, C4)

  # panning and dynamic oscillate based on planetary velocity
  pan = mapValue(sin(i * planetVelocity * speedFactor * 2),
              -1.0, 1.0, PAN_LEFT, PAN_RIGHT)
  dyn = mapValue(cos(i * planetVelocity * speedFactor * 3),
              -1.0, 1.0, 40, 127)

  # create the note and add it the the phrase
  n = Note(pitch, choice(durations), dyn, pan)
  phr.addNote(n)

# now, all notes have been created
```

Notice how the pitch of a planet remains the same for all generated notes—
it depends on its planetary velocity.[*]

The panning position and dynamic level of the notes associated with
each planet are continually modulated using sin() and cos(). The rate of
change is associated with the planetary velocity. This way, different planets
will vary at different speeds.

---

[*] This statement could be moved before the loop, for efficiency. We leave it in the loop to facilitate
   further experimentation, e.g., we may wish to try to somehow alter the pitch of each note for this
   planet.

Since the planetary velocities have harmonic relationships to each other, this musical mapping *sonifies* these harmonies using the panning and dynamic musical dimensions. The effect is breathtaking—dynamic variations give an impression of proximity (loud appears close to the listener, while soft appears far), and panning gives an impression of left to right movement. Two orthogonal (harmonic) oscillations perceivable in one ever-changing sound.

Notice how this program combines the lateral harmonograph model to provide harmonious oscillations for the panning and dynamic (volume) of the notes produced. The effect is quite stunning. Since the planet velocities are related to their orbits, the harmony of the spheres literally emerges. Play with variable `speedFactor` to adjust the speed of the oscillations.

This sonification requires a nice pair of stereo headphones (or speakers).

The comments, and your familiarity by now with Python, should make the rest of the code self-explanatory.

10.6.1 Exercises

1. The two orthogonal oscillations for panning and dynamic have an 1:1 ratio. Modify the above program to use other harmonic ratios, e.g., 1:2, 2:1, 3:2, 5:8, etc. What do you notice? Which ratios do you prefer for panning and dynamic?

2. Modify the above program to use the rotary harmonograph model. Notice that the rotational harmonograph combines two orbits (oscillations). Use the planet periods (in days) around the sun (including Ceres in the asteroid belt, and Pluto*), from inner to outer: 87.969 (Mercury), 224.701 (Venus), 365.256 (Earth), 686.980 (Mars), 1680.1 (Ceres), 4,332.6 (Jupiter), 10,759.2 (Saturn), 30,685 (Uranus), 60,190 (Neptune), and 90,465 (Pluto).

   a. One idea is to sonify the celestial harmonies these planets generate relative to Earth.[†] First try them in the rotary harmonograph program above. Then combine this code with the case study above to generate music.

   b. Another possibility is to try ratios of consecutive planets, from inner to outer.

---

[*] Although Pluto is officially not a planet, it is still out there revolving in a sustainable orbit, so feel free to use it.

[†] That is the harmonic patterns observed by the ancients in the night sky, which most probably inspired the Pythagoreans (and the Egyptians and Babylonians before them) to try and model what they saw through mathematics.

    c. Finally, try different ratios. Some interesting possibilities include Mercury:Venus, Venus:Ceres, Mars:Ceres, Jupiter:Neptune, and Neptune:Pluto (see Martineau, 2002, pp. 54–58).

    d. Consider truncating the ratios (by dividing them, say, by 100, and then converting the result to an int); for example, the ratio Venus:Earth (i.e., 224.701:365.256) will become 2:3. Not accurate, but nice.

Use the rotary harmonograph to first visualize these ratios. Then choose the ones you like better, that is, consider, visualize, and then sonify.

10.7 SUMMARY

Up to this point in the book, we have exposed you to fundamentals in music theory, computer programming, and algorithmic design (that is, finding ways to express musical or other processes through the computer). This chapter focused on the themes introduced in Chapter 1 and explored deeper connections between music, number, and nature.

The last case study was particularly insightful. It demonstrated the types of possibilities that open up when different paths in an exploration converge. In particular, it combined knowledge about math (sin and cos), knowledge about harmonic ratios (the type of ratios that Kepler was exploring when he derived his three laws), principles of music theory and composition (pitch, duration, dynamic, and panning), and, finally, computer programming (how to put all this in a program that works). Now you have a thorough understanding of how to connect music, number, and nature.

Is this the end? Well, it could be... But we hope it is only a beginning. We hope we have inspired you and have provided the foundation for you to continue this exploration on your own. Some of you may even possibly further the state-of-the-art in computer music and (why not?) the knowledge of this new discipline defined by the intersection of computing and the arts. Therefore, we give you one more chapter. In it, we present several powerful ideas that may guide your exploration. Pay close attention!

## CHAPTER 11

Exploring
Powerful Ideas

Topics: Fractals, recursion, Fibonacci numbers, the Golden Ratio, Zipf's Law, top-down design, Python dictionaries, defining Python classes, Python exceptions, animation, color gradients, Python complex numbers, cymatics and dynamical systems (boids).

11.1 OVERVIEW

We have finally reached the last chapter of the book. Throughout this book we have studied fundamentals of music theory, computer programming, sonification, and algorithmic design. Clearly, we have covered a lot of territory in our journey of learning how to make music with Python. But this journey can and, we hope, will continue. There are many creative possibilities for you to explore. This chapter opens the door to some of them.

In this chapter, we present several advanced computational concepts and algorithmic techniques. For some of them, we will show directly how to use them to make music (as we have done throughout the book so far). For others, we will leave it up to you to add a musical dimension to the sample code. Although we could have easily provided you with our musical examples, we wanted to give you the opportunity to find your own, especially for the algorithms and models presented here that are rich with musical (sonification) possibilities. We didn't want to bias you toward a single possibility. Therefore, this chapter presents a few opportunities for you to explore and engage your creative side. When reading the sections that follow, start thinking about how you can combine what you have learned so far to create your own music. This will probably involve musical and algorithmic brainstorming, further readings, and experimentation.

Use your imagination. What musical parameters or sonification ideas would you like to explore? Let the concepts inspire you, and use the visualizations to guide your creativity. Your programs will probably involve most of the knowledge of making music with Python you have accumulated thus far. Some of the ideas may be simple. Others may be more involved. Either way, expect your ideas to evolve, as you are finding your own style and your own musical voice.

Let's begin.

11.2 FRACTALS AND RECURSION

As discussed in Chapter 1, fractals are *self-similar* objects (or phenomena), that is, objects consisting of multiple parts, with the property that the smaller parts are the same shape as the larger parts, but of a smaller size. The field of fractal geometry was developed by Benoit Mandelbrot to study these types of artifacts (Mandelbrot 1982).

In this section we explore an elegant and powerful programming technique called *recursion*. Recursion may be used to perform a variety of computation tasks. It mimics the subdivision process found in nature, for example, the branching (starting with a single sprout) that results in a pine tree, or the branching (starting from a single cell) that resulted in the human reading this text. Therefore, among other things, recursion is especially suited for creating fractal artifacts.

11.3 FIBONACCI NUMBERS AND THE GOLDEN RATIO

This process of repeated branching has been studied extensively. For instance, Leonardo of Pisa, also known as Leonardo Fibonacci (c. 1170–c. 1250) investigated how fast rabbits could breed under ideal conditions. To model this fractal phenomenon, he developed what is now known as the Fibonacci sequence.

Fibonacci numbers appear in many natural objects and processes, including seashells (see Figure 11.1), branching plants, flower petals, flower

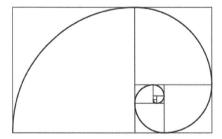

FIGURE 11.1 A nautilus shell and its relationship to Fibonacci numbers.

seeds, leaves, pineapples, and pine cones, among others. They also appear in the formation of tornadoes, hurricanes, and, at the grand scale, of galaxies.

The nautilus shell is a well-known example. As Figure 11.1 shows, its shape can be derived from a geometric process modeled by Fibonacci numbers. This may be seen either as analytical (i.e., holding a shell and seeing how it is made—from large to small), or synthetic (i.e., as a description of the way a nautilus grows its shell—from small to large).

The Fibonacci sequence consists of two numbers, 0 and 1, and a rule on how to generate the next number, namely, "add the previous two numbers." So the sequence goes like this: 0, 1, 1, 2, 3, 5, 8, 13, 21, 34, 55, 89, 144, 233, 377, 610, 987 ... (verify that each number is the sum of the previous two, with the exception, of course, of the two starting numbers, 0 and 1).

Fact: The ratio of two consecutive Fibonacci numbers approximates the golden ratio, φ (phi), which is approximately 0.61803399 ... (or, when the larger of the two Fibonacci numbers is used as the numerator, 1.61803399 ...).

Fact: Artifacts whose proportions incorporate the golden ratio are usually perceived as aesthetically pleasing by humans.

The golden ratio is found in natural and human-made artifacts (Beer 2008; Calter 2008, pp. 46–57; Hemenway 2005, pp. 91–132; Livio 2002; May 1996; Pickover 1991, pp. 203–205). It is also found in the human body (e.g., the bones of our hands, the cochlea in our ears). It has been shown that the golden ratio appears in musical works by Bach, Beethoven, Mozart, and others (Garland and Kahn, 1995).

The Fibonacci numbers can be easily calculated using recursion (where n denotes the nth number in the sequence):

- $fib(0) = 0$

- $fib(1) = 1$

- $fib(n) = fib(n-2) + fib(n-1)$ (i.e., to get the next number you add the previous two)

Definition: A *recursive function* is a function that calls itself.[*]

---

[*] If you are not careful, it is easy to write a recursive function that never terminates. For this reason we make sure we always have a limit (or a simple case), and that we work toward it. When the limit is reached, the function terminates.

Now let's translate this to Python:

```
# fibonacci.py
#
# Find the nth Fibonacci number using recursion.
#

def fib(n):

    if n == 0:          # simple case 1
        result = 0
    elif n == 1:        # simple case 2
        result = 1
    else:               # recursive case

        result = fib(n-2) + fib(n-1)

    return result
```

Notice the two simple cases (or base cases) in the `if` statement inside the function. If we call `fib()` with n equal to 0, result is assigned a 0, and the function terminates (after it returns result). Same for n equal to 1.

However, if we call `fib()` with n equal to 2, it makes two calls to itself with smaller n's (i.e., n-2 and n-1). In other words, it recursively calls itself with n equal to 0 (i.e., n-2) and with n equal to 1 (i.e., n-1). When the two calls return with values 0 and 1, respectively, it adds the two values (0 + 1) and returns the result. That's it.*

We can test the function as follows:

```
# now, let's test it
for i in range(10):
    print "fib(" + str(i) + ") =", fib(i)
```

This outputs:

```
fib(0) = 0
fib(1) = 1
fib(2) = 1
fib(3) = 2
fib(4) = 3
fib(5) = 5
fib(6) = 8
fib(7) = 13
```

---

* Now, try calling `fib()` with n equal to 3. You may need to write intermediate results on paper.

```
fib(8) = 21
fib(9) = 34
```

Since the function is recursive, we are guaranteed that it will work correctly for larger n's.

11.3.1 Case Study: The Golden Tree

The following program generates a fractal tree, also known as a Golden Tree, since it incorporates golden ratio proportions (see Figure 11.2). The golden tree is constructed by dividing a line into two branches, each rotated by 60 degrees (clockwise and counter-clockwise), with a length reduction factor equal to the golden ratio (0.61803399 …). These smaller lines, again, are each subdivided into two lines following the same procedure. This subdivision may go on indefinitely, but (both in nature and computing) there is a practical limit, where further subdivision does not serve any practical purpose and thus it stops (e.g., the brain cavity has filled up with enough grey matter). Similar patterns appear extensively in nature (as they maximize the amount of matter that can fit in a limited space, because surfaces are touching but not overlapping).

In Figure 11.2 there are 14 levels, or 13 subdivision, starting with the main branch.

The program below demonstrates how to code this using recursion.

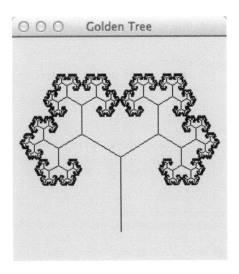

FIGURE 11.2   A fractal (golden) tree with depth 14 (13 subdivisions).

Fact: Recursion occurs when a function calls itself.

Of course, this may result in a never-ending execution of a program. In order for a recursive function to eventually end, we have to make sure that

1. when called recursively, it operates on smaller versions of a problem, and

2. it contains an `if` statement, which checks if we have reached the simplest possible (or smallest practical) case, and if so, it then stops by not calling itself any more.

Let's see how this works:

```
# goldenTree.py
#
# Demonstrates how to draw a golden tree using recursion.
#

from gui import *
from math import *

# create display
d = Display("Golden Tree", 250, 250)
d.setColor(Color.WHITE)

# calculate phi to the highest accuracy Python allows
phi = (sqrt(5) - 1) / 2    # approx. 0.618033988749895

# recursive drawing parameters
depth = 13               # amount of detail (or branching)
rotation = radians(60)   # branch angle is 60 degrees (need radians)
scale = phi              # scaling factor of branches

# initial parameters
angle = radians(90)        # starting orientation is North
length = d.getHeight() / 3 # length of initial branch (trunk)
startX = d.getWidth() / 2  # start at bottom center
startY = d.getHeight() - 33

# recursive function for drawing tree
def drawTree(x, y, length, angle, depth):
    """
    Recursively draws a tree of depth 'depth' starting at 'x', 'y'.
    """
    global d, scale, rotation
```

```
# print "depth =", depth, "x =", x, "y =", y, "length =", length,
# print "angle =", degrees(angle)


# draw this line
newX = x + length * cos( angle ) # calculate run
newY = y - length * sin( angle ) # calculate rise
d.drawLine(int(x), int(y), int(newX), int(newY))


# check if we need more detail
if depth > 1:

    # draw left branch - use line with length scaled by phi,
    # rotated counter-clockwise
    drawTree(newX, newY, length*phi, angle - rotation, depth-1)


    # draw right branch - use line with length scaled by phi,
    # rotated clockwise
    drawTree(newX, newY, length*phi, angle + rotation, depth-1)


# draw complete tree (recursively)
drawTree(startX, startY, length, angle, depth)
```

First, notice how small this program is. This is usually the case with recursive solutions of problems. They tend to be very succinct and elegant.

Notice the calculation of φ, using a mathematical formula (instead of dividing two Fibonacci numbers), for better accuracy:

```
# calculate phi to the highest accuracy Python allows
phi = (sqrt(5) - 1) / 2 # approx. 0.618033988749895
```

Also notice the parameters used to control the recursive process:

```
# recursive drawing parameters
depth    = 14             # amount of detail (or branching)
rotation = radians(60)    # branch angle is 60 degrees (need radians)
scale    = phi            # scaling factor of branches
```

and the parameters for the initial call to function drawTree():

```
# initial parameters
angle  = radians(90)        # starting orientation is North
length = d.getHeight() / 3  # length of initial branch (trunk)
startX = d.getWidth() / 2   # start at bottom center
startY = d.getHeight() - 33
```

The initial call happens at the end of the program, since we first need to define the function `drawTree()`.* This call looks as follows:

```
# draw complete tree (recursively)
drawTree(startX, startY, length, angle, depth)
```

To better understand this recursive solution, uncomment the print statements inside `drawTree()`. Then run the program first with variable depth set to 1, 2, 3, etc.

For example (see Figure 11.3), when depth is 1, the output is

```
depth = 1 x = 125 y = 217 length = 83 angle = 0.0
```

When the depth is 2, the output is

```
depth = 2 x = 125 y = 217 length = 83 angle = 0.0
depth = 1 x = 125.0 y = 134.0 length = 51.30 angle = -60.0
depth = 1 x = 125.0 y = 134.0 length = 51.30 angle = 60.0
```

Here, notice that the first line (i.e., depth = 2) is generated by the very first call. The other two lines are generated by the two recursive calls (one each). And, finally, for the depth 3:

```
depth = 3 x = 125 y = 217 length = 83 angle = 0.0

depth = 2 x = 125.0 y = 134.0 length = 51.30 angle = -60.0
depth = 1 x = 80.58 y = 108.35 length = 31.70 angle = -120.0
depth = 1 x = 80.58 y = 108.35 length = 31.70 angle = 0.0
```

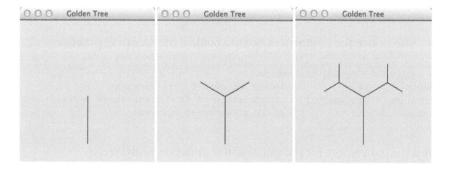

FIGURE 11.3   A fractal (golden) tree with depth 1, 2, and 3, respectively.

---

* In the next case study, we will see a different way to write programs that involve functions, namely, top-down implementation.

```
depth = 2 x = 125.0 y = 134.0 length = 51.30 angle = 60.0
depth = 1 x = 169.42 y = 108.35 length = 31.70 angle = 0.0
depth = 1 x = 169.42 y = 108.35 length = 31.70 angle = 120.0
```

Study this output carefully in conjunction with the code. Actually, the best way to do this is to "play computer," i.e., take a piece of paper and write down the calls to function drawTree() and the parameter values passed each time as if you were the computer executing this program. To help you understand better, answer the following questions:

- What are the input parameters, especially depth?
- What is the value of the condition depth > 1 in the if statement?
- If True, what are the input parameters to the two recursive calls to drawTree()?

Realize that each of these recursive calls executes the same code (i.e., the body of function) as the original call. What changes, for each of these recursive calls, is the value of the input parameters, including depth.

When depth, eventually becomes 1, the corresponding function will draw a line, but will not make any further recursive calls. When that function terminates, Python returns to its parent function (i.e., the function that called it). If that function has more work to do, it does it; otherwise it also returns to whomever called it.

This goes on until we reach the top level of the program the very first call to drawTree(). Then, since that's the last statement in the program, the program terminates.

With the help of the print output, and some paper and pencil, trace this program carefully.[*]

In summary, as with the construction of the golden tree, recursion applies a process (actually a function) to a smaller instance of the task being worked on.

11.3.1.1 Exercises

1. Explore ways to sonify this fractal process. One way is to map coordinates x, y to pitch and velocity (actually negative y or –y), length to duration, and depth to panning. Try to make the recursive process

---

[*] Teaching you how to write recursive programs is beyond the scope of the book. However, if you understand how this example operates, you are halfway there. And remember, nature is full of recursive processes.

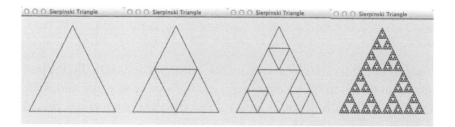

FIGURE 11.4   A Sierpinski fractal triangle with depth 1, 2, 3, and 6, respectively.

audible (e.g., depth to panning). Many other possibilities exist, so experiment.

2. Another fractal shape is the Sierpinski triangle (named after the Polish mathematician who described it in 1915). This fractal triangle consists of three smaller triangles (top, bottom left, bottom right) that have the same shape as the main one (see Figure 11.4). These smaller triangles, again, consist of three even smaller triangles that have the same shape. This repetition or subdivision continues on and on (theoretically) to infinity. Interestingly, similar patterns appear in ancient mosaics, suggesting that self-similarity was a known concept to the ancients. (*Hint:* Consider using the `drawPolygon()` function of GUI displays.)

3. The literature is full of various fractal shapes (Koch curve, Mandelbrot set, etc.). Explore writing recursive programs that generate them.

4. Explore the numerous resources in the literature and on the web on fractals and fractal music. Can you think of any new ideas?

11.4 ZIPF'S LAW

Computers have been used extensively in music to aid humans in analysis, composition, and performance. Is it possible to find algorithmic techniques to help explore and identify aspects of musical aesthetics related to balance, pleasantness, and, why not, beauty?

George Kingsley Zipf (1902–1950) was a linguistics professor at Harvard who studied fractal patterns in language. In his seminal book, *Human Behavior and the Principle of Least Effort* (Zipf 1949), he reports the amazing observation that word proportions in books, as well as notes in musical pieces (among other phenomena) follow the same harmonic proportions first discovered by Pythagoreans on strings (i.e., 1/1, 1/2, 1/3,

1/4, 1/5, etc.).* This means that the most common word appears about twice as many times as the second most common word, three times as the third most common word, four times as the fourth most common word, and so on.

Fact: This type of proportion is called *Zipf's law*.

Zipf proportions (or distributions) have been observed in a wide range of human and naturally occurring phenomena. These include music, city sizes, incomes, computer function calls, earthquake magnitudes, thickness of sediment depositions, clouds, trees, extinctions of species, traffic jams, and visits to websites (e.g., Zipf 1949, Bak 1996, Schroeder 1991).

Moreover, Zipf showed that if we plot the logarithm of the counts of all events in such a phenomenon against the logarithm of the rank of these events, we get a straight line with a slope of approximately –1.0 (i.e., a 45° orientation). For example, Figure 11.5 demonstrates that the book you are currently reading (this book) has Zipfian proportions. Actually, it has a Zipf slope of –1.23.

The second value, R^2, in Figure 11.5 measures how closely the data points fall on the straight line (1.0 means a perfect straight line; 0.0 means data points are scattered around the graph). The data points for this book

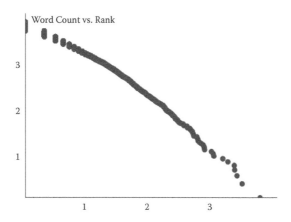

FIGURE 11.5   Zipf plot of word counts in this book (slope is –1.23, and R^2 is 0.98).

---

* Zipf's work had considerable influence on Benoit Mandelbrot, who, being inspired by Zipf's ideas, eventually developed, the field of fractal geometry (Mandelbrot 1982).

fall almost perfectly on a straight line (its R^2 value is 0.98). This is actually quite common for regular, long books (like this one).[*]

Calculating both the Zipf slope and R^2 values is very useful if you do not wish to generate the graph every time. Having an R^2 value near 1.0 (e.g., >0.7) tells you that the data points exhibit harmonic proportions. Then the slope tells you what type of harmonic proportion (or balance) they exhibit. For example, Zipf's ideal slope, –1.0, corresponds to the harmonic series 1/1, 1/2, 1/3, 1/4, 1/5 …, which can also be represented as:

$$Sn = \frac{1}{1} + \frac{1}{2} + \frac{1}{3} + \frac{1}{4} + \frac{1}{5} + \ldots + \frac{1}{n}$$

A more generalized form of this formula is the *Generalized Harmonic Series*, which was explored by Zipf:

$$Sn = \frac{1}{1^p} + \frac{1}{2^p} + \frac{1}{3^p} + \frac{1}{4^p} + \frac{1}{5^p} + \ldots + \frac{1}{n^p}$$

where p is equal to the absolute value of the Zipf slope.[†]

For example, a slope near –2.0 generates the series, 1/1, 1/4, 1/9, 1/16, 1/25 and so on.[‡]

This formula is really nice since it allows us to easily generate the harmonic (or, in the general case, hyperharmonic) proportions encountered in any phenomenon with a Zipf slope other than –1.0.

11.4.1 Zipf's Law and Music

Figure 11.6 shows two examples of measuring the Zipf proportions of pitches in two musical pieces. The left one demonstrates that pitches of Bach's "Air on a G String" exhibit almost ideal Zipf harmonic proportions—notice the 45° slope of the data points.[§] The slope is –1.078 and the R^2 value

---

[*] It has been shown that books written in different human languages (including Esperanto) have different Zipf slopes. This is something that Zipf himself expected, but was statistically verified only recently (Manaris et al. 2006).

[†] Another name for this formula is *Riemann's Zeta function*, which is "probably the most ... mysterious object of modern mathematics, in spite of its utter simplicity" (Watkins 2001, p. 58). Riemann's Zeta function is connected to prime numbers, which shows us that Zipf's law is capturing an essential aspect of the universe we live in (a mathematico-philosophical discussion beyond the scope of this chapter). What is relevant to us is that Zipf's law is connected to music and aesthetics (as discussed below).

[‡] This corresponds to another very common distribution found in nature (known as Brownian motion).

[§] This means that pitches in this piece have approximately harmonic proportions (i.e., 1/1, 1/2, 1/3, etc.).

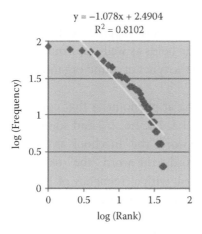
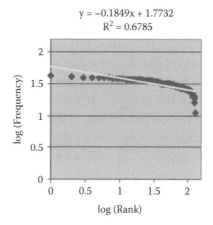

FIGURE 11.6   (Left) Pitch distribution for J. S. Bach's *Orchestral Suite No. 3* in D "Air on a G String" BWV 1068. (Right) Pitch distribution for a piece generated using function random().

is 0.81 (actually an R^2 value over 0.7 is considered significant). So this piece is almost Zipfian in terms of pitch.

On the other hand (Figure 11.6, right), the random piece has an almost horizontal slope (–0.19) and an R^2 of 0.7. This means that pitches have pretty much equal counts, i.e., they have an equal probability of appearing. This is precisely what we expect from using function random() to select pitches, which is how this piece was created using a computer (e.g., see the "Pierre Cage" case study in Chapter 6).

Fact: Zipf's law allows us to measure the balance and proportions of events in music pieces (e.g., pitch, duration, dynamic, etc.).

11.4.2 What Does It Mean?

In the previous chapter, we saw the types of harmonic shapes that can be generated from simple integer ratios. We also discussed how the Pythagoreans worked on sonifying those same integer ratios, and discovered that some of them sounded better than others. These ratios became the basis of modern musical scales (1:2 is octave, 2:3 is fifth, 3:4 is fourth, 4:5 is minor third, 5:6 is major third, and so on).

But what happens when you put many of these musical intervals together in a song? The answer is that you get a complex artifact consisting of numerous harmonic ratios (musical intervals), some of which overlap (e.g., create chords). We call this complex system of harmonic ratios *music*, and our ears and minds are built to easily process and appreciate this world of organized sound. Of course, the opposite holds as well—our music has evolved based on our ears and minds, and the types of sound organization we find harmonious or not.

In the case of Zipf's law, we are looking at phenomena that combine attributes in harmonic ratios 1:1, 1:2, 1:3, 1:4, 1:5, 1:6, 1:7, and so on. For example, think of every appearance of the word "the" in a book as though its appearances were a cycle. In other words, the word "the" oscillates (i.e., appears) in a book every now and then, with a frequency of 1:1. The next most frequent word, say, "to," oscillates (i.e., appears) in a book half as often, with a frequency of 1:2. And so on.

This approach can be used to understand all phenomena that exhibit Zipf's law.

Fact: Zipf's law describes artifacts (music or other) that consist of characteristics that occur according to harmonic ratios which, when taken as a whole, exhibit a specific balance.

Zipf also adapted a Pythagorean approach to explain his observation. He theorized that such harmonic proportions are the result of orthogonal forces. In any environment containing self-adapting agents able to interact with their surroundings, such agents tend to minimize their overall effort associated with this interaction (economy principle). That is, a system of such interacting agents tends to find a global optimum that minimizes overall effort (Bak 1996). This interaction involves some sort of exchange (e.g., information, energy, etc.). We will see this principle at play in the Cymatics section later in this chapter.

Finally, what if we were able to measure music from various music styles (baroque, classical, 20th century, blues, jazz, etc.)? The next section demonstrates how to do this. The technique shown below has been used to help classify music according to composer, style, and pleasantness using computers. Also, it has been used to perform experiments in computer-aided music composition (Manaris et al. 2005, 2007).

11.4.3 Measuring Zipf Proportions

Psychologists have shown that people prefer music, and other experiences, that have a balance of predictability and surprise. The following program reads in a sequence of MIDI files and measures how well the note pitches in each song follow Zipf's law. To do so, it uses functionality from the Zipf library (provided with this book).

It is relatively easy to expand it to measure additional proportions, such as note durations. Even with a single metric, for pitch, we can still get an approximate idea of how a piece may sound in terms of balance and proportion.

Here is the code:

```
# zipfMetrics.py
#
# Demonstrates how to calculate Zipf metrics from MIDI files for
# comparative analysis. It calculates Zipf slopes and R^2 values
# for pitch.
#
# It also demonstrates how to use Python dictionaries to store
# collections of related items.
#
# Finally, it demonstrates how to implement an algorithm in a top-down
# fashion. First function encountered in the program performs the
# highest-level tasks, and any sub-tasks are relegated to lower-level
# functions.
#

from music import *
from zipf import *

# list of MIDI files to analyze
pieces = ["sonifyBiosignals.mid", "ArvoPart.CantusInMemoriam.mid",
          "DeepPurple.SmokeOnTheWater.mid",
          "soundscapeLoutrakiSunset.mid",
          "Pierre Cage.Structures pour deux chances.mid"]

# define main function
def main( pieces ):
    """Calculates and outputs Zipf statistics of all 'pieces'."""

    # read MIDI files and count pitches
    for piece in pieces:

        # read this MIDI file into a score
        score = Score()           # create an empty score
        Read.midi( score, piece )  # and read MIDI file into it

        # count the score's pitches
        histogram = countPitches( score )

        # calculate Zipf slope and R^2 value
        counts = histogram.values()
        slope, r2, yint = byRank(counts)

        # output results
        print "Zipf slope is", round(slope, 4), ", R^2 is", round(r2, 2),
```

```
        print "for", piece
        print

    # now, all the MIDI files have been read into dictionary
    print       # output one more newline

def countPitches( score ):
    """Returns count of how many times each individual pitch appears
       in 'score'.
    """

    histogram = {}        # holds each of the pitches found and its count

    # iterate through every part, and for every part through every
    # phrase, and for every phrase through every note (via nested
    # loops)
    for part in score.getPartList(): # for every part
      for phrase in part.getPhraseList(): # for every phrase in this
          part
        for note in phrase.getNoteList(): # for every note in this
            phrase

          pitch = note.getPitch() # get this note's pitch

          # count this pitch, if not a rest
          if (pitch != REST):

              # increment this pitch's count (or initialize to 1)
              histogram[ pitch ] = histogram.get(pitch, 0) + 1

          # now, all the notes in this phrase have been counted
        # now, all the phrases in this part have been counted
      # now, all the parts have been counted

    # done, so return counts
    return histogram

# start the program
main( pieces )
```

First this program demonstrates a slightly different style of writing Python programs using functions. It demonstrates how to implement an algorithm in a top-down fashion.

11.4.3.1 Top-Down Design (Revisited)

As mentioned in Chapter 3, top-down design is a strategy for constructing programs. It starts from the biggest, highest-level task, and subdivides it into smaller tasks. Each of these tasks is implemented using functions.

By specifying the highest-level pieces of our program first, and then dividing them into successively smaller pieces, we gain perspective, and the structure of the program is clearly defined. Then it is easy to go back and fix problems, or update the program to perform slightly different tasks. Top-down design is very beneficial when developing large programs.

The above program first defines a function `main()` which contains the high-level algorithmic tasks of the program. Then, it defines the function `countPitches()` which, as is explained below, performs a more specific (lower-level) task of the program. Finally, it calls the function `main()` to start the program.

By writing programs this way, we can focus first on the high-level tasks and then on the lower-level tasks. Top-down design is very nice and is used extensively by experienced programmers.

The program reads in and analyzes several MIDI pieces we have created in this book. When you run it, it uses `Read.midi()` to read in MIDI files into scores.

Then it calls the function `countPitches()` to count every pitch in the score. It is interesting to observe the triple-nested loop in this function:

```
for part in score.getPartList(): # for every part
    for phrase in part.getPhraseList(): # for every phrase in part
        for note in phrase.getNoteList(): # for every note in phrase
```

As the comments indicate, the outer loop traverses all parts in the score. For every part, the middle loop is executed. This loop traverses all phrases inside a part. Then, for every phrase, the inner loop is executed. This traverses all notes in a phrase.

For each note in a phrase, we get its pitch, and if it is not a REST, we count it.

To store this count we use Python dictionaries (see next section).

Finally, the program outputs the results. When you run this program, you will notice that these results are interspersed between output from `Read.midi()`. Collected together the results read:

```
Zipf slope is -2.3118 , R^2 is 0.76 for sonifyBiosignals.mid
Zipf slope is -0.8641 , R^2 is 0.79 for ArvoPart.CantusInMemoriam.mid
Zipf slope is -1.2507 , R^2 is 0.94 for DeepPurple.SmokeOnTheWater.mid
Zipf slope is -0.929 , R^2 is 0.55 for soundscapeLoutrakiSunset.mid
Zipf slope is -0.4619 , R^2 is 0.71 for Pierre Cage.Structures pour
  deux chances.mid
```

Analyzing the results:

- It is interesting to note that Deep Purple's "Smoke on The Water" has near Zipfian pitch proportions (slope is –1.25, and R^2 is 0.94). Perhaps it is not an accident that this guitar riff was very popular among rock guitarists for decades.

- The "sonifyBiosignals" piece is measured as monotonous in terms of pitch (slope is –2.3). This means that certain notes predominate.

Which is true, when you listen to it carefully. It does sound monotonous and repetitive (with some variation, of course).

- Arvo Pärt's "Cantus in Memoriam" has near Zipfian pitch proportions (slope is –0.85). Another balanced piece in terms of pitch. Can you hear that?

- The "soundscape Loutraki Sunset" appears to have near Zipfian pitch proportions (slope is –0.929), but the R^2 is 0.5; this means that the data points are kind of scattered, so the slope is not as reliable. Yet, as you listen to it, it does sound well-proportioned in terms of pitch, if not beautiful.

- Finally, the Zipf pitch proportion for "Pierre Cage.Structures pour deux chances" indicates that the piece is near chaotic (i.e., pitches are approaching randomness), with a slope of –0.46.[*] This corresponds with how the piece sounds.

In summary, using a single metric (e.g., Zipf proportion of pitch) can give us an idea about a piece, but it is not definitive in any way with regard to our aesthetic appreciation of it. However, calculating many diverse metrics based on Zipf's law can give us a more accurate idea about the proportions of a piece, as several experiments demonstrate (e.g., Manaris et al. 2005, 2007).

11.4.4 Python Dictionaries

Python has an additional data type, called a *dictionary*. Dictionaries are similar to Python lists. They are used to store collections of related items. The main difference is that, while lists are indexed using increasing integers (0 to the length of the list minus 1), dictionaries are indexed using arbitrary Python values.

A Python dictionary is similar to a real dictionary (which associates words with their definitions). In the case of Python dictionaries, though, the "word" (or key or index) can be almost any Python value, and the "definition" (or value) can also be any Python value. In other words, Python dictionaries are used to store key-value pairs.

To create an empty dictionary, we use this syntax:

```
d = {}
```

---

[*] Ideally, random() should generate a slope closer to 0.0. However, since this piece contains only 100 pitches, with different runs of that program, we may get slightly different results (as it is possible sometimes to get far more heads than tails, when flipping a coin). Also, more notes may be needed for the random behavior to emerge more consistently.

Dictionaries have syntax similar to list indexing. For example, given d, the following statements load into it key-value pairs, where the key is a word in English, and the value is the corresponding word in Spanish.

```
d['hello'] = 'hola'
d['yes'] = 'si'
d['one'] = 'uno'
d['two'] = 'dos'
d['three'] = 'tres'
d['red'] = 'rojo'
d['black'] = 'negro'
d['green'] = 'verde'
d['blue'] = 'azul'
```

Dictionaries have a rich set of built-in operations:

Function	Description
d.keys()	Returns the list of keys in dictionary d.
d.values()	Returns the list of values in dictionary d. This list is ordered to correspond with the list returned by d.keys(), that is, the two lists are parallel.
d.items()	Returns the list of key-value pairs (as sublists) in d.
del d[k]	Deletes item k from the dictionary d.
d.has_key(k)	Returns True if k exists as a key in dictionary d, False otherwise.
d.get(k, default)	Returns the value associated with key k in d. If k does not exist, it returns value default (whatever that may be).

Most Python values will work as indices to dictionaries (e.g., strings, integers, floats, etc.). In the previous case study, we used the pitch MIDI value as the key. Thus we associated a MIDI pitch value with its count. This is done with the statement:

```
# increment this pitch's count (or initialize to 1)
histogram[ pitch ] = histogram.get(pitch, 0) + 1
```

A little more explanation is needed for this. The name of the dictionary is histogram. The expression

```
histogram.get(pitch, 0) + 1
```

gets the current value (i.e., count) for the index pitch and adds 1 to it. If index pitch had not been seen before (i.e., it is not a valid index in the dictionary), get(pitch, 0) will create it and return the second parameter, i.e., 0. To this we add 1. Either way, we store the result in histogram[pitch], either incrementing the previous value (count) stored in this dictionary location, or creating a new dictionary location for the new pitch containing the value 1.

11.4.5 Exercises

1. Listen to the pieces used in the last case study. Can you hear the pitch proportions in the pieces, as analyzed by the Zipf metric computed by the program cited?

2. Search for and download MIDI files for Arnold Schoenberg (or other) 12-tone compositions. Listen to them. What type of Zipf pitch slopes would you expect? Verify your prediction using the cited program.

3. Using the cited program, measure Zipf proportions of popular pieces available in MIDI online. What do you observe?

4. Add more functions, such as `countDurations()`, `countDynamics()`, and `countPitchDurations()`. The first two are self-explanatory. For the third one, extract both pitch and duration values for each note. Combine them into a single value (for counting purposes) using this statement: `pitchDuration = pitch + duration * 0.01` This ensures that, for normal values for pitch and duration, the two values will be safely combined into one, where the pitch forms the integral part of the number (to the left of the decimal point) and duration forms the decimal part of the number. Try some examples to verify this works.

5. Modify the above code to measure Zipf proportions in a text document. *Hint:* Instead of reading MIDI pieces, read a text file (as shown in Chapter 7). Using string library functions, separate the text into words. Then use the histogram method used in `countPitches()` to instead `countWords()`. Finally, go on project Gutenberg (www.gutenberg.org) and download electronic books to measure. What used to take months to do, by Zipf and his research students, you will be able to do in a few seconds. Enjoy.

11.5 PYTHON CLASSES

As mentioned earlier, functions and classes are containers provided by Python to group related functionality. For example, earlier we saw how to create notes:

```
n = Note(C4, QN, 127) # create a middle C quarter note
```

At the time, we focused on the functionality, that is, that variable n gave us an efficient way to store related information about a musical note and eventually play them back through the MIDI synthesizer.

What we left out is that Note is a Python class.

Python classes are a "packaging" mechanism—they allow a programmer to offer functionality, while hiding all the implementation details. This allows others to be more productive and efficient (e.g., to focus on music-making tasks and avoid computational details).

Definition: "Packaging" is done using *encapsulation* and *information hiding* that are programming language properties allowing access to functionality through a well-defined interface, while hiding implementation details.

Fact: Python provides two mechanisms for encapsulation and information hiding: functions and classes.

We have already seen how to define functions (Chapter 7). This section focuses on defining classes. As our code is getting more advanced, we need to start hiding things, for simplicity of use. This way, other programmers can efficiently build on our work (the same way that you have been using this book's music and other libraries).

It is through encapsulation and information hiding (e.g., libraries of Python functions and classes) that a community of creative software developers can collaborate and thrive. It is through encapsulation and information hiding that books can be written on creative programming, while providing useful functionality in well-defined libraries, such as the music, gui, image, midi, and osc libraries.

11.6 CASE STUDY: THE NOTE CLASS

Throughout the book, we have written code that uses Note objects. In this case study, we see how to define (a simplified version of) this class on our own. This class is used to store related information about musical notes (i.e., pitch, duration, dynamic, and panning). It also provides functions to retrieve (get) and update (set) the information encapsulated in a Note object.

Learning how to define Python classes is very useful. It allows us to package related information (data) and, in essence, create new Python data types.

Fact: Python classes allow us to define new data types.

You will have various opportunities to develop new classes in your own coding practice.

As we recall from Chapter 2, the most basic musical structure in the Python music library is a Note. Python notes have the following attributes:

- pitch—an integer from 0 (low) to 127 (high)

- duration—a positive real number (quarter note is 1.0)

- dynamic (or volume)— an integer from 0 (silent) to 127 (loudest)

- panning—a real number from 0.0 (left) to 1.0 (right)

To create a note, we specify its pitch, duration, dynamic, and panning position, as follows:

```
Note(pitch, duration, dynamic, panning)
```

where dynamic and panning are optional. If omitted, dynamic is set to 85 and panning to 0.5 (center).

For example, this Python statement creates a middle C quarter note and stores it in variable n.

```
n = Note(C4, QN)
```

Here we create the same note, but as loud as possible (127) and placed to the left side (in the stereo field):

```
n = Note(C4, QN, 127, 0.0)
```

Definition: The data values encapsulated within a class are called *class attributes* or *instance variables*.

In order to set and retrieve the value of instance variables, we create functions to access them.

Definition: The functions provided to access class attributes are called *class functions*.

In this simplified case study, Note objects will have the following functions[*]:

- `Note(pitch, duration, dynamic, pan)` — create a new note object
- `getPitch()` — return the note's pitch (0–127)
- `setPitch(value)` — updates the note's pitch to value (0–127)
- `getDuration()` — returns the note's duration
- `setDuration(value)` — updates the note's duration to value (a float)

11.6.1 Creating Note Objects

As we saw in Chapter 2, to create a new note, we say:

```
n = Note(C4, QN)
```

This creates a Note object and stores it in variable `n`. Once the note has been created, we can access its data (i.e., data stored inside it) through its class functions:

```
>>> n = Note(C4, QN)
>>> n.getPitch()
60
>>> n.getDuration()
1.0
>>>
```

11.6.2 Defining the Class

Defining a class is similar to defining a function. A function encapsulates variables and statements. A class, on the other hand, encapsulates both variables and functions.

Here is the code to define the Note class:

```
# note.py
#
# It demonstrates how to create a class to encapsulate a musical
# note. (This is a simplified version of Note class found in the
# music library.)
#

from music import *
```

---

[*] Compare these functions to the Note class functions provided in Appendix A.

```
class Note:

    def __init__(self, pitch=C4, duration=QN, dynamic=85, panning=0.5):
        """Initializes a Note object."""

    self.pitch = pitch        # holds note pitch (0-127)
    self.duration = duration  # holds note duration (QN = 1.0)
    self.dynamic = dynamic    # holds note dynamic (0-127)
    self.panning = panning    # holds note panning (0.0-1.0)

    def getPitch(self):
        """Returns the note's pitch (0-127)."""

        return self.pitch

    def setPitch(self, pitch):
        """Sets the note's pitch (0-127)."""

        # first ensure data integrity, then update
        if 0 <= pitch <= 127:  # is pitch in the right range?
            self.pitch = pitch    # yes, so update value
        else:                     # otherwise let them know
            print "TypeError: Note.setPitch(): pitch ranges from",
            print "0 to 127 (got " + str(pitch) + ")"

    def getDuration(self):
        """Returns the note's duration."""

        return self.duration

    def setDuration(self, duration):
        """Sets the note's duration (a float)."""

        if type(duration) == type(1.0): # is duration a float?
            self.duration = float(duration) # yes, so update value
        else:                            # otherwise let them know
            print "TypeError: Note.setDuration(): duration must be",
            print "a float (got " + str(duration) + ")"
```

First, notice how we import the music library to have access to MIDI constants, such as C4, QN, etc.

The first actual line of the code,

```
class Note:
```

is called the *class header*. It simply begins the class definition and states the class name, that is, Note.*

---

* Beware. If you run the above code, the music library's Note will be superseded by our new Note class definition. This only lasts during the execution of the code. No permanent change is made in the music library.

Notice how the class name, Note, begins with an uppercase letter. This is a useful convention, so class names can be easily distinguished from function names. (The latter start, by convention, with lowercase letters.)

Good Style: Class names should begin with an uppercase letter.

Below the class header follow the definitions of the class functions. These need to be indented (again, we use 3 spaces). Class Note has five class functions, as seen above.

Of these functions, __init __ (), is called the *constructor* function.

Definition: The *constructor* function is a special class function which is called automatically every time we create a new instance.

For example, in the following

```
n = Note(60, 1.0)
```

the parameters 60 and 1.0 will be passed to the Note constructor function, __init __ ().

Therefore, when defining the class, we use the constructor to initialize any class attributes or instance variables that need initial values.

Here is the Note constructor function once more:

```
def __init__(self, pitch=C4, duration=QN, dynamic=85, panning=0.5):
   """Initializes a Note object."""

   self.pitch = pitch        # holds note pitch (0-127))
   self.duration = duration  # holds note duration (QN = 1.0)
   self.dynamic = dynamic    # holds note dynamic (0-127)
   self.panning = panning    # holds note panning (0.0-1.0)
```

As seen in the previous section, class Note has four instance variables, pitch, duration, dynamic, and panning. Accordingly, the constructor function creates the four instance variables, self.pitch, self.duration, self.dynamic, and self.panning, and gives them whatever values were passed in as parameters.

Fact: Instance variables in Python classes are always prefixed by the symbol self.

Fact: Instance variables are *global* to the whole class, that is, every class function can access them.
The remaining class functions utilize these instance variables to get their job done.

Fact: When we define a function within a class, Python requires that we use self as the first parameter.*

Fact: When we call a function defined within a class, we do *not* provide a value for the self parameter. We only provide values for the remaining parameters.

For example, it appears as if function `setPitch()` has two parameters, namely, self and pitch:

```
def setPitch(self, pitch):
    """Sets the note's pitch (0-127)."""

    ...
```

However, when we call it, we omit self and only provide a value for pitch:

```
>>> n.setPitch(D4)
```

This is one of the few awkward moments in learning Python. A class function is called with one less parameter than it is defined. (The value for self is provided automatically by the Python interpreter.)

In summary, classes allow us to encapsulate related variables (instance variables) and functions (class functions). This allows us to define new data types (as needed or desired) for data that we may use often in our programming practice.

11.6.2.1 Checking for Data Integrity
Class functions can be used to ensure data integrity. What happens in the Note class if a user (end-programmer, like you in Chapter 2) accidentally tried to set a note pitch to 128?

This would probably cause a significant error when playing the music generated by the program. Most likely, this would generate a cryptic message by your computer's synthesizer (such as "Ill-formed MIDI sequence"

---

* Actually, self is not a reserved word. It is possible to use another prefix, but that would go against Python convention.

or "MIDI message contains illegal data"). Regardless of the message, tracing this problem back to your code would take forever (especially if your program generated hundreds of notes).[*]

Good Style: Class functions used to set internal class values (class attributes) should check for errors, whenever possible. If an error occurs, they should issue an informative error message (and ignore the provided value).[†]

For example, let's see function `setPitch()` once more:

```
def setPitch(self, pitch):
   """Sets the note's pitch (0-127)."""

   # first ensure data integrity, then update
   if 0 <= pitch <= 127:  # is pitch in the right range?
      self.pitch = pitch      # yes, so update value
   else:                      # otherwise let them know
      print "TypeError: Note.setPitch(): pitch ranges from",
      print "0 to 127 (got " + str(pitch) + ")"
```

Notice how the `if` statement is used to ensure that the provided pitch is within the acceptable range. Without this error check, anything may happen (that is, the pitch could be set to a float or even a string!).

Notice that the `else` part outputs an informative error message (which explains what the error is, what the input value was that caused the error, and provides guidance/feedback on how to remedy the error).

Good Style: Error messages should:

1. state where the error occurred in the code,
2. describe the error, in a way that can be understood by someone who cannot see your internal code, and
3. provide guidance on how to fix the error.

For example, the `else` part above uses these statements:

```
print "TypeError: Note.setPitch(): pitch ranges from",
print "0 to 127 (got " + str(pitch) + ")"
```

---

[*] Most likely, you would think there is something wrong with your computer or MIDI synthesizer. You would never think to look back through your code. And you shouldn't have to.

[†] By providing this defense mechanism, we ensure that class data always remains correct. This is a great guarantee to be able to offer, as it leads to error-free software.

11.6.3 Python Exceptions

Python provides *exceptions*, a very useful mechanism for issuing error messages within programs.

For example, the above error message could be implemented as follows:

```
raise TypeError( "Note.setPitch(): pitch range is 0-127 (got " + \
                 str(pitch) + ")" )
```

Notice the difference: instead of using print followed by a string, we use raise followed by the name of an exception, such as TypeError, and the error message in parenthesis.*

Fact: Raising a Python exception stops execution of the code and, in addition to the provided error message, outputs information about where the error occurred during execution.

For example, here is the first error message we ever discussed (see Chapter 2):

```
>>> note2
Traceback (most recent call last):
    File "<stdin>", line 1, in <module>
NameError: name 'note2' is not defined
```

This was generated by the Python interpreter raising an exception (to tell us that it does not recognize the variable note2).

Python has a wide variety of predefined exceptions, including ArithmeticError, ZeroDivisionError, NameError, TypeError, and ValueError. All these can be raised, if needed, from inside our programs (using the syntax shown above).

11.6.4 Exercises

1. Extend the provided Note class with get and set functions for note dynamic and panning. (*Hint:* Use provided functions as models.) Your get functions should ensure data integrity (that is check for

---

* This is actually creating an instance of the class TypeError, with the provided string as the parameter to be passed to the TypeError constructor. This is very similar to Note(C4, QN).

error values). After you define them, run tests to make sure they work as expected.

2. The music library defines musical rests as notes with pitch equal to –2147483648.* Add a function `isRest()`, which reports if a note is a rest. *Hint:* You can use an if statement, or (more succinctly) use the following single line in the function body: `return (self.pitch == -2147483648)` Why does this work?

3. Define a class, called Rest. This class should have the same instance variables and class functions as class Note. However, since rests only have duration, the Rest constructor should accept only one parameter, namely, duration. The other values should be set as follows: pitch = –2147483648, dynamic = 85, panning = 0.5.†

4. Define a class, called Phrase. Just like the one provided by the music library, this class should encapsulate a list of notes. To simplify this exercise, the constructor should accept no parameters (it creates an empty phrase). Also, a Phrase's start time is always 0.0. The following class functions should be defined: `addNote(note)`, `getSize()`, `getNoteList()`, `getStartTime()`. (See Appendix B for precise descriptions of these functions.)

5. Extend the Phrase class to include the function `getEndTime()`. (*Hint:* Using a loop, iterate through all the notes in the internal list and accumulate (add) their durations. Return the result.)

11.7 CASE STUDY: A SLIDER CONTROL

In Chapter 8, we used sliders in conjunction with other controls on GUI displays. Here we define a new class, SliderControl, which can be used to create individual control surfaces (displays). A SliderControl object has a display with a single slider (see Figure 11.7). It can be used to control the value of any program variable interactively. The uses are limitless.‡

The SliderControl class has the following attributes:

---

* This is the smallest integer that can be stored in 32 binary digits (bits) and is clearly outside the valid range for note pitches (0 to 127). A *bit* is a special memory unit, at the hardware (electronics) level, that can store either a 0 or 1. With its long integers, Python theoretically has no limit as to how large an integer may be. However, other languages, such as Java and C/C++, have such limits (with 32 bits being a common one).

† Since `Rest` is also a `Note`, Python provides an elegant way to do this, called *inheritance*. Inheritance is beyond the scope of this chapter.

‡ Actually, we will get to use it in the case studies that follow.

FIGURE 11.7   The GUI display of a SliderControl (here used to control a timer delay).

- title — the control surface (display) title,

- updateFunction — the function to call when the slider position changes (this function should accept a single parameter, the new value of the slider),

- minValue — the slider's min limit,

- maxValue — the slider's max limit,

- startValue — the slider's starting value,

- x — the x coordinate of the top-left corner of the control's display, and

- y — the y coordinate of the top-left corner of the control's display.

11.7.1 Creating SliderControl Objects

Here is a SliderControl object whose update function simply outputs the updated slider value in the standard output:

```
# first, define the update function
def printValue(value):
    print value


# create the SliderControl providing the label and update function
s = SliderControl("Print", printValue)
```

Now for another example, let's create a SliderControl object to update the value of a program variable. This is a very useful application:

```
testValue = -1 # the variable to update interactively

# define the update function
def changeAndPrintValue(value):
```

```
    global testValue

    testValue = value
    print testValue
```

```
# create the SliderControl, providing the update function
s = SliderControl("Change and Print", changeAndPrintValue,
        0, 100, 50, 300, 300)
```

where 0 is the `minValue`, 100 is the `maxValue`, 50 is the `startValue`, and the control display appears at coordinates 300, 300 of the computer screen.

The SliderControl class works only with integer values. Also, its accuracy depends on mouse accuracy (although it is possible to use arrow keys for finer control).* Working with float values is left as an exercise (see below).

11.7.2 Defining the Class

Here is the code for the SliderControl class:

```
# sliderControl.py
#
# It creates a simple slider control surface.
#

from gui import *

class SliderControl:

    def __init__(self, title="Slider", updateFunction=None,
                        minValue=10, maxValue=1000, startValue=None,
                        x=0, y=0):
        """Initializes a SliderControl object."""

        # holds the title of the control display
        self.title = title

        # external function to call when slider is updated
        self.updateFunction = updateFunction

        # determine slider start value
        if startValue == None:  # if startValue is undefined

            # start at middle point between min and max value
            startValue = (minValue + maxValue) / 2
        # create slider
        self.slider = Slider(HORIZONTAL, minValue, maxValue,
                            startValue, self.setValue)

        # create control surface display
        self.display = Display(self.title, 250, 50, x, y)
```

---

* For finer control, it is possible to use arrow keys to control a slider. To do so, put the slider in focus (click on the bubble) and then use the arrow keys to increment (or decrement) the slider value by 1.

```
         # add slider to display
         self.display.add(self.slider, 25, 10)

         # finally, initialize value and title (using 'startValue')
         self.setValue( startValue )

   def setValue(self, value):
       """Updates the display title, and calls the external update
          function with given 'value'.
       """

         self.display.setTitle(self.title + " = " + str(value))
         self.updateFunction(value)
```

This class definition consists of two functions, the constructor and set-Value(). Notice how the class encapsulates a slider and a display. The constructor creates a Slider object and a Display object and adds the former to the latter. Also, it registers the class function setValue() with the Slider object, so that when the slider is interactively changed, this function will be called.

As we create new SliderControl objects, we will see new displays opening up. This is because, whenever a new object is created, Python runs the class __init__() function. In other words, for each SliderControl object created, function __init__() creates a new display window (in addition to all the other instance variables).

Fact: Every object of class has its own copies of instance variables and class functions.

One question is why create a separate class function setValue(), and not register the external updateFunction directly with the Slider object?

The answer is that setValue() also updates the display's title (as seen above). Afterwards, setValue() calls the external updateFunction and gives it the new slider value.

The rest is straightforward.

11.7.3 Exercises

1. Create a SliderControl object that calls the following function to play single notes:

```
def playNote(pitch):
    """Plays a 1-sec note with the provided pitch."""
    Play.note(pitch, 0, 1000)
```

Of course, you will need to import the music library.

2. Add one more SliderControl objects to one of your programs, to control variables of interest. Focus on programs that produce interactive music (for instance, using the AudioSample class). This may lead to impromptu programs for quick audio mixing or playing applications. SliderControl objects work only with integer ranges. Use a SliderControl object together with function `mapValue()` to update a program variable that contains a float value.

3. Create a SliderFloatControl class that works with float values. To do so, separate the range provided by the user (`minValue`, `maxValue`, and `startValue`) from the range used by the slider (e.g., `minSlider`, `maxSlider`, `startSlider`), and use function `mapValue()` to provide mapping between the two ranges.

4. Modify the SliderControl class to work with other GUI widgets of interest, such as Button and DropDownList objects. Explore what types of AudioSample applications you can quickly develop with an arsenal of such independent control surfaces.

5. Consider combining GUI widgets to create a few common, yet generic control surfaces (such as a button plus a slider surface, or two buttons and a slider). Give these classes appropriate names and consider the type of information (parameters) needed for their constructor. Aim for simplicity and usability.

11.8 ANIMATION

We are all familiar with computer animation, as seen in computer-generated cartoons, modern computer games, and CGI (computer-generated imagery) in cinema. Animation is based on the same principle used to make movies; a series of slightly varied still images is replayed in quick succession. This creates the perception of a moving image.

Fact: Animation is produced by showing individual images (frames) in quick succession.

The human eye begins to perceive continuous movement out of individual image sequences at a rate of about 15 frames per second. The standard cinema frame rate is 24 frames per second, although various experiments have been done at higher rates (e.g., 48). The higher the frame rate, the less "choppy" and more natural the experience of continuous movement is.

In order to write an animation program, we need to be able to accurately schedule Python statements (functions) to be executed at a later time, and also in quick (but precisely timed) succession. This is done through a Timer object (which we first used in Chapter 8).

11.8.1 Frame Rate

Timer objects require a callback function and a delay (in milliseconds), which specifies when into the future to call this function. Animation is best expressed in terms of frame rate (i.e., changes per second) rather than in milliseconds, so we need to convert from one to the other.

The formula to convert the delay between consecutive events from milliseconds to frame rate is:

```
frameRate = 1000 / delay # convert delay in milliseconds to frame rate
```

Similarly, the formula to convert frame rate to delay is, of course:

```
delay = 1000 / frameRate # convert frame rate to delay in milliseconds
```

11.8.2 Case Study: A Revolving Musical Sphere

The following case study demonstrates how to create an animation involving a revolving sphere. We model the sphere by using points drawn on a GUI display. As seen on Figure 11.8, we can adjust the animation speed using a slider on a separate control surface (display).* Additionally, as the points rotate, the code plays musical notes (more on that later).

The points are moved together in a synchronized way (in small increments, one per frame) to create the illusion of a rotating sphere. Using mathematical operations (actually, simple trigonometric functions) we calculate the points to always remain on (or trace) the surface of an imaginary sphere.

Additionally, by changing the colors of points, as they rotate, we create an illusion of depth. We use white to make points appear in front (closer to the viewer). We use darker colors to make points appear further back (that is, as points rotate away, they slowly disappear in the distance). The colors for points slowly change from white, to orange, to black. This is accomplished using a list of RGB colors that define a color gradient.

---

* This uses a TimerControl object (as defined above).

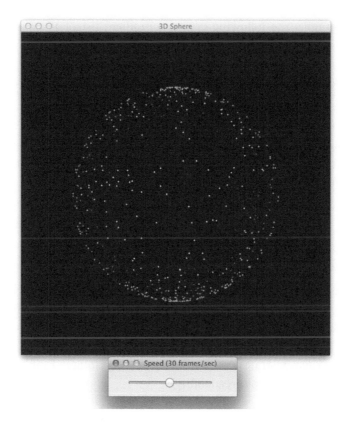

FIGURE 11.8   A revolving musical sphere (with speed slider control).

11.8.2.1 Color Gradients

Definition: A *color gradient* is a smooth color progression from one color to another which creates the illusion of continuity between the two color extremes.

A color gradient may be easily created using the following function (included in the GUI library):

```
colorGradient(color1, color2, steps)
```

This returns a list of RGB colors creating a "smooth" gradient between color1 and color2. The amount of smoothness is determined by steps, which specifies how many intermediate colors to create. The result includes color1, but **not** color2, to allow for connecting one gradient to another (without duplication of colors). The number of steps equals the number of colors returned.

For example, the following creates a gradient list of 12 colors, from black (that is, [0, 0, 0]) to orange (that is, [251, 147, 14]):

```
>>> colorGradient([0, 0, 0], [251, 147, 14], 12)
[[0, 0, 0], [20, 12, 1], [41, 24, 2], [62, 36, 3], [83, 49, 4], [104,
61, 5], [125, 73, 7], [146, 85, 8], [167, 98, 9], [188, 110, 10],
[209, 122, 11], [230, 134, 12]]
>>>
```

Notice how the code above excludes the final color (i.e., [251, 147, 14]). This allows us to create composite gradients (without duplication of colors). For example,

```
black = [0, 0, 0]       # RGB values for black
orange = [251, 147, 14] # RGB values for orange
white = [255, 255, 255] # RGB values for white

cg = colorGradient(black, orange, 12) +
    colorGradient(orange, white, 12) + [white]
```

The code above creates a combined list of gradient colors from black to orange, and from orange to white. Notice how the final color, white, has to be included separately (using list concatenation, i.e., "+"). Now cg contains a total of 25 unique gradient colors.

Thia also demonstrates now to use '\' to divide a long Python statement into two lines (to improve readability).

11.8.3 Defining the Class

The code for the revolving musical sphere is presented below. By encapsulating this code in a Python class, MusicalSphere, we can easily have more than one sphere running at the same time. Each sphere has its own display and speed control surface (see Figure 11.8).

Class MusicalSphere has the following attributes:

- radius — determines the radius of the sphere (in pixels),
- density — how many points to distribute across the surface of the sphere (e.g., 200),
- velocity — how many pixels to move each point by per animation frame (e.g., 0.01), and
- frameRate — how many animation frames per second (e.g., 30).

As we have seen with other classes already (e.g., Note), creating a new sphere will be as simple as:

```
sphere = MusicalSphere(radius, density, velocity, frameRate)
```

where the provided parameters are defined as above.

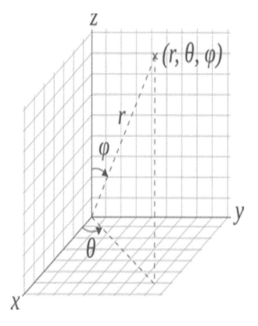

FIGURE 11.9 The spherical coordinate system (math version). (*Note:* In the physics version, the meanings of φ and θ are swapped.)

11.8.3.1 Spherical Coordinate System

A sphere is modeled using the spherical coordinate system. The spherical coordinate system is a 3D coordinate system, where the position of a point is specified by three numbers (see Figure 11.9):

- the distance, r, of that point from the sphere's center (also known as *radial distance*),

- its polar angle, φ (also known as *latitude*), and

- its azimuth angle, θ (also known as *longitude*).

Here is the code[*]:

```
# musicalSphere.py
#
# Demonstrates how to create an animation of a 3D sphere using
# GUI points on a Display.  The sphere is modeled using points on
# a spherical coordinate system
# (see http://en.wikipedia.org/wiki/Spherical_coordinate_system).
# We convert from spherical 3D coordinates to cartesian 2D
# coordinates to position the individual points on the display.
```

---

[*] Based on code by Uri Wilensky (1998), distributed with NetLogo.

```
# The z axis (3D depth) is mapped to color, using an orange gradient,
# ranging from white (front surface of sphere) to black (back
# surface of sphere).  Also, when a point passes the primary meridian
# (the imaginary vertical line closest to the viewer), a note is
# played using the point's latitude for pitch (low to high).
# Also the point turns red momentarily.
#


from gui import *
from music import *
from random import *
from math import *
from sliderControl import *


class MusicalSphere:
    """Creates a revolving sphere that plays music."""

    def __init__(self, radius, density, velocity=0.01, frameRate=30):
        """
        Construct a revolving sphere with given 'radius', 'density'
        number of points (all on the surface), moving with 'velocity'
        angular (theta / azimuthal) velocity, at 'frameRate' frames
        (or movements) per second.  Each point plays a note when
        crossing the zero meridian (the sphere's meridian (vertical
        line) closest to the viewer).
        """

        ### musical parameters ####################################
        self.instrument = XYLOPHONE
        self.scale = PENTATONIC_SCALE
        self.lowPitch = C2
        self.highPitch = C7
        self.noteDuration = 100    # milliseconds

        Play.setInstrument(self.instrument, 0)    # set the instrument

        ### visual parameters #####################################
        # create display to draw sphere (with black background)
        self.display = Display("3D Sphere", radius*3, radius*3)
        self.display.setColor( Color.BLACK )

        self.radius = radius        # how wide sphere is
        self.numPoints = density    # how many points on sphere surface
        self.velocity = velocity    # how far sphere rotates per frame
        self.frameRate = frameRate  # how many frames to do per second

        # place sphere at display's center
        self.xCenter = self.display.getWidth() / 2
        self.yCenter = self.display.getHeight() / 2

        ### sphere data structure (parallel lists) ################
        self.points      = []  # holds all the points
        self.thetaValues = []  # holds point rotation (azimuthal angle)
        self.phiValues   = []  # holds point latitude (polar angle)
```

```
### timer to drive animation ###############################
delay = 1000 / frameRate   # convert frame rate to delay (ms)
self.timer = Timer(delay, self.movePoints)   # create timer

### control surface for animation frame rate ###############
xPosition = self.display.getWidth() / 3      # position control
yPosition = self.display.getHeight() + 45
self.control = SliderControl(title="Frame Rate",
                             updateFunction=self.setFrameRate,
                             minValue=1, maxValue=120,
                             startValue=self.frameRate,
                             x=xPosition, y=yPosition)

   ### color gradient (used to display depth) ##################
   black = [0, 0, 0]            # RGB values for black (back)
   orange = [251, 147, 14]      # RGB values for orange (middle)
   white = [255, 255, 255]      # RGB values for white (front)

   # create list of gradient colors from black to orange, and from
   # orange to white (a total of 25 colors)
   self.gradientColors = colorGradient(black, orange, 12) + \
                         colorGradient(orange, white, 12) + \
                         [white]  # include the final color

   self.initSphere()         # create the circle

   self.start()              # and start rotating!

def start(self):
   """Starts sphere animation."""
   self.timer.start()

def stop(self):
   """Stops sphere animation."""
   self.timer.stop()

def setFrameRate(self, frameRate=30):
   """Controls speed of sphere animation (by setting how many
      times per second to move points).
   """

   # convert from frame rate to delay between each update
   delay = 1000 / frameRate     # (in milliseconds)
   self.timer.setDelay(delay)   # and set timer delay
 def initSphere(self):
   """Generate a sphere of 'radius' out of points (placed on the
      surface of the sphere).
   """

   for i in range(self.numPoints):      # create all the points

       # get random spherical coordinates for this point
       r = self.radius                      # placed *on* the surface
       theta = mapValue( random(), 0.0, 1.0, 0.0, 2*pi) # rotation
       phi = mapValue( random(), 0.0, 1.0, 0.0, pi)     # latitude

       # remember this point's spherical coordinates by appending
       # them to the two parallel lists (since r = self.radius
```

```
                    # for all points, no need to save that)
                    self.thetaValues.append( theta )
                    self.phiValues.append( phi )

                    # project spherical to cartesian 2D coordinates (z is depth)
                    x, y, z = self.sphericalToCartesian(r, phi, theta)

                    # convert depth (z) to color
                    color = self.depthToColor(z, self.radius)

                    # create point at these x, y coordinates, with this color
                    # and thickness 1
                    point = Point(x, y, color, 1)

                    # remember point by appending it to the third parallel list
                    self.points.append( point )        # this point

                    # now, display this point
                    self.display.add( point )
         def sphericalToCartesian(self, r, phi, theta):
             """Convert spherical to cartesian coordinates."""

             # adjust rotation so that theta is 0 at max z (near viewer)
             x = r * sin(phi) * cos(theta + pi/2)      # horizontal axis
             y = r * cos(phi)                          # vertical axis
             z = r * sin(phi) * sin(theta + pi/2)      # depth axis

             # move sphere's center to display's center
             x = int( x + self.xCenter )  # pixel coordinates are integer
             y = int( y + self.yCenter )
             z = int( z )

             return x, y, z

         def depthToColor(self, depth, radius):
             """Create color based on depth."""

             # map depth to gradient index (farther away less luminosity)
             colorIndex = mapValue(depth, -self.radius, self.radius,
                                   0, len(self.gradientColors))

             # get corresponding gradient (RBG value), and create the color
             colorRGB = self.gradientColors[colorIndex]
             color = Color(colorRGB[0], colorRGB[1], colorRGB[2])

             return color

         def movePoints(self):
             """Rotate points on y axis as specified by angular velocity."""

             for i in range(self.numPoints):  # for every point

                 point = self.points[i]           # get this point
                 theta = self.thetaValues[i]      # get rotation angle
                 phi = self.phiValues[i]          # get latitude (altitude)

                 # animate by incrementing angle to simulate rotation
                 theta = theta + self.velocity
```

```
# wrap around at the primary meridian (i.e., 360 degrees
# become 0 degrees) - needed to decide when to play note
theta = theta % (2*pi)

# convert from spherical to cartesian 2D coordinates
x, y, z = self.sphericalToCartesian(self.radius, phi, theta)

# check if point crossed the primary meridian (0 degrees),
# and if so, play musical note, and change its color to red
if self.thetaValues[i] > theta:    # did we just wrap around?

    # yes, so play note
    pitch = mapScale(phi, 0, pi,
                     self.lowPitch, self.highPitch,
                     self.scale)  # phi is latitude
    dynamic = randint(0, 127)     # get random dynamic

    Play.note(pitch, 0, self.noteDuration, dynamic)

    # set point's color to red
    color = Color.RED

else:   # otherwise, not at the primary meridian, so

    # set point's color based on depth, as usual
    color = self.depthToColor(z, self.radius)

# now, we have the point's new color and x, y coordinates

self.display.move(point, x, y)  # move point to new position
point.setColor(color)           # and update its color

# finally, save this point's new rotation angle
self.thetaValues[i] = theta
```

This program is a little more challenging than what we have seen so far. This is because it combines classes, top-down design, and a new physical model (simulating a sphere through individual, animated points) using trigonometric functions. It also makes music (resembling a music box).

Fact: Good programs are made to be read.[*]

Advice: When confronted with a new, large program, start slowly from the top, and familiarize yourself in with its components—its class(es), instance variables, and its functions.

For the above program, a good strategy is to focus on the __init__(), initSphere(), and movePoints() functions. You may need to use a paper

---

[*] Also, further explore Donald Knuth's concept of Literate Programming (http://www.literate programming.com).

and pencil to write down some notes or questions. This way you can write important things down for later consideration (e.g., what do the various instance variables hold, or what function does the Timer object call repeatedly), while keeping your mind clear to absorb new things. If you approach things systematically, everything will soon make sense.

The __init__() function (constructor) starts, as usual, setting up the instance variables. First, notice the musical parameters, which define the type of sounds generated by the points, as they rotate through the primary meridian (the imaginary vertical line closest to the viewer).

```
self.instrument = XYLOPHONE
self.scale = PENTATONIC_SCALE
self.lowPitch = C2
self.highPitch = C7
self.noteDuration = 100          # milliseconds
```

Of particular interest is the way we model a sphere:

```
### sphere data structure (parallel lists) ##################
self.points      = [] # holds the points
self.thetaValues = [] # holds the points' rotation (azimuthal angle)
self.phiValues   = [] # holds the points' latitude (polar angle)
```

As mentioned earlier, we simulate a sphere using rotating points. For each of these points, we create a GUI point object (to be placed on the display) and its spherical coordinates (see previous section).

Since all points are placed on the sphere's surface, we only need to store the θ and φ values (as the r value is always equal to the sphere's radius — see self.radius).

Function initSphere() creates the individual points and updates the three parallel lists. (This will be discussed in more detail below.)

Also, of particular interest are the Timer and the SliderControl objects:

```
### timer to drive animation ###############################
delay = 1000 / frameRate   # convert frame rate to delay (ms)
self.timer = Timer(delay, self.movePoints)   # create timer


### control surface for animation frame rate ###############
xPosition = self.display.getWidth() / 3    # position control
yPosition = self.display.getHeight() + 45
self.control = SliderControl(title="Frame Rate",
                     updateFunction=self.setFrameRate,
                     minValue=1, maxValue=120,
                     startValue=self.frameRate,
                     x=xPosition, y=yPosition)
```

First, we convert the frame rate (measured in frames per second, e.g., 30) to a timer delay (that is, how often should the timer call the animation function). The timer delay is measured in milliseconds, hence the formula (`delay = 1000 / frameRate`).

Then we create the Timer object providing the delay and, most importantly, the animation function, `movePoints()`. This function is very important and is discussed later.

Notice how, when creating the SliderControl object, we specify its initial parameters (see previous case study). An important parameter is the update function, `setFrameRate()`. As you can see in the code, this function does one thing—it updates the timer delay. So, by having the slider control call it, when the slider is moved, in essence, we have connected the slider to the timer delay.

Another important set of parameters for the slider control is its initial x and y position. The goal here is to place the control surface below the main sphere display. So we use the main display's (`self.display`) width and height to calculate a good initial position for the timer control surface (approximately right below the main display).[*]

Function `initSphere()` uses a `for` loop (controlled by the number of points needed) to create all the points and store them (and their spherical coordinates) into the three parallel lists described above. Notice how the θ (azimuthal) angle, for each point, is set to a random value between 0 and 2*pi (that is, 360 degrees):

```
theta = mapValue(random(), 0.0, 1.0, 0.0, 2*pi) # rotation
```

Similarly, the ϕ (polar) angle is set to a random value between 0 and pi (that is, 180 degrees):

```
phi = mapValue(random(), 0.0, 1.0, 0.0, pi) # latitude
```

We use pi (180 degrees) for ϕ (as opposed to 2*pi) because ϕ controls the placement of the point on the meridian (vertical half circle). The other angle, θ, moves this imaginary meridian around 360 degrees, thus covering the complete surface of the sphere. If we had used 360 degrees for ϕ, we would be covering the sphere surface twice. This would also work, but it is inelegant and wasteful.

---

[*] Ideally, this calculation should be relative to the main display's position (here we are assuming that it is always created at the top-left screen corner. To be more correct, we should be using `self.display.getPosition()`, which returns the display's x and y coordinates in a list. This is left as an exercise.

Function `sphericalToCartesian()`, as its name indicates, converts from spherical to cartesian 3D coordinates, x, y, and z. We use the x and y coordinates to place the point on the GUI display. The math behind the conversion guarantees that the x and y coordinates will change proportionally to how the point is placed on the imaginary (simulated) sphere. This bit of math creates a quite stunning effect.[*]

The third coordinate, z, corresponds to depth. As mentioned earlier, to simulate depth of a point, we use a color gradient (from white, to orange, to black). Function `depthToColor()` performs the necessary conversion, using the `self.gradientColors` list. This creates an effective illusion of visual depth.[†]

Finally, function `movePoints()` advances the θ angle of each point, as follows:

```
theta = self.thetaValues[i]    # get rotation angle

# animate by incrementing angle to simulate rotation
theta = theta + self.velocity

# wrap around at the primary meridian (i.e., 360 degrees
# become 0 degrees) - needed to decide when to play note
theta = theta % (2*pi)
```

This adds the velocity (e.g., 0.01) specified by the end-programmer to the current θ value and ensures that when we reach 360° (2*pi), we zero out the θ angle. This is done by using the modulo, %, operator. (Recall that modulo 2*pi is the remainder of the division by 2*pi.) If the value in question is larger than 2*pi, it is replaced by the remainder, that is, it "wraps around" to the corresponding value smaller than 2*pi.[‡]

Function `movePoints()` also checks for when a point crosses the primary meridian. This happens every time the modulo 2*pi statement above returned a value that's smaller than the original θ value:

```
if self.thetaValues[i] > theta:    # did we just wrap around?

        # yes, so play note
    pitch = mapScale(phi, 0, pi,
                    self.lowPitch, self.highPitch,
```

---

[*] Similar math is used in modern computer games to project 3D objects onto the 2D computer screen. If you are interested in computer games, this little math demystifies how its all done.

[†] Explore using different colors in the gradient. Is white, to orange, to black the best combination? What other colors can you use?

[‡] Wrapping around ever increasing (or decreasing) values, using the modulo operator with the limiting value, is a common technique in computer science. Try it out to see that it works.

```
                    self.scale)   # phi is latitude
        dynamic = randint(0, 127)       # get random dynamic

        Play.note(pitch, 0, self.noteDuration, dynamic)

        # set point's color to red
        color = Color.RED

else:   # otherwise, not at the primary meridian, so

        # set point's color based on depth, as usual
        color = self.depthToColor(z, self.radius)
```

If so, we temporarily change the color of the point to red (to suggest a "plucking" effect, similar to a musical box). We also play a note with pitch corresponding to the φ angle (low to high)[*] and random dynamic.

Otherwise, if the point has not crossed the primary meridian, we simply assign it a color based on its depth and the selected color gradient.

Finally, function movePoints() moves the point on the display, using the updated information.

Again, notice how the following line drives the whole simulation:

```
sphere = MusicalSphere(radius=200, density=200, velocity=0.01, frameR-
        ate=30)
```

This concludes this case study. When you get this program running think of ways that you can map the data to sound and music and then implement those processes to "hear" your animation.

11.8.4 Exercises

1. Modify the musical parameters to create a different effect. The original parameters create a happier, more cheerful musical effect (aided by the use of the pentatonic scale and the xylophone sound). Switch to the following:

```
# musical parameters (alternate)
self.instrument = PIANO
self.scale = MAJOR_SCALE
self.lowPitch = C1
self.highPitch = C6
self.noteDuration = 2000  # milliseconds (2 seconds)
```

---

[*] Ideally this should be inverted so that a low-placed point produces a low pitch. But, given the number of points, this subtlety is lost (so we aimed for a more efficient solution – one less operation). Sometimes efficiency (there are hundreds of points to be calculated per animation frame, and many animation frames per second) trumps completeness.

Notice how these create a more solemn, meditative mood. This demonstrates the importance of design in music composition, that is, to carefully consider the effect produced by a particular pitch set, register choice, and orchestration (in addition to melody and harmony).

2. Explore different musical parameters to create different musical effects. What do you observe?

3. Experiment with creating more than one musical sphere, each having different musical parameters. Explore creating a set of complementary musical spheres. Explore performance possibilities with adjusting their rotation speeds.

4. Create SliderControl objects to control musical parameters in real time. Explore the performance possibilities opened by this modification.

5. (*Advanced*) Modify the MusicalSphere code to create other rotating geometrical shapes.

11.9 CYMATICS

As discussed in Chapter 1, Cymatics (from the Greek κύμα, "wave") is the study of visible (visualized) sound and vibration in 1-, 2-, and 3-dimensional artifacts. It was influenced significantly by the work of the 18th century physicist and musician Ernst Chladni, who developed a technique to visualize modes of vibration on mechanical surfaces, known as Chladni plates (see Figure 11.10).

In the previous case study, we introduced the idea of simulating a sphere using individual points placed on a GUI display. These points were controlled by the simulation algorithm to move on the surface of an

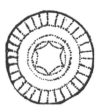

FIGURE 11.10 Chladni plates, vintage engraving. Old engraved illustration of Chladni plates isolated on a white background. (From Charton, É. and Cazeaux, E., eds. (1874), Magasin Pittoresque.)

imaginary sphere. The algorithm changed their colors and played sounds when they crossed a certain threshold.

The concept of cymatics takes this idea a step up.

What if, instead of having passive points, the points were actively moving on their "own." That is, instead of each point being controlled by an external (to it) algorithm, what if each point encapsulated its own set of rules on how to react to its environment? This idea leads to simulations that "behave" more naturally, as they model interactions with various phenomena that surround them.

Since its creation, the universe, as we perceive it, consists of innumerable individual elements, at different scales of size and time. These elements interact, converge, and diverge to create systems of organized behavior for example, an electron, a water molecule, the human brain and its neurons, a sandpile and its sand particles, the solar system and its planets and asteroids, and so on (Bak 1996). Anything we perceive can potentially be described in terms of more atomic elements that somehow interact, and, in some cases, balance out with each other, following the principle of least effort (that is, finding a place of apparent rest or a place of quasi-periodic movement).

The universe is filled with such systems.

This case study demonstrates how to simulate such a system. It is based on "Boids," a program developed by Craig Reynolds in 1986, modeling the flocking behavior of birds (Reynolds 1987).*

Boids are self-adapting entities (agents) which are capable of perceiving their environment and acting upon the environment. This idea is used in developing artificial intelligence applications and simulations. This is a quite powerful concept, full of creative musical potential.

The following program demonstrates the types of behaviors that emerge from self-adapting agents (boids), which sense their surroundings and modify their behavior (movement) accordingly (see Figure 11.11). It is hard to appreciate the liveliness of the resulting animation through a static figure. Imagine each of these boids (points) moving around the display, staying in close proximity to other boids, yet avoiding them, while trying to collectively move in a certain direction. The resulting movement, in some cases, resembles the movement of birds, fish, and other animal collectives.

---

* "Boid" refers to a bird-like object (a bird-oid), and, in a stereotypical New York accent, sounds like "bird."

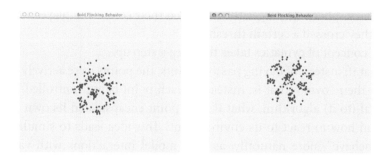

FIGURE 11.11   Two snapshots of the boid animation (200 boids, min separation = 47, flock threshold = 267, separation factor = 0.01, alignment factor = 0.01, cohesion factor = 0.15, friction = 1.1). (*Note:* These parameters will be explained soon.)

These boids follow three simple rules:

1. **Rule of Separation:** Each boid keeps separate from other boids; it avoids being at the same place with another boid at the same time.

2. **Rule of Alignment:** Each boid moves toward the average heading of its local flock.

3. **Rule of Cohesion (or Attraction):** Each boid moves toward a certain point. This may be the center of the universe, the mouse pointer, or the center of a local flock.

An additional rule (left as an exercise) is the following:

4. **Rule of Avoidance:** Each boid moves away from a certain point (e.g., an obstacle). This could be a particular point in the universe or the mouse pointer.

A boid applies the above rules to decide what to do next. Each of these rules contributes a direction and a distance in which to move. The boid combines these directions and distances to generate a single direction and distance in which to move (in each animation frame). Sound simple? It is.

As discussed in the section on Zipf's law, such agents tend to minimize their overall effort associated with this interaction (economy principle). That is, a system of such interacting agents tends to find a global optimum that minimizes overall effort (Bak 1996). This interaction models some sort of exchange (e.g., information, energy, etc.).

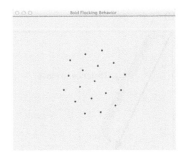

FIGURE 11.12 Boids have converged to a hexagon resembling the molecular structure of benzene, or water molecules in an ice crystal, among others (19 boids, min separation = 30, flock threshold = 100, separation factor = 0.13, alignment factor = 0.16, cohesion factor = 0.01, friction = 4.2).

Given the simulation parameters, the collective movements of each and all boids, over time, can result in three possibilities:

1. **Divergence:** The boids move away from each other and outside of the universe window (never to be seen again). This divergence resembles a slow explosion—particles moving away from each other.

2. **Convergence:** The boids find a place of balance, which, given the initial conditions, may be a single collapsed point (resembling a black hole), or several collapsed points (resembling several black holes), or a static pattern (see Figure 11.12).

3. **Orbital Behavior:** The boids oscillate in a quasi-periodic movement that resembles bees flying around a flower, or complex planetary systems revolving around one, or more suns (points of attraction).[*]

11.9.1 Vectors and Python Complex Numbers

As mentioned above, each of the simple rules (followed by a boid) contributes a direction and a distance to move toward. A boid combines these directions and distances to generate a single direction and distance to move toward.

Each direction and distance is simply modeled by x and y values. The larger the values, the larger the movement. This information is called a *vector* (see Figure 11.13).

---

[*] This behavior is quite mesmerizing to watch (and tinker with, by adjusting the program parameters).

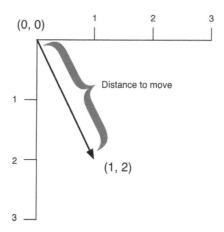

FIGURE 11.13 A 2D vector corresponds to x and y displacement (per animation frame).

Definition: A *vector* is a pair of x and y values, which indicate how far to move from a given position (in x and y).*

Python provides a convenient data structure for storing x and y coordinates, called *complex*. Python complex numbers consist of two values, the *real* value (which we will use to store x coordinates) and the *imaginary* value (which we will use to store y coordinates).†

For example, the following creates a complex number:

```
>>> c = complex(1, 2)
>>> c
(1+2j)
>>> c.real
1.0
>>> c.imag
2.0
```

Notice how, after the complex number c is created, we can easily retrieve its real (*x*) and imaginary (*y*) parts.

Python complex numbers are great, since they implement addition and multiplication (or division).

---

* Actually, this is the definition of a 2D vector (a vector in two dimensions, like a GUI display). For a vector in 3D space, we add one more dimension (z). And so on.
† Do not let the terms "real" and "imaginary" confuse you. For our purposes, "real" means "x coordinate" and "imaginary" means "y coordinate."

The addition of two complex numbers simply adds the two x values together and two y values together. For example:

```
>>> c = complex(1, 2)
>>> d = complex(4, 6)
>>> e = c + d
>>> e.real
5.0
>>> e.imag
8.0
```

Multiplication (or division) of a complex number with a regular (single) number simply multiplies each of the two coordinates by the regular number (or divides them by, for division). For example:

```
>>> c = complex(1, 2)
>>> f = c * 3
>>> f.real
3.0
>>> f.imag
6.0
```

In the next three sections, we present the code for the Boid Universe simulation. These three pieces of code were placed in the same Python file ("boids.py"). They have been separated here for convenience in explaining their key points.

11.9.2 Defining the Boid Universe

Here is the code for the boid universe, together with the global simulation parameters:

```
# boids.py
#
# This program simulates 2D boid behavior.
#
# See http://www.red3d.com/cwr/boids/ and
# http://www.vergenet.net/~conrad/boids/pseudocode.html
#

from gui import *
from math import *
from random import *

# universe parameters
universeWidth  = 1000   # how wide the display
universeHeight = 800    # how high the display

# boid generation parameters
numBoids    = 200         # from 2 to as much as your CPU can handle
boidRadius  = 2           # radius of boids
boidColor   = Color.BLUE  # color of boids
```

```
# boid distance parameters
minSeparation  = 30    # min comfortable distance between two boids
flockThreshold = 100   # boids closer than this are in a local flock

# boid behavior parameters (higher means quicker/stronger)
separationFactor = 0.01# how quickly to separate
alignmentFactor = 0.16 # how quickly to align with local flockmates
cohesionFactor = 0.01  # how quickly to converge to attraction point
frictionFactor = 1.1   # how hard it is to move (dampening factor)

### define boid universe, a place where boids exist and interact ####
class BoidUniverse:
    """This is the boid universe, where boids exist and interact. It is
       basically a GUI Display, with boids (moving Circles) added to
       it.  While boids are represented as circles, they have a little
       more logic to them - they can sense where other boids are, and
       act accordingly.  While individual boids have simple rules for
       sensing their environment and reacting, very intricate, complex,
       naturally-looking patterns of behavior emerges, similar to those
       of birds flying high in the sky (among others).  The rules of
       behavior are the same for all boids, and are defined in the Boid
       class (a sister class to this one).
    """

    def __init__(self, title = "", width = 600, height = 400,
                    frameRate=30):

        self.display = Display(title, width, height) # universe display

        self.boids = []            # list of boids

        # holds attraction point for boids (initially, universe center)
        self.attractPoint = complex(width/2, height/2)

        # create timer
        delay = 1000 / frameRate # convert frame rate to delay (ms)
        self.timer = Timer(delay, self.animate)     # animation timer

        # when mouse is dragged, call this function to set the
        # attraction point for boids
        self.display.onMouseDrag( self.moveAttractionPoint )

    def start(self):
        """Starts animation."""
        self.timer.start() # start movement!

    def stop(self):
        """Stops animation."""
        self.timer.stop() # stop movement!

    def add(self, boid):
        """Adds another boid to the system."""

        self.boids.append( boid )         # remember this boid
        self.display.add( boid.circle )  # add a circle for this boid

    def animate(self):
        """Makes boids come alive."""
```

```
### sensing and acting loop for all boids in the universe !!!
for boid in self.boids:   # for every boid

    # first observe other boids and decide how to adjust movement
    boid.sense(self.boids, self.attractPoint)

    # and then, make it so (move)!
    boid.act(self.display)

def moveAttractionPoint(self, x, y):
    """Update the attraction point for all boids."""
    self.attractPoint = complex(x, y)
```

Notice how the code starts with various parameters that control the boid simulation. These parameters are as follows:

```
# boid generation parameters
numBoids   = 200        # from 1 to as much as your CPU can handle
boidRadius = 2          # radius of boids
boidColor  = Color.BLUE # color of boids

# boid distance parameters
minSeparation  =  30    # min comfortable distance between any two boids
flockThreshold = 100    # boids closer than this are in a local flock

# boid behavior parameters (higher means quicker/stronger)
separationFactor = 0.01 # how quickly to separate
alignmentFactor  = 0.16 # how quickly to align with local flockmates
cohesionFactor   = 0.01 # how quickly to converge to attraction point
frictionFactor   = 1.1  # how hard it is to move (dampening factor)
```

These parameters are very important. Modifying them may drastically change the system's behavior. Again, boids may converge (to a single point), diverge (and disappear), or engage in quasi-periodic oscillations. Run the code and experiment with different settings (also see the exercises at the end of the chapter).

Next we have the class defining the boid universe, called (what else?), BoidUniverse. Notice how its constructor, __init__(), encapsulates the GUI display, a list of boids to be used in the simulation, and the attraction point for all the boids. Also, notice how it creates the timer that runs the animation (probably the most important line of code in the whole program, if there is one):

```
# create timer
delay = 1000 / frameRate    # convert frame rate to delay (ms)
self.timer = Timer(delay, self.animate)      # animation timer
```

This causes function animate() to be called repeatedly by the timer, once every delay milliseconds (e.g., if frameRate is 30, then animate() will be called approx. once every 33 milliseconds).

Finally, the BoidUniverse constructor assigns function moveAttrac-
tionPoint() to the display's mouse-dragging event. This function simply
updates the attraction point, when the mouse is dragged. This allows the
end-user to control boid flocking behaviors interactively.

The next interesting function is add(). Notice how, when called with a
boid, it simply adds it to the BoidUniverse's list of boids, and also, adds the
boid's internal representation, a circle, to the GUI display. This is necessary, because the GUI display knows nothing about boids—it only knows
about GUI widgets, like a circle.

```
def add(self, boid):
   """Adds another boid to the system."""

   self.boids.append( boid )       # remember this boid
   self.display.add( boid.circle ) # add a circle for this boid
```

Finally, the most important function of the BoidUniverse class is ani-
mate(). As its name indicates, this function creates the animation (or
makes boids come alive). Notice how it uses a for loop to iterate through
the BoidUniverse's list of boids. Then, for each boid, it calls its sense() and
act() functions.

```
def animate(self):
   """Makes boids come alive."""

   ### sensing and acting loop for all boids in the universe !!!
   for boid in self.boids:   # for every boid

      # first observe other boids and decide how to adjust movement
      boid.sense(self.boids, self.attractPoint)

      # and then, make it so (move)!
      boid.act(self.display)
```

Notice how the boid sense() function expects the list of boids, as well as
the attraction point. As we will see soon, this is because it will examine
the locations of all the other boids, as well as the location of the attraction
point. Then, using the boid behavior rules, it will determine a new direction and distance for each boid to travel, in the current animation frame.[*]
This movement is accomplished by function act(), which only requires the
GUI display, since the rest of the information is maintained in each Boid
object's instance variables.

Next, let's examine the Boid class.

---

[*] Recall that function animate() is called repeatedly by the BoidUniverse timer, according to
the frame rate (e.g., 30 times a second).

11.9.3 Defining the Boids

Here is the code for the individual boids, followed by an explanation of key points:

```
### define the boids, individual agents who can sense and act #######
class Boid:
    """This a boid.  A boid is a simplified bird (or other species)
       that lives in a flock.  A boid is represented as a circle,
       however, it has a little more logic to it - it can sense where
       other boids are, and act, simply by adjusting its direction of
       movement.  The new direction is a combination of its reactions
       from individual rules of thumb (e.g., move towards the center of
       the universe, avoid collisions with other boids, fly in the same
       general direction as boids around you (follow the local flock,
       as you perceive it), and so on.  Out of these simple rules
       intricate, complex behavior emerges, similar to that of real
       birds (or other species) in nature.
    """

    def __init__(self, x, y, radius, color,
                    initVelocityX=1, initVelocityY=1 ):
        """Initialize boid's position, size, and initial
           velocity (x, y).
        """

        # a boid is a filled circle
        self.circle = Circle(x, y, radius, color, True)

        # set boid size, position
        self.radius = radius                 # boid radius
        self.coordinates = complex(x, y)     # boid coordinates (x, y)

        # NOTE: We treat velocity in a simple way, i.e., as the
        # x, y displacement to add to the current boid coordinates,
        # to find where to move its circle next.  This moving is done
        # once per animation frame.

        # initialize boid velocity (x, y)
        self.velocity = complex(initVelocityX, initVelocityY)

    def sense(self, boids, center):
        """
        Sense other boids' positions, etc., and adjust velocity
        (i.e., the displacement of where to move next).
        """

        # use individual rules of thumb, to decide where to move next

        # 1. Rule of Separation - move away from other flockmates
        #                         to avoid crowding them
        self.separation = self.rule1_Separation(boids)
```

```
        # 2. Rule of Alignment - move towards the average heading
        #                        of other flockmates
        self.alignment = self.rule2_Alignment(boids)

        # 3. Rule of Cohesion - move toward the center of the universe
        self.cohesion = self.rule3_Cohesion(boids, center)

        # 4. Rule of Avoidance: move to avoid any obstacles
        #self.avoidance = self.rule4_Avoidance(boids)

        # create composite behavior
        self.velocity = (self.velocity / frictionFactor) + \
                        self.separation + self.alignment + \
                        self.cohesion

    def act(self, display):
        """Move boid to a new position using current velocity."""

        # Again, we treat velocity in a simple way, i.e., as the
        # x, y displacement to add to the current boid coordinates,
        # to find where to move its circle next.

        # update coordinates
        self.coordinates = self.coordinates + self.velocity

        # get boid (x, y) coordinates
        x = self.coordinates.real   # get the x part
        y = self.coordinates.imag   # get the y part

        # act (i.e., move boid to new position)
        display.move( self.circle, int(x), int(y) )

    ##### steering behaviors ####################
    def rule1_Separation(self, boids):
        """Return proper velocity to keep separate from other boids,
           i.e., avoid collisions.
        """

        newVelocity = complex(0, 0)   # holds new velocity

        # get distance from every other boid in the flock, and as long
        # as we are too close for comfort, calculate direction to
        # move away (remember, velocity is just an x, y distance
        # to travel in the next animation/movement frame)
        for boid in boids:              # for each boid

            separation = self.distance(boid)   # how far are we?

            # too close for comfort (excluding ourself)?
            if separation < minSeparation and boid != self:
                # yes, so let's move away from this boid
                newVelocity = newVelocity - \
                              (boid.coordinates - self.coordinates)

        return newVelocity * separationFactor  # return new velocity
```

```python
def rule2_Alignment(self, boids):
    """Return proper velocity to move in the same general direction
       as local flockmates.
    """

    totalVelocity = complex(0, 0) # holds sum of boid velocities
    numLocalFlockmates = 0        # holds count of local flockmates

    # iterate through all the boids looking for local flockmates,
    # and accumuate all their velocities
    for boid in boids:

        separation = self.distance(boid)     # get boid distance

        # if this a local flockmate, record its velocity
        if separation < flockThershold and boid != self:
            totalVelocity = totalVelocity + boid.velocity
            numLocalFlockmates = numLocalFlockmates + 1

    # average flock velocity (excluding ourselves)
    if numLocalFlockmates > 0:
        avgVelocity = totalVelocity / numLocalFlockmates
    else:
        avgVelocity = totalVelocity

    # adjust velocity by how quickly we want to align
    newVelocity = avgVelocity - self.velocity

    return newVelocity * alignmentFactor  # return new velocity

def rule3_Cohesion(self, boids, center):
    """Return proper velocity to bring us closer to center of the
       universe.
    """

    newVelocity = center - self.coordinates

    return newVelocity * cohesionFactor  # return new velocity

    ##### helper function ####################
    def distance(self, other):
        """Calculate the Euclidean distance between this and
           another boid.
        """

    xDistance = (self.coordinates.real - other.coordinates.real)
    yDistance = (self.coordinates.imag - other.coordinates.imag)

    return sqrt( xDistance*xDistance + yDistance*yDistance )
```

The Boid constructor, __init__(), creates the boid's visual representation (a GUI circle) and initializes the boid's size (self.radius), position (self.coordinates), and initial velocity (self.velocity). Again, keep in mind that we model velocity in a simple, meaningful way—as the distance and direction a boid needs to travel in one animation frame (that is, a relative x, y displacement from its current position). Thus, moving a boid is as simple as adding `self.velocity`'s x component to `self.coordinate`'s x component. The same is done for y. That's all!

The Boid class has two main functions, `sense()` and `act()`.

11.9.3.1 Boid Sensing

The `sense()` function calls a sequence of simple functions that implement the steering (or behavior) rules. Each of these functions contributes a suggested velocity (again, modeled simply as an x and y displacement). We say "suggested" because this velocity is this contribution of one only rule and needs to be combined with the suggested velocities of all other rules. Combining velocities is simply done by **adding** their x and y components, respectively:

```
# create composite behavior
self.velocity = (self.velocity / frictionFactor) + \
                self.separation + self.alignment + \
                self.cohesion
```

It is as simple as addition. (If some of the velocities are pointing in opposite directions, the corresponding x and y components simply cancel out—their signs will be opposite. Addition does this.)

All the rules are computed similarly. So here we will study more carefully a representative one, the rule for alignment. As mentioned earlier,

- **Rule of alignment:** Each boid moves toward the average heading of its local flock.

This means that each boid needs to adjust its velocity to match the average velocity of all other boids in its local flock. The local flock is determined by distance. That is, the local flock of a boid consists of boids that are closer to it than a certain distance (flockThreshold).[*]

This function is called by a single boid (`self`). It is given access to the list of all boids.

```
def rule2_Alignment(self, boids):
    """Return proper velocity to move in the same general direction
       as local flockmates.
    """
```

---

[*] Recall that `flockThreshold` is one of the system's adjustable parameters.

Fact: Variable `self` has the same value as the value returned when we create a new boid by `Boid(x, y, boidRadius, boidColor, 1, 1)`.

That is, inside the boid's code, we use self to determine who we are (when needed by the algorithm/code).

Inside the function, we first initialize needed variables:

```
totalVelocity = complex(0, 0) # holds sum of boid velocities
numLocalFlockmates = 0        # holds count of local flockmates
```

Next we loop through all the boids. For every boid that is close enough, we add up its velocity to the total velocity of other flockmates, so far:

```
for boid in boids:

    separation = self.distance(boid)    # get boid distance

    # if this a local flockmate, record its velocity
    if separation < flockThershold and boid != self:
       totalVelocity = totalVelocity + boid.velocity
       numLocalFlockmates = numLocalFlockmates + 1
```

Notice how we use function `self.distance()` to get the Euclidean distance between the two boid positions. Also, notice how, as we loop through all the boids, we exclude the current boid from the calculation:

```
if separation < flockThershold and boid != self:
```

Notice how addition of Python complex numbers takes care of adding the x and y components of the velocities:

```
    totalVelocity = totalVelocity + boid.velocity
```

And, of course, we keep track of how many boids are in the local flock (to use later, for calculating the average):

```
    numLocalFlockmates = numLocalFlockmates + 1
```

Then we simply divide `totalVelocity` by the number of flockmates.

```
# average flock velocity vector (excluding ourselves)
if numLocalFlockmates > 0:
    avgVelocity = totalVelocity / numLocalFlockmates
else:
    avgVelocity = totalVelocity
```

Finally, we return the difference between the flock's average velocity and our velocity (`self.velocity`), adjusted by the global parameter, `alignment-Factor`. This parameter controls the strength of this contribution to the overall steering behavior of this boid (see previous section).

```
# adjust velocity by how quickly we want to align
newVelocity = avgVelocity - self.velocity

return newVelocity * alignmentFactor # return new velocity
```

11.9.3.2 Boid Acting

The `act()` function simply updates the position of the current boid, given the new composite velocity, which was calculated by function `sense()`.

```
# update coordinates
self.coordinates = self.coordinates + self.velocity

# get boid (x, y) coordinates
x = self.coordinates.real      # get the x part
y = self.coordinates.imag      # get the y part
```

Then it updates the boid's visual representation on the GUI display.

```
# act (i.e., move boid to new position)
display.move(self.circle, int(x), int(y))
```

This concludes the explanation of the `Boid` class.

11.9.4 Creating the Simulation

The last section of the code uses the above class definitions to create one `BoidUniverse` object and several `Boid` objects:

```
# start boid simulation universe = BoidUniverse(title="Boid Flocking
Behavior",
                      width=universeWidth, height=universeHeight,
                      frameRate=30)

# create and place boids
for i in range(0, numBoids):
    x = randint(0, universeWidth) # get a random position for this boid
    y = randint(0, universeHeight)

    # create a boid with random position and velocity
    boid = Boid(x, y, boidRadius, boidColor, 1, 1)
    universe.add(boid)

# animate boids
universe.start()
```

Notice how boids are placed randomly across the boid universe. The rest of the code is self-explanatory. The exercises below will help you explore the code better and appreciate its creative musical (and other) potential.

11.10 EXERCISES

1. Experiment with the above code. Try the following sets of simulation parameters and explore the different boid behaviors they generate. Get a feeling about what each parameter contributes to the simulation. Make notes. Realize that you are dealing with a complex chaotic system (like the ones found in nature). Under the right conditions, a small change can have an immense effect. Some of the parameters are interfering/interacting with others. Derive your own combinations.

2. Add a few SliderControl (or SliderFloatControl) objects to adjust the various boid parameters. Explore the different types of flocking behavior you can create. You are dealing with a chaotic system with various places of balance and behavior. See how many different behaviors you can create. Experiment.

Parameter	Set 1	Set 2	Set 3	Set 4	Set 5	Set 6	Set 7	Set 8	Set 9
boids	200	200	200	49	19	2	3	4	6
min separation	30	47	47	30	30	70	30	30	47
flock threshold	100	267	100	100	100	100	100	100	100
separation factor	0.01	0.01	0.01	0.13	0.13	0.1	0.01	0.01	0.03
alignment factor	0.16	0.01	0.16	0.16	0.16	1.6	1.6	1.6	0.16
coherence factor	0.01	0.15	0.01	0.01	0.01	0.001	0.001	0.001	0.01
friction factor	1.1	1.1	4.3	4.2	4.2	1.1	1.1	1.1	1.1

3. Experiment with the number of boids and their sizes (radius). Notice how, under certain conditions, boids will derive a geometric shape that maximizes balance. This demonstrates how geometric shapes emerge in the universe. Euclidean geometry is then an abstraction of structures and properties that emerge in any universe (or system) given enough freely moving and interacting particles.

4. Using a prime number (2, 3, 5, 7, 11) forces the boids to keep looking for a place of rest, balance—they never find it.

   a. Explore the different geometric shapes that emerge using 2, 3, 4, 5, 6, 7, 8, 9, and 10 boids. What do you see?

   b. At larger numbers, for example, 250, you see universe-like properties emerging. Some boids are forced to form local systems, being pushed closer than the rule of separation alone would dictate. They are "pushed" by the collective "mass" (or force) of the global system. A real universe (as in cosmos) emerges. Experiment with larger numbers of boids.

5. Modify the code to make the boids be attracted to the center of the universe, but also *avoid* the mouse pointer. How can you accomplish this? *Hint:* Add an avoidance rule. This is very similar (actually, opposite) to the cohesion rule. What types of behaviors can you create now?

6. Add a rule for boids to be attracted to the center of their local flock. To do so, average the coordinates of all other local flockmates and make that the center of attraction for a given boid. This will create even more natural flocking behaviors, in that different subsystems will emerge with their own local behavior (e.g., direction of movement). This can be very interesting. Experiment.

7. What other rules can you think of? How about adding control of boids, either local (controlling a single boid) or global (controlling an attraction or avoidance point), using a MIDI instrument or OSC controller (e.g., a smartphone). How about having several boids be controlled by such instruments/controllers. How about using such a system to create a collaborative music performance? How can you incorporate/combine other techniques seen in this book? Brainstorm. Experiment. Create. Replace the boid visual representation (currently, a Circle object) with a Icon object (see Chapter 8 and Appendix C). Pick an interesting icon (a bird or something more abstract/creative). Make use of Icon's functions for rotation (and scaling/resizing, if needed).* Convert each boid's vector information (x and y displacement, per animation frame) to a direction, or orientation. This can be done easily with Python's atan2(y, x)

---

* For instance, scaling would be helpful if you implemented three dimensions (x, y, z). In that case, you would map the z coordinate (depth) to Icon scaling/resizing (via the setSize() function).

trigonometric function. This function returns the angle of a vector with this y displacement (or rise) and this x displacement (or run).[*] Use the resultant angle to orient the boid's visual representation, via Icon's `rotate()` function. Notice that `atan()` returns an angle in radians (in terms of π units), whereas Icon's `rotate()` expects the angle in degrees. (Use Python's `radians()` function.) All the parts are there. Assemble them.

8. Add a third dimension to the boid universe. Unfortunately, the Python complex data type works only with two dimensions. Replace it with three individual coordinates (x, y, z), as done in MusicalSphere. (Alternatively, you could create a Coord class, which encapsulates all three coordinates.) Adding 3D coordinates (x, y, z) follows the same approach as adding 2D coordinates. The same holds for multiplication (or division). Map the z coordinate (depth) to a color gradient (as in MusicalSphere).[†]

9. And, finally, music making! The various boid systems can drive music generation in amazing ways. This creative space is vast, so we will provide only a few seed ideas. This section invites you to apply everything you have learned so far about music generation/performance in this book (and beyond). Ask yourself what types of musical parameters would you like to drive/control through the position/motion of boids. Here are some ideas:

   a. Perhaps each boid plays a MIDI note, via `Play.note()`, where the boid's x coordinate determines the pitch, and the y coordinate determines the volume.

   b. Perhaps each boid is connected with an AudioSample object, where the boid's velocity controls the sample's frequency. What are some other AudioSample control possibilities?

   c. Explore playing with timbre, using the AudioSample class to introduce looped recordings of different sounds and of different

---

[*] Notice how `atan2()` provides the inverse calculation of `sin()` and `cos()`. In other words, `atan2(sin(theta), cos(theta)) == theta`.

[†] Another possibility is to (also) change the size of the `Circle` object (but to do this requires creating a new `Circle` every time a boid changes depth). This is costly, but interesting to try out, especially for few boids. Using an `Icon` object instead allows you to easily do rescaling via the `setSize()` function. Explore.

lengths.* Clearly, once you open this door, the possibilities are endless. Some great compositions wait to be discovered.

d. Explore using `mapScale()` to derive musical outcomes that sound harmonious.

e. Modify some of the boids to be controlled by input from MIDI instruments and/or OSC controllers (e.g., smartphones).† Simply create a MIDIBoid and/or OSCBoid class, which use slightly modified rules of behavior. They may use a single rule, which simply uses MIDI pitch (or some OSC parameter) to set boid's new velocity (or position). You could still use some of the other rules (e.g., separation) to let the boid system constrain the musical (or OSC) input. This can help guide your musical creativity/ performance in natural, unexpected ways.

f. Brainstorm. Experiment. Create. Perform.

11.11 SUMMARY

Different people throughout history believed that exploring, studying, classifying, and modeling the types of patterns and structures that emerge in nature holds the key to understanding the mystery of life and the universe we live in. The development of music and mathematics (and, eventually, computer science) was fueled by this human need to explain nature and the observable phenomena that surround us. The mathematical concepts of the golden ratio, Fibonacci numbers, Zipf's law, cymatics, fractals, and boids are all part of the same theme that interweaves music, nature, and numbers.

In this book, we have underscored the connections among music, number, and nature. We also explored how to represent useful mathematical knowledge and processes in Python. It is interesting to realize that programming is also a representation. What sets programming apart is that it represents not only knowledge (i.e., data) but also ways to act on that knowledge (i.e., process these data with algorithms). Also, and most importantly, this programmed representation is "runnable" on everyday

---

* It is relatively easy to create a loopable audio sample. Start with a sound you like. Using an audio editor (like Audacity), split the audio sample in half. Switch the two halves (that is, put the second half at the beginning, and the first half at the end), making sure they overlap a little. Using cross-fade mix the two switched halves into one sample. Now the end and the beginning of the new sample match perfectly. An audio loop is born!

† This way, you could engage your audience in a musical performance/happening.

computing devices. To the best of our knowledge, this is the first time in the history of our species that the everyday person (e.g., you and your friends) has access to such powerful (magical?) representations (such as Python) *and* also access to cheap devices that can execute Python programs.

Our hope is that you have been inspired by this intersection of music, mathematics, and nature to hopefully continue this exploration on your own. And remember, the universe is a mysterious and magical place, which you may explore through music, numbers, and computing. Ars longa, vita brevis ...

References

Aristotle (1992). *Complete Works*, vol. 10. "Metaphysics I," Hatzopoulos, O. (ed.). Kaktos, Athens, Greece (in Greek).

Arnheim, R. (1971). *Entropy and Art: An Essay on Disorder and Order*. London: University of California Press.

Ashton, A. (2003). *Harmonograph: A Visual Guide to the Mathematics of Music*. New York: Wooden Books.

Bak, P. (1996). *How Nature Works: The Science of Self-Organized Criticality*. New York: Springer-Verlag.

Beer, M. (2008). "Mathematics and music: Relating science to arts?" *Mathematical Spectrum*, 41(1): 36–42.

Berendt, J.-E. (1991). "Before we make music, the music makes us." In *The World Is Sound*, chap. 4, pp. 57–75. Rochester: Destiny Books.

Calter, P.A. (2008). *Squaring the Circle: Geometry in Art and Architecture*. New York: John Wiley & Sons.

Capra, F. (1991). "Foreword." In J.-E. Berendt, *The World Is Sound*, pp. xi–xiii. Rochester: Destiny Books.

Charton, É. and Cazeaux, E., eds. (1874). Magasin Pittoresque.

Chomsky, N. (1957). *Syntactic Structures*. The Hague, Holland Mouton.

Clark, M. A. (2005). "Genetic Music: An Annotated Source List." Published online November 2, 2005, http://www.whozoo.org/mac/Music/Sources.htm.

Cook, N. (1990). *Music, Imagination and Culture*. Oxford: Oxford University Press.

Cooke, R. L. (2005). *The History of Mathematics: A Brief Course*, 2nd ed. Hoboken, NS: Wiley.

Cope, D. (2004). *Virtual Music: Computer Synthesis of Musical Style*. Cambridge, MIT Press.

Cope. D. (2013). "Experiments in Musical Intelligence." Accessed online—http://artsites.ucsc.edu/faculty/cope/experiments.htm.

Eco, U. (1988). "The aesthetics of proportion." In *Art and Beauty in the Middle Ages*, chap. 3, pp. 28–42. New Haven, CT: Yale University Press.

Gardner, H. (1985). *The Mind's New Science: A History of the Cognitive Revolution*. New York: Basic Books.

Garland, T. H. and Kahn, C. V. (1995). *Math and Music: Harmonious Connections*. Palo Alto: Dale Seymour Publications.

Grant, E. (2009). "Making Sound Visible through Cymatics." *TED Talk* [Video file]. Retrieved from https://www.ted.com/talks/lang/en/evan_grant_cymatics.html.

Greenberg, S. et al. (2011). *Sketching User Experiences: The Workbook*. Morgan Kaufmann.

Hemenway, P. (2005). *Divine Proportion: Φ (Phi) in Art, Nature, and Science*. Sterling Publishing.

Hofstadter, D. (1979). *Gödel, Escher, Bach: An Eternal Golden Braid*. New York: Basic Books.

James, J. (1995). *The Music of the Spheres: Music, Science, and the Natural Order of the Universe*. Copernicus.

Kepler, J. (1619). *Harmonices Mundi* (trans. by Charles Wallis, "Harmonies of the World," Hong Kong: Forgotten Books, 2008).

Levine, M. (1995). *The Jazz Theory Book*. Petaluma, CA: Sher Music Co.

Livio, M. (2002). *The Golden Ratio*. New York: Broadway Books.

Magnusson, T. (2010). "Designing constraints: Composing and performing with digital musical systems." *Computer Music Journal*, 34(4): 62–73. MIT Press, Winter.

Manaris, B. et al. (2005). "Zipf's law, music classification and aesthetics." *Computer Music Journal*, 29(1): 55–69. MIT Press, Spring.

Manaris, B. et al. (2006). "Investigating Esperanto's statistical proportions relative to other languages using neural networks and Zipf's law," *Proceedings of the 2006 IASTED International Conference on Artificial Intelligence and Applications (AIA 2006)*. Innsbruck, Austria, February.

Manaris, B. et al. (2007). "A corpus-based hybrid approach to music analysis and composition." *Proceedings of 22nd Conference on Artificial Intelligence (AAAI-07)*, Vancouver, BC, pp. 839–845, July.

Mandelbrot, B. B. (1982). *The Fractal Geometry of Nature*. New York: W. H. Freeman and Company.

Martineau, J. (2002). *A Little Book of Coincidence*. New York: Wooden Books.

May, M. (1996). "Did Mozart Use the Golden Section?" *American Scientist*, 84(1): 118–119.

Miranda, E. R. (2001). *Composing Music with Computers*. Oxford: Focal Press.

Moggridge, B. (2007). *Designing Interactions*. Cambridge: The MIT Press, pp. 704–706.

Nielsen, J. (2000). Why You Only Need to Test with 5 Users. Nielsen Norman Group. Retrieved from http://www.nngroup.com/articles/why-you-only-need-to-test-with-5-users/

Pickover, C. A. (1991). *Computers and the Imagination*. New York: St. Martin's Press.

Pickover, C. A. (2002). *The Zen of Magic Squares, Circles, and Stars: An Exhibition of Surprising Structures across Dimensions*. Princeton NS: Princeton University Press.

Rushkoff, D. (2010). *Program or Be Programmed—Ten Commands for a Digital Age*. New York: OR Books.

Reynolds, C.W. (1987). Flocks, Herds and Schools: A distributed behavioral model. *Computer Graphics*, 21(4), 25–34.

Scaletti, C. (1993). "Sonification—an ancient idea made feasible by new technology." *ACM SIGRAPH '93—Course Notes 81*, August 1993, p. 4.2.

Schwanauer, S. and Levitt, D. (1993). *Machine Models of Music*. Cambridge, MA: MIT Press.

Scruton, R. (1983). Understanding Music. *Ratio*, 25, 97–120.

Smith, T. A. (1996). "Fourteen Canons on the First Eight Notes of the Goldberg Ground (BWV 1087)." Accessed online—http://www2.nau.edu/tas3/fourteencanonsgg.html.

Schroeder, M. (1991). *Fractals, Chaos, Power Laws: Minutes from an Infinite Paradise*. New York: W. H. Freeman.

Stone, D. et al. (2005). *User Interface Design and Evaluation*. Morgan Kaufmann.

Surmani, A., Surmani, K. F., and Manus, M (1999). *Alfred's Essentials of Music Theory*. Van Nuys, CA: Alfred Publishing Co.

Takahashi, R. and Miller, J. H. (2007). "Conversion of Amino-Acid Sequence in Proteins to Classical Music: Search for Auditory Patterns." *Genome Biology*, 8(5): 405, published online May 3, 2007, http://www.ncbi.nlm.nih.gov/pmc/articles/PMC1929127.

Valle, A. et al. (2010). "In a concrete space—reconstructing the spatialization of Iannis Xenakis' Concret PH on a multichannel setup." *Proceedings of the Sound and Music Computing Conference (SMC-2010)*. Barcelona, Spain, 21–24 July.

Vallianatos, E. (2012). "Deciphering and appeasing the heavens: The history and fate of an ancient Greek computer." *Leonardo*, 45(3): 250–257.

Watkins, M. (2001). *Useful Mathematical and Physical Formulae*. Virginia: Walker & Company.

Xenakis, I. (1971). *Formalized Music: Thought and Mathematics in Composition*. Indiana University Press.

Zelle, J. M. (2004). *Python Programming: An Introduction to Computer Science*. Franklin, Beedle & Associates, Inc.

Zbikowski, L. M. (2005). *Conceptualizing Music: Cognitive Structure, Theory, and Analysis*. Oxford, UK: Oxford University Press.

Zipf, G. K. (1949). *Human Behavior and the Principle of Least Effort: An Introduction to Human Ecology*. Eastford, CT: Martino Fine Books (2012 reprint of 1949 edition).

Appendix A: MIDI Constants

A.1 OVERVIEW

To simplify your coding, the **Python Music Library** defines many musical constants for you. These constants (words and abbreviations) can be used in place of numbers in your code. In many cases, each value is represented by several alternate constants (words). For example, the rhythm value 1.0 is equivalent to the constants QUARTER_NOTE and QN.

Notice that most of the constants are in upper case, which makes them easier to recognize in Python code. Python is a case-sensitive language, which means that these constants must be written using the case that appears here. Their numerical values are shown here, but you can also find them easily by typing the constants in the interpreter.

You should use these constants whenever possible instead of the actual (literal) values, for two reasons[*]:

1. To improve the readability of your programs

2. To ensure that your programs continue working as intended, in case some of these values change in subsequent releases of the music library (unlikely, but possible).

This is an important rule to remember.

A.2 PITCH CONSTANTS

Note pitches are represented as in the MIDI specification, using integers from 0 (lowest pitch) to 127 (highest pitch). That's a total of 128 pitches

---

[*] These reasons are true for all programs and constants, regardless of whether these constants come from the music library or elsewhere.

(i.e., 10 octaves). There is some disagreement on how to number these octaves. A common way is to number the highest octave as 9, the second to lowest octave as 0, and the lowest octave as −1.

The music library defines the following pitch constants for convenience. For example, the highest possible pitch is G9; it is equal to 127. The lowest possible note is C_1 (C negative one); it is equal to 0. Use these constants whenever possible. They improve the readability of the code.

Additionally, there is a special pitch corresponding to rests. A rest is a note whose pitch is:

```
REST
```

Rests are notes that last as long as you specify (see rhythm value constants), but produce no sound.

Here are the regular pitch constants in decreasing pitch value:

```
G9 = 127, GF9 = 126, FS9 = 126, F9 = 125, FF9 = 124, ES9 = 125, E9 =
124, EF9 = 123, DS9 = 123, D9 = 122, DF9 = 121, CS9 = 121, C9 = 120,
CF9 = 119, BS8 = 120, B8 = 119, BF8 = 118, AS8 = 118, A8 = 117, AF8 =
116, GS8 = 116, G8 = 115, GF8 = 114, FS8 = 114, F8 = 113, FF8 = 112,
ES8 = 113, E8 = 112, EF8 = 111, DS8 = 111, D8 = 110, DF8 = 109, CS8 =
109, C8 = 108, CF8 = 107, BS7 = 108, B7 = 107, BF7 = 106, AS7 = 106,
A7 = 105, AF7 = 104, GS7 = 104, G7 = 103, GF7 = 102, FS7 = 102, F7 =
101, FF7 = 100, ES7 = 101, E7 = 100, EF7 = 99, DS7 = 99, D7 = 98, DF7
= 97, CS7 = 97, C7 = 96, CF7 = 95, BS6 = 96, B6 = 95, BF6 = 94, AS6 =
94, A6 = 93, AF6 = 92, GS6 = 92, G6 = 91, GF6 = 90, FS6 = 90, F6 = 89,
FF6 = 88, ES6 = 89, E6 = 88, EF6 = 87, DS6 = 87, D6 = 86, DF6 = 85,
CS6 = 85, C6 = 84, CF6 = 83, BS5 = 84, B5 = 83, BF5 = 82, AS5 = 82, A5
= 81, AF5 = 80, GS5 = 80, G5 = 79, GF5 = 78, FS5 = 78, F5 = 77, FF5 =
76, ES5 = 77, E5 = 76, EF5 = 75, DS5 = 75, D5 = 74, DF5 = 73, CS5 =
73, C5 = 72, CF5 = 71, BS4 = 72, B4 = 71, BF4 = 70, AS4 = 70, A4 = 69,
AF4 = 68, GS4 = 68, G4 = 67, GF4 = 66, FS4 = 66, F4 = 65, FF4 = 64,
ES4 = 65, E4 = 64, EF4 = 63, DS4 = 63, D4 = 62, DF4 = 61, CS4 = 61, C4
= 60, CF4 = 59, BS3 = 60, B3 = 59, BF3 = 58, AS3 = 58, A3 = 57, AF3 =
56, GS3 = 56, G3 = 55, GF3 = 54, FS3 = 54, F3 = 53, FF3 = 52, ES3 =
53, E3 = 52, EF3 = 51, DS3 = 51, D3 = 50, DF3 = 49, CS3 = 49, C3 = 48,
CF3 = 47, BS2 = 48, B2 = 47, BF2 = 46, AS2 = 46, A2 = 45, AF2 = 44,
GS2 = 44, G2 = 43, GF2 = 42, FS2 = 42, F2 = 41, FF2 = 40, ES2 = 41, E2
= 40, EF2 = 39, DS2 = 39, D2 = 38, DF2 = 37, CS2 = 37, C2 = 36, CF2 =
35, BS1 = 36, B1 = 35, BF1 = 34, AS1 = 34, A1 = 33, AF1 = 32, GS1 =
32, G1 = 31, GF1 = 30, FS1 = 30, F1 = 29, FF1 = 28, ES1 = 29, E1 = 28,
EF1 = 27, DS1 = 27, D1 = 26, DF1 = 25, CS1 = 25, C1 = 24, CF1 = 23,
BS0 = 24, B0 = 23, BF0 = 22, AS0 = 22, A0 = 21, AF0 = 20, GS0 = 20, G0
= 19, GF0 = 18, FS0 = 18, F0 = 17, FF0 = 16, ES0 = 17, E0 = 16, EF0 =
15, DS0 = 15, D0 = 14, DF0 = 13, CS0 = 13, C0 = 12, CF0 = 11, BS_1 =
12, B_1 = 11, BF_1 = 10, AS_1 = 10, A_1 = 9, AF_1 = 8, GS_1 = 8, G_1 =
7, GF_1 = 6, FS_1 = 6, F_1 = 5, FF_1 = 4, ES_1 = 5, E_1 = 4, EF_1 = 3,
DS_1 = 3, D_1 = 2, DF_1 = 1, CS_1 = 1, C_1 = 0
```

A.3 RHYTHM VALUE CONSTANTS

Note durations are represented using real numbers (floats). For example, 4.0 stands for a whole note, 2.0 for a half note, 1.0 for a quarter note, and so on. The music library defines constants for common duration values.

Here are the common durations in decreasing duration value:

```
WHOLE_NOTE, WN = 4.0
DOTTED_HALF_NOTE, DHN = 3.0
DOUBLE_DOTTED_HALF_NOTE, DDHN = 3.5
HALF_NOTE, HN = 2.0
HALF_NOTE_TRIPLET, HNT = 4.0/3.0
QUARTER_NOTE, QN = 1.0
QUARTER_NOTE_TRIPLET, QNT = 2.0/3.0
DOTTED_QUARTER_NOTE, DQN = 1.5
DOUBLE_DOTTED_QUARTER_NOTE, DDQN = 1.75
EIGHTH_NOTE, EN = 0.5
DOTTED_EIGHTH_NOTE, DEN = 0.75
EIGHTH_NOTE_TRIPLET, ENT = 1.0/3.0
DOUBLE_DOTTED_EIGHTH_NOTE, DDEN = 0.875
SIXTEENTH_NOTE, SN = 0.25
DOTTED_SIXTEENTH_NOTE, DSN = 0.375
SIXTEENTH_NOTE_TRIPLET, SNT = 1.0/6.0
THIRTYSECOND_NOTE, TN = 0.125
THIRTYSECOND_NOTE_TRIPLET, TNT = 1.0/12.0
```

A.4 DYNAMIC CONSTANTS

Dynamic or volume of notes (also known as MIDI velocity) is represented as in the MIDI specification using integers from 0 (silent) to 127 (loudest). That's a total of 128 volume settings. The music library defines the following dynamic constants for convenience.

Here are the dynamic constants in decreasing loudness value:

```
FFF = 120
FORTISSIMO, FF = 100
FORTE, F = 85
MEZZO_FORTE, MF = 70
MEZZO_PIANO, MP = 60
P = 50
PIANISSIMO, PP = 25
PPP = 10
SILENT = 0
```

A.5 PANNING CONSTANTS

Panning (or panoramic) constants are represented in the music library using real numbers (floats) from 0.0 (left) to 1.0 (right). Given that human ears are very sensitive to placement of sound in a stereo field, using floats allows for much greater detail than the following constants allow:

Here are the panning constants in increasing value:

```
PAN_LEFT = 0.0
PAN_CENTER = 0.5
PAN_RIGHT = 1.0
```

A.6 GENERAL MIDI INSTRUMENT CONSTANTS

Instruments (timbre options) (also known as MIDI program changes) are represented as in the MIDI specification using integers from 0 to 127. That's a total of 128 timbre settings. The music library defines the following instrument constants for convenience.

Below are the instrument constants in increasing value (and grouped in terms of instrument family).

A.6.1 Piano Family

```
ACOUSTIC_GRAND, PIANO = 0
BRIGHT_ACOUSTIC = 1
ELECTRIC_GRAND = 2
HONKYTONK_PIANO, HONKYTONK = 3
EPIANO1, RHODES_PIANO, RHODES = 4
EPIANO2, DX_PIANO, DX = 5
HARPSICHORD = 6
CLAVINET = 7
```

A.6.2 Pitched Percussion Family

```
CELESTA = 8
GLOCKENSPIEL = 9
MUSIC_BOX = 10
VIBRAPHONE, VIBES = 11
MARIMBA = 12
XYLOPHONE = 13
TUBULAR_BELLS = 14
DULCIMER = 15
```

A.6.3 Organ Family

```
DRAWBAR_ORGAN, ORGAN = 16
PERCUSSIVE_ORGAN, JAZZ_ORGAN = 17
ROCK_ORGAN = 18
CHURCH_ORGAN = 19
REED_ORGAN = 20
ACCORDION = 21
HARMONICA = 22
TANGO_ACCORDION, BANDONEON = 23
```

A.6.4 Guitar Family

```
NYLON_GUITAR, GUITAR = 24
STEEL_GUITAR = 25
```

```
JAZZ_GUITAR = 26
CLEAN_GUITAR, ELECTRIC_GUITAR = 27
MUTED_GUITAR = 28
OVERDRIVEN_GUITAR = 29
DISTORTION_GUITAR = 30
GUITAR_HARMONICS = 31
```

A.6.5 Bass Family

```
ACOUSTIC_BASS = 32
BASS, ELECTRIC_BASS, FINGERED_BASS = 33
PICKED_BASS = 34
FRETLESS_BASS = 35
SLAP_BASS1 = 36
SLAP_BASS2 = 37
SYNTH_BASS1 = 38
SYNTH_BASS2 = 39
```

A.6.6 Strings and Timpani Family

```
VIOLIN = 40
VIOLA = 41
CELLO = 42
CONTRABASS = 43
TREMOLO_STRINGS = 44
PIZZICATO_STRINGS = 45
ORCHESTRAL_HARP, HARP = 46
TIMPANI = 47
```

A.6.7 Ensemble Family

```
STRING_ENSEMBLE1, STRINGS = 48
STRING_ENSEMBLE2 = 49
SYNTH_STRINGS1, SYNTH = 50
SYNTH_STRINGS2 = 51
CHOIR_AHHS, CHOIR = 52
VOICE_OOHS, VOICE = 53
SYNTH_VOICE, VOX = 54
ORCHESTRA_HIT = 55
```

A.6.8 Brass Family

```
TRUMPET = 56
TROMBONE = 57
TUBA = 58
MUTED_TRUMPET = 59
FRENCH_HORN, HORN = 60
BRASS_SECTION, BRASS = 61
SYNTH_BRASS1 = 62
SYNTH_BRASS2 = 62
```

A.6.9 Reed Family

```
SOPRANO_SAX, SOPRANO_SAXOPHONE = 64
ALTO_SAX, ALTO_SAXOPHONE = 65
TENOR_SAX, TENOR_SAXOPHONE, SAX, SAXOPHONE = 66
```

```
BARITONE_SAX, BARITONE_SAXOPHONE = 67
OBOE = 68
ENGLISH_HORN = 69
BASSOON = 70
CLARINET = 71
```

A.6.10 Pipe Family

```
PICCOLO = 72
FLUTE = 73
RECORDER = 74
PAN_FLUTE = 75
BLOWN_BOTTLE, BOTTLE = 76
SHAKUHACHI = 77
WHISTLE = 78
OCARINA = 79
```

A.6.11 Synth Lead Family

```
LEAD_1_SQUARE, SQUARE = 80
LEAD_2_SAWTOOTH, SAWTOOTH = 81
LEAD_3_CALLIOPE, CALLIOPE = 82
LEAD_4_CHIFF, CHIFF = 83
LEAD_5_CHARANG, CHARANG = 84
LEAD_6_VOICE, SOLO_VOX = 85
LEAD_7_FIFTHS, FIFTHS = 86
LEAD_8_BASS_LEAD, BASS_LEAD = 87
```

A.6.12 Synth Pad Family

```
PAD_1_NEW_AGE, NEW_AGE = 88
PAD_2_WARM, WARM_PAD = 89
PAD_3_POLYSYNTH, POLYSYNTH = 90
PAD_4_CHOIR, SPACE_VOICE = 91
PAD_5_GLASS, BOWED_GLASS = 92
PAD_6_METTALIC, METALLIC = 93
PAD_7_HALO, HALO = 94
PAD_8_SWEEP, SWEEP = 95
```

A.6.13 Synth Effects Family

```
FX_1_RAIN, ICE_RAIN = 96
FX_2_SOUNDTRACK, SOUNDTRACK = 97
FX_3_CRYSTAL, CRYSTAL = 98
FX_4_ATMOSPHERE, ATMOSPHERE = 99
FX_5_BRIGHTNESS, BRIGHTNESS = 100
FX_6_GOBLINS, GOBLINS = 101
FX_7_ECHOES, ECHO_DROPS = 102
FX_8_SCI_FI, SCI_FI = 103
```

A.6.14 Ethnic Family

```
SITAR = 104
BANJO = 105
SHAMISEN = 106
KOTO = 107
```

```
KALIMBA = 108
BAGPIPE = 109
FIDDLE = 110
SHANNAI = 111
```

A.6.15 Percussive Family

```
TINKLE_BELL, BELL = 112
AGOGO = 113
STEEL_DRUMS = 114
WOODBLOCK = 115
TAIKO_DRUM, TAIKO = 116
MELODIC_TOM, TOM_TOM = 117
SYNTH_DRUM = 118
REVERSE_CYMBAL = 119
```

A.6.16 Sound Effects Family

```
GUITAR_FRET_NOISE, FRET_NOISE = 120
BREATH_NOISE, BREATH = 121
SEASHORE, SEA = 122
BIRD_TWEET, BIRD = 123
TELEPHONE_RING, TELEPHONE = 124
HELICOPTER = 125
APPLAUSE = 126
GUNSHOT = 127
```

A.7 GENERAL MIDI DRUM AND PERCUSSION CONSTANTS

In addition to the above MIDI instruments, when using MIDI channel 9 (the MIDI protocol has 16 channels, numbered 0–15), each pitch corresponds to a different drum sound. The music library defines the following drum sound constants for convenience.

Here are the drum sound constants in increasing value:

```
ACOUSTIC_BASS_DRUM, ABD = 35
BASS_DRUM, BDR = 36
SIDE_STICK, STK = 37
SNARE, SNR = 38
HAND_CLAP, CLP = 39
ELECTRIC_SNARE, ESN = 40
LOW_FLOOR_TOM, LFT = 41
CLOSED_HI_HAT, CHH = 42
HIGH_FLOOR_TOM, HFT = 43
PEDAL_HI_HAT, PHH = 44
LOW_TOM, LTM = 45
OPEN_HI_HAT, OHH = 46
LOW_MID_TOM, LMT = 47
HI_MID_TOM, HMT = 48
CRASH_CYMBAL_1, CC1 = 49
HIGH_TOM, HGT = 50
RIDE_CYMBAL_1, RC1 = 51
CHINESE_CYMBAL, CCM = 52
```

```
RIDE_BELL, RBL = 53
TAMBOURINE, TMB = 54
SPLASH_CYMBAL, SCM = 55
COWBELL, CBL = 56
CRASH_CYMBAL_2, CC2 = 57
VIBRASLAP, VSP = 58
RIDE_CYMBAL_2, RC2 = 59
HI_BONGO, HBG = 60
LOW_BONGO, LBG = 61
MUTE_HI_CONGA, MHC = 62
OPEN_HI_CONGA, OHC = 63
LOW_CONGA, LCG = 64
HIGH_TIMBALE, HTI = 65
LOW_TIMBALE, LTI = 66
HIGH_AGOGO, HAG = 67
LOW_AGOGO, LAG = 68
CABASA, CBS = 69
MARACAS, MRC = 70
SHORT_WHISTLE, SWH = 71
LONG_WHISTLE, LWH = 72
SHORT_GUIRO, SGU = 73
LONG_GUIRO, LGU = 74
CLAVES, CLA = 75
HI_WOOD_BLOCK, HWB = 76
LOW_WOOD_BLOCK, LWB = 77
MUTE_CUICA, MCU = 78
OPEN_CUICA, OCU = 79
MUTE_TRIANGLE, MTR = 80
OPEN_TRIANGLE, OTR = 81
```

A.8 SCALE AND MODE CONSTANTS

Scales in the music library are defined as pitch class sets (i.e., chromatic scale degrees within one octave). The first pitch class sets is often 0, denoting the starting (root) note. There is wisdom in this representation. It allows us to use any MIDI note (pitch) as the starting note offset. For example, the following piece of code outputs the pitches for a major scale starting at C4:

```
root = C4        # starting pitch
scale = MAJOR_SCALE # scale
# output the pitches in this scale
for interval in scale:
        print root+interval # add the interval to the root
```

Change the C4 to another pitch, or change MAJOR_SCALE to another scale, and you will get the actual pitches of any possible scale. This demonstrates the power, flexibility, and economy of the music library's representation for scales.

Here are the music library scale definitions in alphabetical order:

```
AEOLIAN_SCALE = [0, 2, 3, 5, 7, 8, 10]
BLUES_SCALE = [0, 2, 3, 4, 5, 7, 9, 10, 11]
CHROMATIC_SCALE = [0, 1, 2, 3, 4, 5, 6, 7, 8, 9, 10, 11]
DIATONIC_MINOR_SCALE = [0, 2, 3, 5, 7, 8, 10]
DORIAN_SCALE = [0, 2, 3, 5, 7, 9, 10]
HARMONIC_MINOR_SCALE = [0, 2, 3, 5, 7, 8, 11]
LYDIAN_SCALE = [0, 2, 4, 6, 7, 9, 11]
MAJOR_SCALE = [0, 2, 4, 5, 7, 9, 11]
MELODIC_MINOR_SCALE = [0, 2, 3, 5, 7, 8, 9, 10, 11]
MINOR_SCALE = [0, 2, 3, 5, 7, 8, 10]
MIXOLYDIAN_SCALE = [0, 2, 4, 5, 7, 9, 10]
NATURAL_MINOR_SCALE = [0, 2, 3, 5, 7, 8, 10]
PENTATONIC_SCALE = [0, 2, 4, 7, 9]
TURKISH_SCALE = [0, 1, 3, 5, 7, 10, 11]
```

Appendix B: Music Library Functions

B.1 OVERVIEW

To simplify making music, the Python Music Library defines many musical functions for you. These functions are organized in terms of four fundamental classes (Note, Phrase, Part, Score) and a few additional classes (Mod, Read, Write, View). Each of these functions may have several flavors, giving you increasing control and expression over musical parameters. These functions are presented below in terms of class.

B.2 NOTE FUNCTIONS

Note objects contain the simplest possible musical events, consisting of pitch, duration, etc. The music library provides several functions to create musical notes. It also provides several others to retrieve or modify attributes (e.g., pitch) or existing notes. You can often choose to either modify an existing note or create a new one with the proper attributes. The decision depends on the context. Pick the option that appears more natural or efficient (to you, to the computer, or both).

Here are functions used to create musical notes. Each of these functions creates a new Note object, so you need to save it in a variable (or other memory location).

Function	Description
Note(pitch, duration)	Creates a new note, where pitch is 0–127, and duration is a float (e.g., 1.0 is a quarter note).
Note(pitch, duration, dynamic)	Creates a new note, where pitch is 0–127, duration is a float, and dynamic is 0–127.
Note(pitch, duration, dynamic, pan)	Creates a new note, where pitch is 0–127, duration is a float, dynamic is 0–127, and pan is 0.0 (left) to 1.0 (right).

Below are functions used to retrieve and modify attributes of existing notes. It is assumed that a note has already been created, for example, as follows:

```
n = Note(C4, HN)
```

Function	Description
n.getPitch()	Retrieves the pitch (0–127) of note n.
n.setPitch(pitch)	Sets the pitch (0–127) of note n.
n.getDuration()	Retrieves the duration (a float) of note n.
n.setDuration(duration)	Sets the duration (a float) of note n.
n.getDynamic()	Retrieves the dynamic (0–127) of note n.
n.setDynamic(dynamic)	Sets the dynamic (0–127) of note n.
n.getPan()	Retrieves the pan (0.0–1.0) of note n.
n.setPan(pan)	Sets the pan (0.0–1.0) of note n.
n.isRest()	Checks if note n is a rest or a note with pitch. Returns a Boolean value (True or False).
n.copy()	Returns a new note with the same attributes as note n. This is used to create a copy to be modified, while the original note is not affected.

Finally, here are some helper note functions:

Function	Description
Note.freqToMidiPitch(freq)	Converts freq (a float) in Hertz (e.g., 440.0) to the equivalent MIDI pitch (e.g., 69).
Note. midiPitchToFreq(pitch)	Converts pitch (0–127) from a MIDI value to the equivalent frequency in Hertz (a float).

B.3 PHRASE FUNCTIONS

Phrase objects contain a sequence of Note objects. These Note objects are played sequentially (i.e., one after the other). If a gap is desired between two notes, then a Rest note should be introduced. Phrases may also contain chords (i.e., sets of concurrent notes).

Phrases have start times. If no start time is specified, then the phrase starts at the end of the previous phrase (or at the beginning of the piece, if this is the first phrase). The music library provides several functions to create musical phrases. It also provides several others to retrieve or modify existing phrases.

Here are functions used to create phrases. Each of these functions creates a new Phrase object, so you need to store it in a variable (or other memory location) to be able to use it later.

Function	Description
Phrase()	Creates an empty phrase. This phrase starts at the end of the previous phrase (or at the beginning of the piece, if this is the first phrase).
Phrase(startTime)	Creates an empty phrase starting at specified time (e.g., 0.0 is beginning of piece, 1.0 is a quarter note into the piece, etc.).
Phrase(note)	Creates a phrase containing the specified note.

Below are functions used to retrieve and modify attributes of existing phrases. It is assumed that a phrase has already been created, for example, as follows:

```
phr = Phrase()
```

Function	Description
phr.addNote(note)	Appends the given note to the phrase.
phr.addNote(pitch, duration)	Appends a new note of given pitch (0–127) and duration (a float) to the phrase.
phr.addNoteList(listOfPitches, listOfDurations, listOfDynamics, listOfPanoramics)	Appends the notes specified in terms of pitches (a list), durations (a list), dynamics (a list), and panning values (a list) to the phrase. The lists are parallel. Dynamics and panoramics lists are optional.
phr.addChord(listOfPitches, duration, dynamics, panoramics)	Appends a chord containing the specified pitches (a list) and having the specified duration (a float), dynamics (0–127), and panoramics (0.0–1.0). Dynamics and panoramics values are optional.
phr.getEndTime()	Returns the phrase's end time (a float).
phr.getNoteList()	Returns the phrase's notes (a list).
phr.getSize()	Returns the number of notes in phrase phr.
phr.getStartTime()	Returns the phrase's start time (a float).
phr.setStartTime(startTime)	Sets the phrase's start time (a float).
phr.setDynamic(dynamic)	Sets the dynamic (0–127) of the phrase.
phr.copy()	Returns a new phrase with the same notes and attributes as phrase phr. This is used to create a copy to be modified, while the original phrase is not affected.
phr.empty()	Removes all notes from phrase phr.

B.4 PART FUNCTIONS

A Part object contains a set of Phrase objects to be played by a particular instrument. These phrase objects are played in parallel (i.e., simultaneously), and thus may overlap (according to their start times and durations). Even if the particular instrument does not allow for polyphony (e.g., a flute), a part using this instrument can have different simultaneous melodies. In other words, a part can be thought of as a group of several instruments of the same type (e.g., flutes), each playing a different melody (a phrase).

There are 128 different instruments to pick from (see Appendix A).

Parts may have a title (a string).

Parts may also be assigned to one of 16 MIDI channels (0–15) available on a standard computer's audio system. Each MIDI channel is capable of playing any of the 128 different instruments possible, but only one at a time. So it is important to keep parts using different instruments different MIDI channels. If two parts are using the same instrument, they may be assigned to the same MIDI channel.

It should be noted that MIDI channel 9 is reserved for percussion sounds. Regardless of a part's selected instrument, if that part is assigned to MIDI channel 9, its notes will generate percussion sounds, based on the notes' pitches (see Appendix A).

The music library provides several functions to create musical parts. It also provides several others to retrieve or modify existing parts.

Here are functions used to create parts. Each of these functions creates a new Part object, so you need to save it in a variable (or other memory location).

Function	Description
Part()	Creates an empty part.
Part(instrument)	Creates an empty part with the timbre of the specified instrument (0–127).
Part(instrument, channel)	Creates an empty part with the timbre of the specified instrument (0–127) and using the specified MIDI channel (0–15).
Part(title, instrument, channel)	Creates an empty part with the specified title (a string), with the timbre of the specified instrument (0–127), and using the specified MIDI channel (0–15).
Part(phrase)	Creates a part containing the specified phrase.

Below are functions used to retrieve and modify attributes of existing parts. It is assumed that a part has already been created, for example, as follows:

```
part = Part("An example flute part", FLUTE, 0)
```

This creates a part with a descriptive title, using instrument FLUTE (see Appendix A for a complete list of instruments), and assigned to MIDI channel 0 of the computer's audio system. Again, you should assign parts with different instruments to different MIDI channels (0–8, and 10–15). Remember that MIDI channel 9 is dedicated to percussive sounds, as explained above.

Function	Description
part.addPhrase(phrase)	Add a phrase to this part. If the phrase does not have a specific start time, it is added to the end of the part.
part.addPhraseList(listOfPhrases)	Adds the specified phrases (a list) to the part. If a phrase does not have a specific start time, it is added to the end of the part.
part.copy()	Returns a new part with the same phrases and attributes as part. This is used to create a copy to be modified, while the original part is not affected.
part.empty()	Removes all phrases from this part.
part.getChannel()	Returns the channel for this part.
part.getEndTime()	Returns the part's end time (a float).
part.getInstrument()	Returns the part's instrument number (MIDI program change).
part.getPhraseList()	Returns the part's phrases (a list).
part.getSize()	Returns the number of phrases in this part.
part.getTempo()	Returns the part's tempo (a float).
part.getTitle()	Returns the part's title (a string).
part.getVolume()	Returns the part's volume (0–127).
part.setChannel(channel)	Sets the MIDI channel for this part (0–15).
part.setInstrument(instrument)	Sets the instrument number (MIDI program change) for this part (0–127).
part.setPan(pan)	Sets the pan position for all notes in this part (0.0–1.0).
part.setTempo(tempo)	Sets the part's tempo (a float).
part.setTitle(title)	Gives the part a new title (a string).
part.setDynamic(dynamic)	Sets the part's dynamic (0–127).

B.5 SCORE FUNCTIONS

A Score object contains a set of Part objects. Score contents (parts, phrases, notes) are algorithmically generated or read from a standard MIDI file (see Read.midi()). Scores can be written to standard MIDI files (see Write. midi()).

The music library provides several functions to create musical scores. It also provides several others to retrieve or modify existing scores.

Here are functions used to create scores. Each of these functions creates a new Score object, so you need to save it in a variable (or other memory location).

Function	Description
Score()	Creates an empty score.
Score(title)	Creates an empty score with the specified title (a string).
Score(tempo)	Creates an empty score with the specified tempo (in beats-per-minute, e.g., 120.0).
Score(title, tempo)	Creates an empty score with the specified title (a string) and with the specified tempo (in beats-per-minute, e.g., 120.0).
Score(part)	Creates a score containing the specified part.

Below are functions used to retrieve and modify attributes of an existing Score object. It is assumed that a score has already been created, for example, as follows:

```
score = Score("Morning Glory", 135.0)
```

This creates a score with the descriptive title "Morning Glory," with a tempo of 135 beats per minute.

Function	Description
score.addPart(part)	Adds a part to this score.
score.addPartList(listOfParts)	Adds the specified parts (a list) to the score.
score.copy()	Returns a new score with the same parts and attributes as the score. This is used to create a copy to be modified, while the original score is not affected.
score.empty()	Removes all parts from this score.
score.getDenominator()	Returns the time signature denominator for this score.
score.getEndTime()	Returns the score's end time (a float).
score.getKeyQuality()	Returns the score's key quality (0 is major, 1 is minor).
score.getKeySignature()	Returns the score's key signature (as an integer). Zero (0) means the score is in the key of C. A positive (+) integer indicates the number of sharps, whereas a negative (–) integer indicates the number of flats.
score.getNumerator()	Returns the time signature numerator for this score.

Continued

Function	Description
`score.getPartList()`	Returns the score's parts (a list).
`score.getSize()`	Returns the number of parts in this score.
`score.getTempo()`	Returns the score's tempo (a float).
`score.getTitle()`	Returns the score's title (a string).
`score.getVolume()`	Returns the score's volume (0–127).
`score.setDenominator(denominator)`	Sets the time signature denominator for this score.
`score.setKeyQuality(quality)`	Sets the score's key quality (0 is major, 1 is minor).
`score.setKeySignature(signature)`	Sets the score's key signature (as an integer). Zero (0) means the score is in the key of C. A positive (+) integer indicates the number of sharps, whereas a negative (–) integer indicates the number of flats.
`score.setNumerator(numerator)`	Sets the time signature's numerator for this score.
`score.setPan(pan)`	Sets the pan position for all notes in this score (0.0–1.0).
`score.setTempo(tempo)`	Sets the score's tempo (in beats-per-minute, e.g., 120.0).
`score.setTimeSignature(num, den)`	Specifies the score's time signature (i.e., num/den).
`score.setTitle(title)`	Gives the part a new title (a string).

B.6 VIEW FUNCTIONS

There are a number of functions to help you visually display music in Python. These are part of the View class. Of these functions, notation(), sketch(), and pianoRoll() also allow adding new notes to the musical material being displayed. However, these notes cannot be saved back to the original program (that called the View functions), only to an external MIDI file.

Function	Description
`View.notation(material)`	Displays the phrase as staff notation, where material may be a Score, Part, or Phrase.
`View.internal(material)`	Prints out the music data to the screen, where material may be a Score, Part, or Phrase.
`View.sketch(material)`	Displays the music as a small piano-roll display, where material may be a Score, Part, or Phrase.
`View.pianoRoll(material)`	Displays the music as a piano-roll display, where material may be a Score, Part, or Phrase.

B.7 MOD FUNCTIONS

The Mod class contains many functions for modifying, or varying, phrases, parts, and scores. Each of these functions modifies the data passed to them. For example, the `Mod.repeat()` function creates repetitions of the given musical material. It is called like this:

```
Mod.repeat(phrase, 41)
```

This will modify phrase to contain a total of 41 copies of the original musical material.

Below are the available Mod functions. You may apply these functions repeatedly to achieve a desired effect.

Function	Description
`Mod.accent(material, beats)`	Accents the first beat of each measure in `material`. `Beats` (a float) is the number of beats per measure. Material may be a `Phrase`, `Part`, or `Score`.
`Mod.accent(material, meter, accentedBeats, accentAmount)`	Accents by `accentAmount` (an int) at the `accentedBeats` locations (a list of floats, denoting time in quarter notes). `Meter` (a float) is the number of beats per measure. `Material` may be a `Phrase`, `Part`, or `Score`.
`Mod.append(material1, material2)`	Appends `material2` to `material1`. Works with notes, phrases, parts, and scores. In the case of notes, it extends the duration value of the first note (without changing its pitch).
`Mod.bounce(material)`	Adjusts the pan values of all notes in material to alternate between extreme left and right from note to note. `Material` may be a `Phrase`, `Part`, or `Score`.
`Mod.changeLength(phrase, newLength)`	Alters the `phrase` so that its notes are stretched or compressed until the phrase is the specified length (number of beats—a float). It works in the same manner as `elongate()`, except it takes an absolute length parameter, not a ratio.
`Mod.compress(material, ratio)`	Compresses (or expands) the `material`. Compression ratio numbers are between 0 and 1. Values larger than 1 expand. Negative values invert the volume about the mean. This compression applies only to the volume of the notes. It will multiply the difference between each note's volume and the volume, by the compression `ratio`. Thus, a ratio of 0 will change every note's volume to the average volume; 1 will leave every note unchanged; and 2 will make every note's volume twice as far from the mean. Negative values will have a similar affect but leave the volume of each note on the other side of the mean. Material may be a `Phrase`, `Part`, or `Score`.

Continued

Function	Description
`Mod.consolidate(part)`	Merges `part`'s phrases into one phrase. This works well with `View.notate(part)`, so that `notate()` can display all the notes in the part (`View.notate()` works only with a single phrase).
`Mod.cycle(material, numberOfNotes)`	Repeats the `material` until it contains the specified number of notes. The repetitions work in the same manner as `Mod.repeat()`, except that the final repetition will not be a complete copy of the original if the note count is reached before the repetition is completed. `Material` may be a `Phrase`, `Part`, or `Score`.
`Mod.elongate(material, scaleFactor)`	Stretches the time of each note in the phrase by `scaleFactor` (a float). `Material` may be a `Phrase`, `Part`, or `Score`.
`Mod.fadeIn(material, fadeLength)`	Linearly fades in the `material` (fadeLength is quarter notes). `Material` may be a `Phrase`, `Part`, or `Score`.
`Mod.fadeOut(material, fadeLength)`	Linearly fades out the `material` (fadeLength is quarter notes). `Material` may be a `Phrase`, `Part`, or `Score`.
`Mod.fillRests(material)`	Lengthens notes followed by a rest in `material` by creating one longer note and deleting the rest. This will reduce the overall note count. Material may be a Phrase, Part, or Score.
`Mod.invert(material)`	Mirrors the pitch of notes in the material around the first note's pitch. The order of the notes is not affected; it is only the pitches that are mirrored. That is, notes which are n semitones above the first pitch will be changed to be n semitones below. `Material` may be a `Phrase`, `Part`, or `Score`.
`Mod.mutate(phrase)`	Mutate `phrase` by changing one pitch and one duration value. The new pitch is selected randomly between the lowest and the highest note of `phrase`. The random duration is selected from those in the existing notes.
`Mod.palindrome(material)`	Extends the phrase by adding all notes backwards, repeating the last note of the material. Material may be a Phrase, Part, or Score.
`Mod.randomize(material, pitchVariation, durationVariation, dynamicVariation)`	Randomly adjusts every note's pitch, duration, and dynamic values to a random value within the range plus or minus the specified amount; `pitchVariation` is an int (make sure result stays within 0–127); `durationVariation` is a float (e.g., 1.0 is a quarter note); `dynamicVariation` is an int (make sure result stays within 0–127). Material may be a `Phrase`, `Part`, or `Score`.

Continued

Function	Description
`Mod.repeat(material, number)`	Repeats the `material` a number of times. Material may be a `Phrase`, `Part`, or `Score`. For example, `Mod.repeat(phrase, 2)` will play phrase two times.
`Mod.retrograde(material)`	Reverses the order of notes in the `material`. Material may be a `Phrase`, `Part`, or `Score`.
`Mod.rotate(material)`	Moves the notes around the `material`, first becoming second, second becoming third, … and last becoming first. Material may be a `Phrase`, `Part`, or `Score`.
`Mod.shake(material)`	Randomly adjusts the volume of notes to create uneven loudness. `Material` may be a `Phrase`, `Part`, or `Score`.
`Mod.shake(material, amount)`	Randomly adjusts the volume of notes to create uneven loudness. Amount (an int) denotes how strong the effect will be, e.g., 5 will be +/–5 from the current volume. `Material` may be a `Phrase`, `Part`, or `Score`.
`Mod.shuffle(material)`	Randomizes the order of notes in the `material` without repeating any note. `Material` may be a `Phrase`, `Part`, or `Score`.
`Mod.spread(material)`	Randomly adjusts the pan values of all notes in `material` to create an even spread across the stereo spectrum. `Material` may be a `Phrase`, `Part`, or `Score`.
`Mod.tiePitches(material)`	Joins consecutive notes in `material` that have the same pitch, creating one longer note. This is similar to the musical function of a tie. This will reduce the overall note count. `Material` may be a `Phrase`, `Part`, or `Score`.
`Mod.tieRests(material)`	Joins consecutive rests in `material`, creating a longer note. This is similar to the musical function of a tie. This will reduce the overall note count. `Material` may be a `Phrase`, `Part`, or `Score`.
`Mod.transpose(material, semitones)`	Chromatic transposition. It shifts the pitch of every note in the material by the given number of semitones. Material may be a Phrase, Part, or Score.
`Mod.transpose(material, steps, scale, key)`	Diatonic transposition. It shifts the pitch of every note in the `material` by the given number of scale degrees (`steps`), given the `scale` and key. `Material` may be a `Phrase`, `Part`, or `Score`. Key is an integer (0 means the key of C, a positive integer indicates the number of sharps, whereas a negative integer indicates the number of flats). See Appendix A for a list of available scales.

Appendix C: GUI Library Functions

C.1 OVERVIEW

Graphical user interfaces (GUIs) may be developed in Python using the GUI library.[*] Unlike many other GUI libraries, this library keeps simple things simple and makes complicated things possible.

C.2 GUI DISPLAY

To build a GUI, you have to create at least one display (window).

- Display objects are application windows. They contain other GUI objects (widgets and graphics objects). A program may have several displays open. Displays may contain any number of GUI objects, but they cannot contain another display.

Once a display has been created, you populate it by placing various GUI widgets and graphics objects on it. The library provides various GUI widgets and graphics objects.

The origin (0, 0) of a display is at the top-left corner.

The following function creates a new Display, so you need to save it in a variable (in order to add GUI objects to it later).

Function	Description
`Display(title, width, height)`	Creates a display window with the specified `title` (string — default is blank), width (default is 600 pixels), and height (default is 400 pixels).
`Display(title, width, height, x, y, color)`	Same as above, but also initial x and y position on screen (default is (0, 0) at top-left) and background color (default is Color.WHITE).

---

[*] The GUI library is based on Java's Swing library. It provides a clean, simpler API to use for building graphical user interfaces. For advanced users, existing Swing functionality is also available, but not advertised. Here, as in the rest of the book, the target audience is beginning programmers.

For example, a display may be created as follows:

```
d = Display("Simple GUI", 120, 60)
```

Once a display has been created, you can add GUI widgets and other graphical objects, using the following function:

```
d.add(object, x, y)
```

where object is a GUI widget or graphical object (presented below). The coordinates x, y are optional and specify where in the display to place the object. If omitted, e.g.,

```
d.add(object)
```

the object is placed using its own coordinates (e.g., a line or a circle) specified when it was created (more below).

Once a display has been created, the following functions are available:

Function	Description
d.show()	Shows display d. This happens automatically when a new display is created.
d.hide()	Hides display d.
d.add(object, x, y)	Adds a GUI object on display d, at coordinates (x, y). It aligns the object's top-left corner (for Circle, its center) with these coordinates. The origin (0, 0) is at the display's top-left corner. Any GUI object may be placed, except another display. A GUI object may appear only on one display. Placing a GUI object on another display removes it from the first display.
d.place(object, x, y)	Same as add().
d.move(object, x, y)	Moves a GUI object to the new coordinates (x, y).
d.reposition(object, x, y)	Same as move().
d.remove(object)	Removes a GUI object from display d.
d.removeAll()	Removes all GUI objects from display d. Provides a convenient way to clear a display.
d.setToolTipText(text)	Sets the tooltip of display d to the provided text.
d.setColor(color)	Changes the background color of display d (e.g., Color.RED). If no color is provided, use the dialog box to select.

Continued

Function	Description
`d.getColor()`	Returns the current background color of display d.
`d.setSize(width, height)`	Sets the `width` and `height` of display d.
`d.getHeight()`	Returns the height of display d.
`d.getWidth()`	Returns the width of display d.
`d.setPosition(x, y)`	Sets the position of display d on the screen, where (0, 0) is at top-left.
`d.getPosition()`	Returns the position of display d on the screen.
`d.showMouseCoordinates()`	Shows mouse coordinates using the tooltip of display d. This is useful for discovering coordinates to place widgets at GUI design/ exploration time.
`d.hideMouseCoordinates()`	Stops showing mouse coordinates using the tooltip of display d.

C.2.1 Drawing on Display

Display objects support drawing of various graphics objects. This is done with the following functions.

Function	Description
`d.drawLine(x1, y1, x2, y2)`	Draw a Line on display d between points `(x1, y1)` and `(x2, y2)`. Additional optional parameters (in this order) include `color` (default is `Color.BLACK`) and `thickness` (default is 1).
`d.drawCircle(x, y, radius)`	Draw a Circle on display d with center at `(x, y)` and with the specified `radius`. Additional optional parameters (in this order) include `color` (default is `Color.BLACK`), `fill` (default is `False`), and `thickness` (default is 1).
`d.drawPoint(x, y)`	Draws a Point on display d with center at `(x, y)`. Additional optional parameter is `color` (default is `Color.BLACK`).
`d.drawOval(x1, y1, x2, y2)`	Draws an Oval on display d using the coordinates of its enclosing rectangle, `(x1, y1)` and `(x2, y2)`. Additional optional parameters (in this order) include `color` (default is `Color.BLACK`), `fill` (default is `False`), and `thickness` (default is 1).
`d.drawRectangle(x1, y1, x2, y2)`	Draws a Rectangle on display d using the provided coordinates, `(x1, y1)` and `(x2, y2)`. Additional optional parameters (in this order) include `color` (default is `Color.BLACK`), `fill` (default is `False`), and `thickness` (default is 1).

Continued

Function	Description
d.drawPolygon(xPoints, yPoints)	Draws a Polygon on display d using the provided coordinates, list of x points, and list of y points (parallel lists). Additional optional parameters (in this order) include color (default is Color.BLACK), fill (default is False), and thickness (default is 1).
d.drawIcon(filename, x, y)	Draws an Icon (image) on display d the image from the provided external file (a string, ending in ".jpg" or ".png") at the given coordinates (top-left). Additional optional parameters (in this order) include width and height — if provided, image is rescaled accordingly.
d.drawImage(filename, x, y)	Same as drawIcon().
d.drawLabel(text, x, y)	Draws a Label on display d containing text (a string) at the provided coordinates. Additional optional parameters (in this order) include color (default is Color.BLACK) and font (e.g., Font("Serif", Font.ITALIC, 16)).
d.drawText(text, x, y)	Same as drawLabel().

For convenience, each of the above functions returns the corresponding graphics object (e.g., drawLabel() returns a Label), which can be ignored or saved for further interaction, such as animation.

C.3 GUI WIDGETS

Widgets are used to present information and receive user input. The following widgets are available:

- Label objects present textual information.

- Button objects can be pressed by the user to perform an action.

- Checkbox objects can be selected (or deselected) by the user.

- Slider objects can be adjusted by the user to input a value.

- DropDownList objects contain items which can be selected by the user.

- TextBox objects allow the user to enter a single line of text.

- TextArea objects allow the user to enter multiple lines of text.

- Icon objects allow displaying of external images (.jpg or .png).

- Menu objects contain items which can be selected by the user. Menu objects are fixed at the menu bar (top), whereas DropDownList objects are placed anywhere on a display.

Below we present each of these objects in more detail.

C.3.1 Label

Label objects are used to present labels and other permanent text on displays.

The following function creates a new Label, so you need to save it in a variable (so you can use it later).

Function	Description
Label(text, alignment)	Creates a new label with specified text (string) and alignment (LEFT, CENTER, or RIGHT — default is LEFT).

For example, a label may be created as follows:

```
label1 = Label("Hello World!")
```

Once a label has been created, it cannot be resized. So you should create it with the widest (possibly blank) string necessary, e.g.,

```
label1 = Label("               ") # up to 16 characters
```

Once a label has been created, it may be added to a Display, specifying where to place its top-left corner.

```
d.add(label1, 50, 50)
```

Also, the following functions are available:

Function	Description
label1.setText(text)	Updates the contents of label1 to text (a string).
label1.getText()	Returns the text contained in label1 (as a string).
label1.setFont()	Changes the font used in label1, e.g., Font("Dialog", Font.PLAIN, 12) or Font("Serif", Font.ITALIC, 16).

C.3.2 Button

Button objects can be pressed by the user.

Pressing a Button calls a function. Which function to call is specified when the button is created.

The following function creates a new Button, so you need to save it in a variable (so you can use it later).

Function	Description
`Button(text, function)`	Creates a new button containing `text` (a string). Every time the button is pressed, `function` is called automatically. This function should expect **zero** parameters.

For example, a button may be created as follows:

```
button1 = Button("Play music", playMusic)
```

where playMusic is a function with zero parameters. This function will be called automatically when the user presses this button.

Once a Button has been created, it may be added to a Display, specifying where to place its top-left corner.

```
d.add(button1, 50, 50)
```

C.3.3 Checkbox

Checkbox objects can clicked (i.e., selected or deselected) by the user.

The following function creates a new Checkbox, so you need to save each in a unique variable (so you can use them later).

Function	Description
`Checkbox(text, function)`	Creates a new checkbox with the specified `text` label (a string) and a `function` (optional) to be called every time the checkbox changes state. If provided, this function should expect **one** parameter (Boolean—signifies the changed state of the checkbox—True means checkbox was just checked, False means checkbox was just unchecked).

For example, a checkbox may be created as follows:

```
checkbox1 = Checkbox()
```

Once a Checkbox has been created, it may be added to a Display, specifying where to place its top-left corner.

```
d.add(checkbox1, 50, 50)
```

If you create a Checkbox without using a callback function, then it is a passive GUI element. In other words, a different part of your program needs to check the state (selected, deselected) of the Checkbox. This can be done using the following functions:

Function	Description
`checkbox1.isChecked()`	Returns `True` if `checkbox1` is checked, `False` if unchecked.
`checkbox1.check()`	Sets `checkbox1` (i.e., makes it appear checked).
`checkbox1.uncheck()`	Clears `checkbox1` (i.e., makes it appear unchecked).

If you create a Checkbox with a callback function, then that function will be called any time the changes state (checked, unchecked) by the user. The function should accept one parameter.

C.3.4 Slider

Slider objects contain an indicator which can be moved by the user to set a value.

The function below creates a new Slider, so you need to save it in a variable (so you can use it later).

Function	Description
`Slider(orientation, lower, upper, start, function)`	Creates a new `Slider` with `orientation` (`HORIZONTAL` or `VERTICAL`—default is `HORIZONTAL`), `lower` value (integer—default is 0), `upper` value (integer—default is 100), placing the indicator at `start` value (integer—default is half-way). When the indicator is moved, `function` (optional) is called automatically. If provided, this function should expect **one** parameter (i.e., the new value of the slider).

For example, a slider may be created as follows:

```
slider1 = Slider(VERTICAL, 0, 127, 50, changeVolume)
```

where `changeVolume` is a function which expects one parameter, the new value of the slider. When the function is called, it may use this value to update the volume of some musical material, for instance.

Once a Slider has been created, it may be added to a Display, specifying where to place its top-left corner.

```
d.add(slider1, 50, 50)
```

Additionally, you may use the following functions to get its current value:

Function	Description
`slider1.getValue()`	Returns the current value of the slider (an integer between `lower` and `upper`).
`slider1.setValue(value)`	Sets the current value of the slider to `value` (an integer between `lower` and `upper`).

C.3.5 DropDownList

DropDownList objects contain items which can be selected by the user.[*]

The following function creates a DropDownList, so you need to save it in a variable (so you can use it later).

Function	Description
`DropDownList(items, function)`	Creates a drop-down list containing the provided items (list of strings, e.g. `["item1", "item2", "item3"]`). When an item gets selected, `function` (optional) is called automatically. If provided, the function should expect **one** parameter (string—the selected item).

For example, a drop-down list may be created as follows:

```
 dd1 = DropDownList(["item1", "item2", "item3"], itemSelected)
```

where `itemSelected` is a function which expects one parameter, the selected item (a string).

Once a DropDownList has been created, it may be added to a Display, specifying where to place its top-left corner.

```
d.add(dd1, 50, 50)
```

C.3.6 TextField

TextField objects are used for entering text on a user interface.

The following function creates a TextField object, so you need to store it in a variable (or other memory location) to be able to use it later.

---

[*] Whereas `Menu` objects (seen later in the appendix) are fixed at a display's menu bar (top), `DropDownList` objects can be placed anywhere on a display.

Function	Description
TextField(text, columns, function)	Creates a text field containing specified text (a string—optional), with specified number of columns width (optional—default is 8), with function to call when the ENTER key is pressed in the text field. This function should expect **one** parameter (string—the contents of the text field).

If you create a TextField with a callback function, then that function will be called any time the enter key is typed inside the box. (Presumably, the user will change the text and then press enter.)

For example, a text field may be created as follows:

```
text = TextField("type and hit <ENTER> ", 18, process-Entry)
```

where processEntry is a function which expects one parameter, the updated text (a string).

Once a TextField has been created, it may be added to a Display, specifying where to place its top-left corner.

```
d.add(textfield1, 50, 50)
```

If you create a TextField without using a callback function, then it is a passive GUI element. In other words, a different part of your program needs to manage (e.g., check for change in) the text content. This can be done using the following functions:

Function	Description
textField.getText()	Returns the text contained in the text field (as a string).
textField.setText()	Sets the text contained in the text field (as a string).
textField.setFont()	Changes the font used in the text field, e.g., Font("Dialog", Font.PLAIN, 12) or Font("Serif", Font.ITALIC, 16).

C.3.7 TextArea

TextArea objects are used for entering text that may span several lines.

Function	Description
TextArea(text, rows, columns)	Creates a text area containing the given text (string) with the given rows (default 5) and columns (default 400). If the text exceeds the text area dimensions, a slider bar will appear on the right.

Once a TextArea has been created, it may be added to a Display, specifying where to place its top-left corner.

```
d.add(textarea1, 50, 50)
```

A TextArea is a passive GUI element. You can access its contents with the following functions:

Function	Description
textArea.getText()	Returns the text contained in the text area (a string).
textArea.setText()	Sets the text contained in the text area (a string).
textArea.setFont()	Changes the font used in the text area, e.g., Font("Dialog", Font.PLAIN, 12) or Font("Serif", Font.ITALIC, 16).

C.3.8 Icon

Icon objects contain external images (.jpg or .png). They are created using the following functions.

Function	Description
Icon(filename)	Imports an image from the given filename (e.g. "apple.jpg" or "apple.png").
Icon(filename, width)	Imports an image from the given filename and resizes it (proportionally) using the provided width (in pixels).
Icon(filename, width, height)	Imports an image from the given filename and resizes (stretches) it using the provided width and height (in pixels).

Once an Icon has been created, it may be added to a Display, specifying where to place its top-left corner point.

```
d.add(icon1, 50, 50)
```

Additionally, you may use the following functions:

Function	Description
icon.setSize(width, height)	Changes/stretches the width and height of an image (in pixels).
icon.getWidth()	Returns the width of an image (in pixels).
icon.getHeight()	Returns the height of an image (in pixels).
icon.rotate(angle)	Rotates the image angle degrees.
icon.crop(x, y, width, height)	Crops the image starting from point x, y up to width and height (from the point x, y).

C.3.9 Menu

The GUI library simplifies creation of Menu objects. A display has a menu bar at the top. This is initially empty. Menu items may be added to it (e.g., "File", "Edit", etc.). Menus can also be added to a display (or any other graphical object, for that matter), as pop-up menus (i.e., menus that come up when you press the right mouse button).

Menu objects are created using the following function:

Function	Description
Menu(menuName)	Creates a new Menu with the specified name.

Once a menu has been created, it can be populated with items using the following functions:

Function	Description
menu.addItem(item, functionName)	Adds item (string) to the menu and specifies which function to call when item is selected.
menu. addItemList(itemList, functionNameList)	Adds a list of items (a list of strings) to the menu and specifies the corresponding functions to call (one function per item). The two lists are parallel (and thus need to be of equal length).
menu.addSeparator()	Adds a separator line to the menu.
menu.addSubmenu(menu)	Adds a submenu to the menu. Used for creating hierarchical menus.
menu.enable()	Enables the menu (active).
menu.disable()	Disables/grays out the menu (inactive).

Once a menu has been created, it can be added to a display as follows:

Function	Description
d.addMenu(menu)	Adds menu to the menu bar (left to right) of display d (e.g., "File", "Edit", etc.)
d.addPopupMenu(menu)	Adds a popup menu (e.g. a right-click menu) on display d.

Again, the second function is also available for every GUI object (e.g., a circle). This opens the door for many interesting GUI applications.

The following outlines the process of creating menu objects:

C.3.9.1 Drop-down Menus

Every display has its own menu bar. To add menus to it, follow these steps:

1. Create a Menu, as follows:
   ```
   menu = Menu(name)
   ```
   where name is a string.

2. Add menu items to a menu, as follows*:

```
menu.addItem(name, function)
```

where `name` is a string, and function is a function to be called when the user selects this menu item. This function should expect no parameters.

3. Finally, add the menu to a Display's menu bar:

```
display.addMenu(menu)
```

C.3.9.2 Pop-Up Menus

Pop-up menus are menus displayed when the user right-clicks on a display or other GUI object.

Pop-up menus are created the same way as drop-down menus. The only difference is that pop-up menus are added to a GUI object using the object's `addPopupMenu()` function, as follows:

```
object.addPopupMenu(menu)
```

C.4 GRAPHICS OBJECTS

Graphics objects are used to draw various geometric shapes on a display. The following graphics objects are available:

- Line objects are used for drawing solid lines with a specified color and thickness.

- Circle objects are used for drawing circles with a specified color and thickness. The circles may be filled or not.

- Point objects are used for drawing points with a specified color.

- Oval objects are used for drawing ovals with a specified color and thickness. The ovals may be filled or not.

- Rectangle objects are used for drawing rectangles with a specified color and thickness. The rectangles may be filled or not.

- Polygon objects are used for drawing polygons with a specified color and thickness. The polygons may be filled or not.

- Below we present each of these objects in more detail. Notice that Display has shortcut functions which create and add such objects in one step (see section "Drawing on Display" above).

---

* You may also add separators and submenus, similarly.

C.4.1 Line

Line objects are created using the following functions. Lines are drawn between a starting point (x1, y1) and an ending point (x2, y2).

Function	Description
`Line(x1, y1, x2, y2)`	Creates a line from point `x1, y1` to point `x2, y2`.
`Line(x1, y1, x2, y2, color, thickness)`	Creates a line from point `x1, y1` to point `x2, y2`. Additional optional parameters include `color` (e.g. `Color.BLACK` (default), `Color.ORANGE`, or `Color(255, 0, 255)`, using specific RGB values), and `thickness` (default is 1 pixel).

Once a Line has been created, it may be added to a Display, specifying where to place its leftmost point.

```
line1 = Line(100, 100, 200, 200)
display1.add(line1)
```

C.4.2 Circle

Circle objects are created using the following functions. Circles are drawn specifying their center point (x, y), and their radius.

Function	Description
`Circle(x, y, radius)`	Creates a circle at the given `x, y` coordinates and `radius`.
`Circle(x, y, radius, color, filled, thickness)`	Creates a circle at the given `x, y` coordinates, `radius`, `color` (e.g. `Color.BLACK` (default), `Color.ORANGE`, or `Color(255, 0, 255)`, using specific RGB values), `filled` (boolean — default is `False`), and `thickness` (default is 1 pixel).

Once a Circle has been created, it may be added to a Display, specifying where to place its center point.

```
circle1 = Circle(50, 50, 5)
d.add(circle1)
```

C.4.3 Point

Point objects are created using the following functions. Points are drawn specifying their center point (x, y).

Function	Description
Point(x, y)	Creates a new Point at the given x, y coordinates.
Point(x, y, color)	Creates a new Point at the given x, y coordinates and color (e.g. Color.BLACK (default), Color.ORANGE, or Color(255, 0, 255), using specific RGB values).

Once a Point has been created, it may be added to a Display, specifying where to place its center point.

```
point1 = Point(50, 50)
d.add(point1)
```

C.4.4 Oval

Oval objects are created using the following functions. Ovals are drawn by specifying the top-left corner point (x1, y1) and the bottom-right corner point (x2, y2) of the box that encloses them.

Function	Description
Oval(x1, y1, x2, y2)	Creates an Oval with top-left corner at x1, y1 and bottom-right corner at x2, y2.
Oval(x1, y1, x2, y2, color, filled, thickness)	Creates an Oval with top-left corner at x1, y1, bottom-right corner at x2, y2, color (e.g. Color.BLACK (default), Color.ORANGE, or Color(255, 0, 255), using specific RGB values), filled (boolean — default is False), and thickness (default is 1 pixel).

Once an Oval has been created, it may be added to a Display, specifying where to place its top-left corner point (x1, y1).

```
oval1 = Oval(50, 30, 100, 150)
d.add(oval1)
```

C.4.5 Rectangle

Rectangle objects are created using the following functions. Rectangles are drawn by specifying the top-left corner point (x1, y1) and the bottom-right corner point (x2, y2).

Function	Description
Rectangle(x1, y1, x2, y2)	Creates a Rectangle with top-left corner at x1, y1 and bottom-right corner at x2, y2.
Rectangle(x1, y1, x2, y2, color, filled, thickness)	Creates a Rectangle with top-left corner at x1, y1, bottom-right corner at x2, y2, color (e.g. Color. BLACK (default), Color.ORANGE, or Color(255, 0, 255), using specific RGB values), filled (boolean — default is False), and thickness (default is 1 pixel).

Once a Rectangle has been created, it may be added to a Display, specifying where to place its top-left corner point.

```
rec1 = Rectangle(50, 30, 100, 150)
d.add(rec1)
```

C.5 ADDITIONAL COLOR FUNCTIONS OF GRAPHICS OBJECTS

Once a graphics object has been created, the following functions are available:

Function	Description
object.setColor(color)	Changes the object color to the specified color (e.g. Color.BLACK or Color(255, 0, 255), using specific RGB values). If the color parameter is omitted, **a color selection dialog box will be presented**.
object.getColor()	Returns the current object color.

C.6 EVENT FUNCTIONS

Event functions are used to receive and process user actions at the GUI interface. Every GUI object listens for user actions (e.g., mouse click, mouse drag, typing a key, etc.), and allows you to specify a function to be called if and when a user action occurs.

You do not have to specify a function for every user action on every object on your GUI, only for the ones that you want. For example, you could make a program that draws a circle when the mouse is clicked on a particular location and cleans up all circles when the mouse exits the window.

The following event functions are available for **all** GUI library objects (i.e., Display, Label, Button, Checkbox, DropDownList, Slider, TextBox, Line, Circle, Rectangle, and Icon) except menus.*

Note: In the case of overlapping objects (e.g., a label and display), user events are handled by the object on top (front-most).

C.6.1 Keyboard Events

Keyboard events are divided into "key typed" and "key pressed/released" events. (For more information, see the Java API documentation on KeyEvents.)

C.6.1.1 Key Typed Events

"Key typed" events are higher-level and generally are independent of platform (and keyboard layout). These are generated when a character is typed on the keyboard (typing means both pressing and releasing the character key(s) on the keyboard).

The following function is provided to handle "key typed" events. It is available for all GUI objects (e.g., Display, Circle, etc.).

Function	Description
`object.` `onKeyType(function)`	When the user types a key (i.e., presses it and releases it), the system calls the provided `function`.* This function should accept one parameter (a string), which is the key typed, e.g., "a", "A", "b", "B", "/", etc.). Lower and uppercase characters are distinguished.

* Such functions are known as callback functions, because they are "called back" by the system if and when the specific event happens.

C.6.1.2 Key Down/Up Events

"Key down" and "key up" events are lower-level events and are generated whenever a key is pressed or released. As a result, these events may be used for various gaming applications (e.g., when pressing and holding a key does one thing, and when releasing the key does another).

These events are specific to the platform and keyboard layout (i.e., some keys may not work the same on all platforms). So test on all desired platforms to make sure keystroke controls work as intended.

---

* Menus handle user events (i.e., selection) by definition, as seen in the Menu section.

"Key down" and "key up" events are the only way to find out about keys that do not generate character input (e.g., action keys, modifier keys, etc.).

Function	Description
`object.` `onKeyDown(function)`	When the user presses a key (i.e., pushes a key down), the system calls the provided `function`. This function should accept **one** parameter (an integer), which is the virtual key pressed, e.g., VK_SHIFT or VK_A. NOTE: This function may be called many times, if a key is held down (according to the keyboard's key repeat rate). This is similar to pressing a key and having it repeat many times (e.g., in an editor window).
`object.` `onKeyUp(function)`	When the user releases a key, the system calls the provided `function`. This function should accept **one** parameter (an integer), which is the virtual key released, e.g., VK_SHIFT or VK_A.

The following functions are provided to handle "key down" and "key up" events. They are available for all GUI objects (e.g., Display, Circle, etc.). "Key down" and "key up" events use **virtual key codes** to report which keyboard key has been pressed, rather than a character generated by the combination of one or more keystrokes (such as "A", which comes from shift and "a").

For example, pressing the Shift key will cause a "key down" event with a VK_SHIFT key code whereas pressing the "a" key will result in a VK_A key code. After the "a" key is released, a "key up" event will be fired with VK_A.

Here is a list of the most important virtual key codes:

- VK_0 through VK_9 are for keys "0" thru "9".

- VK_A through VK_Z are for keys "A" thru "Z" (regardless of case - upper/lower).

- VK_LEFT, VK_RIGHT, VK_UP, VK_DOWN are for the arrow keys.

- VK_F1 through VK_F12 are for the function keys.

- other keys, such as VK_AMPERSAND, VK_CAPS_LOCK, VK_COMMA, VK_CONTROL, VK_ENTER, VK_MINUS, VK_PLUS, VK_SPACE, and so on.

For a complete list see the Java API documentation on KeyEvent.

C.6.2 Mouse Events

The following functions handle various mouse events.

The first group handles mouse events which happen **inside** a specific GUI object (e.g., display, circle, etc.):

Function	Description
`object.onMouseClick(function)`	When the user clicks the mouse (left button),* the system calls the provided `function`. This function should accept **two** parameters, x and y (i.e., the coordinates of the mouse cursor).
`object.onMouseDown(function)`	When the user presses the left mouse button, the system calls the provided `function`. This function should accept **two** parameters, x and y (i.e., the coordinates of the mouse cursor).
`object.onMouseUp(function)`	When the user releases the left mouse button, the system calls the provided `function`. This function should accept **two** parameters, x and y (i.e., the coordinates of the mouse cursor).
`object.onMouseMove(function)`	When the user moves the mouse within the object, the system calls the provided `function`. This function should accept **two** parameters, x and y (i.e., the coordinates of the mouse cursor).
`object.onMouseDrag(function)`	When the user drags the mouse within the object (i.e., moves the mouse while clicking), the system calls the provided `function`. This function should accept **two** parameters, x and y (i.e., the coordinates of the mouse cursor).

* The mouse right button is reserved for pop-up menus.

The following functions handle movement of a mouse that crosses the borders of a GUI object (i.e., entering or exiting the object boundaries):

Function	Description
`object.onMouseEnter(function)`	When the user moves the mouse into the borders of an object (from outside), the system calls the provided `function`. This function should accept two parameters, x and y (i.e., the coordinates of the mouse cursor).
`object.onMouseExit(function)`	When the user moves the mouse from inside to outside the borders of an object, the system calls the provided `function`. This function should accept two parameters, x and y (i.e., the coordinates of the mouse cursor).

C.6.3 Display Events

In addition to all the above, Display objects also have an `onClose()` event handling function.

Function	Description
d.onClose(function)	When display d is closed, the system calls the provided function. This function should have **zero** parameters. It may be used to perform cleanup, play a sound, update other displays, etc.

C.7 SCHEDULING TASKS – THE TIMER CLASS

The GUI library supports scheduling tasks (e.g., animation) through Timer class. Timer objects are used to schedule how often to perform a certain task (i.e., how often to call a given function).

C.7.1 Creating Timers

Timer objects are used to to schedule functions to be executed after a given time interval, repeatedly or once. The following function creates a new Timer, so you need to save it in a variable (so you can use it later).

Function	Description
Timer(delay, function, parameters, repeat)	Creates a new `Timer` to execute `function` after `delay` time interval (in milliseconds). The optional parameter parameters is a list of `parameters` to pass to the function (when called). The optional parameter `repeat` (boolean—default is `True`) determines if the timer will go on indefinitely. Note: The list of parameters is fixed at timer creation time and cannot be modified.

For example, the following:

```
t = Timer(500, Play.noteOn, [A4], True)
```

creates a Timer t, which will call function `Play.noteOn(A4)` repeatedly every 500 milliseconds (i.e., half second). In order for a timer to operate, it needs to get started:

```
t.start()
```

Once a Timer t has been created, the following functions are available:

Function	Description
t.start()	Starts timer t.
t.stop()	Stops timer t.
t.getDelay()	Returns the delay time interval of timer t (in milliseconds).
t.setDelay(delay)	Sets a new delay time interval for timer t (in milliseconds). This allows us to change the speed of the animation, after some event occurs.
t.isRunning(delay)	Returns True if timer t is running (has been started), False otherwise.
t.setFunction(function, parameters)	Sets the function to execute. The optional parameter parameters is a list of parameters to pass to the function (when called).
t.getRepeat()	Returns True if timer t is set to repeat, False otherwise.
t.setRepeat(flag)	If flag is True, timer t is set to repeat (this also starts the timer, if stopped). Otherwise, if flag is False, timer t is set to not repeat (this stops the timer, if running).

Appendix D: Other Functions

I N ADDITION TO THE FUNCTIONS covered in Appendices B and C, there are a number of more advanced functions mainly intended for offering convenient shortcuts (i.e., providing useful/common functionality) and for building interactive musical instruments and installations. This appendix provides a reference guide to them in the order they are introduced in the book. They include:

- Mapping functions
- Image functions
- Play functions
- AudioSample functions
- MIDI Sequence functions
- Timer functions
- MIDI input/output functions
- Open Sound Control functions
- Zipf functions
- ColorGradient function

D.1 MAP FUNCTIONS

These functions convert a value from one range to another, maintaining the relative position within the range. They are used to expand, contract, or offset data values.

These functions are included in the music library, so in order to use them, you need the following in your program:

```
from music import *
```

Function	Description
mapValue(value, minValue, maxValue, minResult, maxResult)	Takes a number within one range and returns its equivalent within another range. The arguments are: value–the number to be mapped minValue–the lowest possible number to be mapped (inclusive) maxValue–the highest possible number to be mapped (inclusive) minResult–the lowest value of the destination range (inclusive) maxResult–the highest value of the destination range (inclusive)
mapScale(value, minValue, maxValue, minResult, maxResult, scale)	Takes a number (i.e., MIDI pitch) within one range and returns its equivalent within another range, quantized to the pitch class value in scale. The arguments are: value–the number to be mapped minValue—the lowest possible number to be mapped (inclusive) maxValue—the highest possible number to be mapped (inclusive) minResult—the lowest value of the destination range (inclusive) maxResult—the highest value of the destination range (inclusive) scale—(optional) the musical scale (a list of pitch classes between 0 and 11) to be used in the destination range (see Appendix A for common scale constants)

D.2 IMAGE FUNCTIONS

The Image class provides useful functions for displaying and manipulating jpeg and png graphic files. This class is similar to the Icon class (seen in Appendix C), except that an Icon is intended to be used as a component for an interactive (GUI) application. On the other hand, an Image:

- has its own window,

- can be manipulated at the individual pixel level, and

- can be written/saved to an external file.

The `Image` class is included in its own library, so you need the following in your program:

```
from image import *
```

Function	Description
`Image(filename)`	Reads in a .jpg or .png file called `filename` (a string) and shows an image. It returns the image, so it should be stored in a variable, e.g., `img = Image("sunset.jpg")`.
`Image(width, height)`	Returns an empty (blank) image with provided `width` and `height`. It returns the image, so it should be stored in a variable, e.g., `img = Image(200, 300)`.
`img.getWidth()`	Returns the width of image `img`.
`img.getHeight()`	Returns the height of image `img`.
`img.getPixel(col, row)`	Returns this pixel's RGB values (a list, e.g., [255, 0, 0]), where `col` is the image column, and `row` is the image row. The image origin (0, 0) is at top left.
`img.setPixel(col, row, RGBlist)`	Sets this pixel's RGB values, e.g., [255, 0, 0], where `col` is the image column, and `row` is the image row. The image origin (0, 0) is at top left.
`img.show()`	Displays the image `img` in a window.
`img.hide()`	Hides the image window (if any).
`img.write(filename)`	Writes image `img` to the .jpg or .png `filename`.

D.3 PLAY FUNCTIONS

Earlier in the book we used the `Play.midi()` function to render musical compositions stored in `Note`, `Phrase`, `Part`, and `Score` objects. The `Play` functions discussed below are more advanced, as they are intended for building interactive musical instruments and installations.

These functions are included in the music library, so you need the following in your program:

```
from music import *
```

Function	Description
`Play.noteOn(pitch, volume, channel)`	Starts `pitch` sounding. Specifically, it sends a NOTE_ON message with `pitch` (0–127), at given `volume` (0–127—default is 100), to played on `channel` (0–15—default is 0) through the Java synthesizer.

Continued

Function	Description
`Play.noteOff(pitch, channel)`	Stops `pitch` from sounding. Specifically, it sends a NOTE_OFF message with `pitch` (0–127), on given `channel` (0–15—default is 0) through the Java synthesizer. If the pitch is not sounding on this channel, this has no effect.
`Play.note(pitch, start, duration, volume, channel)`	Schedules a note with `pitch` (0–127) to be sounded after `start` milliseconds, lasting `duration` milliseconds, at given `volume` (0–127, default is 100), on `channel` (0–15, default is 0) through the Java synthesizer.
`Play.allNotesOff()`	Stops all notes from sounding on all channels.
`Play.setInstrument(instrument, channel)`	Sets a MIDI `instrument` (0–127—default is 0) for the given `channel` (0–15—default is 0). Any notes played through `channel` will sound using `instrument`.
`Play.getInstrument(channel)`	Returns the MIDI `instrument` (0–127) assigned to `channel` (0–15—default is 0).
`Play.setVolume(volume, channel)`	Sets the global (main) `volume` (0–127) for this `channel` (0–15). This is different from the velocity level of individual notes—see `Play.noteOn()`.
`Play.getVolume(channel)`	Returns the global (main) `volume` (0–127) for this `channel` (0–15).
`Play.setPanning(position, channel)`	Sets the global (main) panning `position` (0–127) for this `channel` (0–15). The default position is in the middle (64).
`Play.getPanning(channel)`	Returns the global (main) `position` (0–127) for this `channel` (0–15).
`Play.setPitchBend(bend, channel)`	Sets the pitch `bend` for this `channel` (0–15—default is 0) to the Java synthesizer object. Pitch bend ranges from −8192 (max downward bend) to 8191 (max upward bend). No pitch bend is 0 (which is the default). If you exceed these values, the outcome is undefined (it may wrap around or it may cap, depending on the system.)
`Play.getPitchBend(channel)`	Returns the current pitch bend for this `channel` (0–15—default is 0).
`Play.frequencyOn(frequency, volume, channel)`	Starts a note sounding at the given `frequency` and `volume` (0–127—default is 100) on `channel` (0–15—default is 0). **Warning**: You should play only one frequency per channel. (Since this uses pitch bend indirectly, it will affect the pitch of all other notes sounding on this channel.)
`Play.frequencyOff(frequency, channel)`	Stops a note sounding at the given `frequency` on `channel` (0–15—default is 0).

Continued

Function	Description
	Warning: You should play only one frequency per channel. (Since the frequency gets translated to a pitch and a pitch bend, this will also affect notes with nearby frequencies on this channel.)
`Play.allFrequenciesOff()`	Same as `Play.allNotesOff()`. Stops all notes from sounding on all channels.

D.4 AUDIOSAMPLE FUNCTIONS

The `AudioSample` class includes functions related to playing audio samples in real time. An audio sample is a sound object created from an external audio file (supported formats are WAV and AIF—16-, 24-, and 32- bit PCM and 32-bit float), which can be played, looped, paused, resumed, and stopped. The functions below are intended for building interactive musical instruments and installations.

An application may have several `AudioSample` objects active at the same time. It is even possible to create complex timbres by loading several simple sound objects and manipulating them (e.g., changing their frequency and/or volume) in real time. `AudioSample` objects open endless timbral possibilities for interactive applications.

This class is included in the music library, so you need the following in your program:

```
from music import *
```

Use the following function to create an `AudioSample` object:

Function	Description
`AudioSample(filename, pitch, volume)`	Creates an audio sample from the audio file specified in `filename` (supported formats are WAV and AIF—16-, 24-, and 32-bit PCM and 32-bit float). Parameter `pitch` (optional) specifies a MIDI note number to be used for playback (default is A4). Parameter `volume` (optional) specifies a MIDI note velocity to be used for playback (default is 127).

Once an audio sample, `a`, has been created, the following functions are available:

Function	Description
`a.play()` `a.play(start, size)`	Play the sample once. If `start` and `size` are provided, the sample is played from milliseconds `start` until milliseconds `start+size` (default is 0 and –1, respectively, meaning from beginning to end).

Continued

Function	Description
`a.loop()` `a.loop(times, start, size)`	Repeats the sample indefinitely. Optional parameter times specifies the number of times to repeat (default is –1, indefinitely). If start and size are provided, looping occurs between milliseconds start and milliseconds start+size (default is 0 and –1, respectively, meaning from beginning to end).
`a.stop()`	Stops sample playback immediately.
`a.pause()`	Pauses sample playback (remembers current position for resume).
`a.resume()`	Resumes sample playback (from the paused position).
`a.isPlaying()`	Returns True if the sample is still playing, False otherwise.
`a.setPitch(pitch)`	Sets the sample pitch (0–127) through pitch shifting from sample's base pitch.
`a.getPitch()`	Returns the sample's current pitch (it may be different from the default pitch).
`a.setFrequency(freq)`	Sets the sample pitch frequency (in Hz). This is equivalent to setPitch(), except it provides more granularity (accuracy). For instance, pitch A4 is the same as frequency 440 Hz.
`a.getFrequency()`	Returns the current playback frequency.
`a.setVolume(volume)`	Sets the volume (amplitude) of the sample (volume ranges from 0 to 127).
`a.getVolume()`	Returns the current volume (amplitude) of the sample (volume ranges from 0 to 127).
`a.setPanning(panning)`	Sets the panning of the sample (panning ranges from 0–127).
`a.getPanning()`	Returns the current panning of the sample (panning ranges from 0–127).
`a.getFrameRate()`	Returns the sample's recording rate (e.g., 44100.0 Hz).

D.5 MIDISEQUENCE FUNCTIONS

The MidiSequence class includes functions related to playing external MIDI files (as well as Note, Phrase, Part, and Score objects) in real time. A MIDI sequence provides playback features that are similar to the functionality described above for audio samples. Again, these functions are intended for building interactive musical instruments and installations.

This class is included in the music library, so you need the following in your program:

```
from music import *
```

Use the following function to create a MidiSequence object:

Function	Description
MidiSequence(material, pitch, volume)	Creates a MIDI sequence from the MIDI material specified in material (this may be a filename of an external MIDI file or a Note, Phrase, Part, and Score object. Parameter pitch (optional) specifies a MIDI note number to be used for playback (default is A4). Parameter volume (optional) specifies a MIDI note velocity to be used for playback (default is 127).

Once a MIDI sequence, m, has been created, the following functions are available:

Function	Description
m.play()	Plays the MIDI sequence once.
m.loop()	Repeats the MIDI sequence indefinitely.
m.stop()	Stops MIDI sequence playback immediately.
m.pause()	Pauses MIDI sequence playback (remembers current position for resume).
m.resume()	Resumes MIDI sequence playback (from the paused position).
m.isPlaying()	Returns True if the MIDI sequence is still playing, False otherwise.
m.setPitch(pitch)	Sets the MIDI sequence's playback pitch (0–127) by transposing the MIDI material.
m.getPitch()	Returns the MIDI sequence's playback pitch (0–127).
m.setTempo(tempo)	Sets the MIDI sequence's playback tempo in beats per minute (e.g., 60).
m.getTempo()	Returns the MIDI sequence's playback tempo (in beats per minute).
m.getDefaultTempo()	Returns the MIDI sequence's default tempo (in beats per minute).
m.setVolume(volume)	Returns the volume of the MIDI sequence (0–127).
m.getVolume()	Returns the current volume of the MIDI sequence (0–127).

D.6 TIMER FUNCTIONS

The Timer class allows us to schedule events to happen after a precise time interval has passed and/or to be repeated at precise time intervals in the future. An event is anything that can be specified/included in the body of a Python function. In essence, a Timer object is given a function to execute in the future. This function you may include one or more sound events (such as a note or an audio sample), altering parameters of an on-going process (such as changing the frequency or volume of an audio sample, or changing the color of a graphics object on a GUI display), and many other possibilities.

`Timer` objects are useful building blocks for more advanced interactive musical instruments and installations. Among other things, `Timer` objects can be used for creating animation.

The `Timer` class is included in its own library, so you need the following in your program:

```
from timer import *
```

Use the following function to create a `Timer` object:

Function	Description
`Timer(delay, function, parameters, repeat)`	Creates a new `Timer` to execute `function` after `delay` time interval (in milliseconds). The optional parameter `parameters` is a list of parameters to pass to the function (when called). The optional parameter `repeat` (boolean — default is `True`) determines if the timer will go on indefinitely. Note: The list of parameters is fixed at timer creation time and cannot be modified.

Once a timer object, `t`, has been created, the following functions are available:

Function	Description
`t.start()`	Starts timer `t`.
`t.stop()`	Stops timer `t`.
`t.getDelay()`	Returns the delay time interval of timer `t` (in milliseconds).
`t.setDelay(delay)`	Sets a new `delay` time interval for timer `t` (in milliseconds). This allows us to change the speed of the animation, after some event occurs.
`t.isRunning()`	Returns `True` if timer `t` is running (has been started), `False` otherwise.
`t.setFunction(function, parameters)`	Sets the `function` to execute. The optional parameter `parameters` is a list of parameters to pass to the function (when called).
`t.getRepeat()`	Returns `True` if timer `t` is set to repeat, `False` otherwise.
`t.setRepeat(flag)`	If `flag` is `True`, timer `t` is set to repeat (this also starts the timer, if stopped). Otherwise, if `flag` is `False`, timer `t` is set to not repeat (this stops the timer, if running).

D.7 MIDI INPUT OUTPUT, AND FUNCTIONS

The `MidiIn` and `MidiOut` classes enable communication, from within your programs, to various input and output MIDI devices connected to your computer.

D.7.1 The MidiIn Class

`MidiIn` objects may be used in your programs to get input from MIDI devices that generate input events (e.g., a MIDI guitar, keyboard, or control surface).

The `MidiIn` class is included in the MIDI library, so you need the following in your program:

```
from midi import *
```

Use the following function to create a `MidiIn` object:

Function	Description
MidiIn()	Creates a new `MidiIn` object to connect to an input MIDI device. When called, it presents the user with a GUI to select one from the available MIDI devices (see Figure 9.3).

Once a MIDI input object, `mInput`, has been created, the following functions are available:

Function	Description
mInput.onNoteOn (function)	When a NOTE_ON event happens on the `mInput` device (i.e., the user starts a note), the system calls the provided `function`. This `function` should expect four parameters, eventType, channel, data1, data2. For NOTE_ON events, the eventType is always 144, the channel ranges from 0 to 15, data1 is the note pitch (0–127), and data2 is the volume of the note (0–127).
mInput.onNoteOff (function)	When a NOTE_OFF event happens on the `mInput` device (i.e., the user ends a note), the system calls the provided `function`. This `function` should expect four parameters, eventType, channel, data1, data2. For NOTE_OFF events, the eventType is always 128, the channel ranges from 0 to 15, data1 is the note pitch (0–127), and data2 is ignored.

Continued

Function	Description
mInput.onSetInstrument (function)	When a SET_INSTRUMENT (also known as CHANGE_PROGRAM) event happens on the mInput device (i.e., the user selects a different timbre), the system calls the provided function. This function should expect four parameters, eventType, channel, data1, data2. For SET_INSTRUMENT events, the eventType is always 192, the channel ranges from 0 to 15, data1 is the MIDI instrument (0–127), and data2 is ignored.
mInput.onInput (eventType, function)	Associates an incoming eventType with a callback function. When the specified eventType event happens on the mInput device, the system calls the provided function. This function should expect four parameters, eventType, channel, data1, data2. Can be used repeatedly to associate different event types (128–224) with different callback functions (one function per event type). If eventType is ALL_EVENTS, then function will be called for **all incoming events that have not yet been assigned** callback functions.

D.7.2 The MidiOut Class

MidiOut objects may be used in your programs to send output to MIDI devices that accept output events (e.g., an external MIDI synthesizer).

The MidiOut class is included in the MIDI library, so you need the following in your program:

```
from midi import *
```

Use the following function to create a MidiOut object:

Function	Description
MidiOut()	Creates a new MidiOut object to connect to an output MIDI device. When called, it presents the user with a GUI to select one from the available MIDI devices (see Figure 9.4).

Once a MIDI output object, mOutput, has been created, the following functions are available:

Function	Description
mOutput.noteOn(pitch, velocity, channel)	Sends a NOTE_ON message with pitch (0–127), at a given velocity (0–127—default is 100), to channel (0–15—default is 0) on the mOutput device.

Continued

Function	Description
`mOutput.noteOff(pitch, channel)`	Sends a NOTE_OFF message with `pitch` (0–127), on given `channel` (0–15—default is 0) on the `mOutput` device. If the pitch is not sounding on this channel, this has no effect.
`mOutput. setInstrument(instrument, channel)`	Sets a MIDI `instrument` (0–127—default is 0) for the given `channel` (0–15—default is 0) on the `mOutput` device. Any notes played through `channel` will sound using `instrument`.
`mOutput.playNote(pitch, start, duration, velocity, channel)`	Schedules playing of a note with pitch at the given `start` time (in milliseconds from now), with `duration` (in milliseconds from start time), `velocity` (0–127—default is 100), to `channel` (0–15—default is 0) on the `mOutput` device.
`mOutput.play(material)`	Plays music library material (`Score`, `Part`, `Phrase`, `Note`) on the `mOutput` device.

Note: For more information, see Chapter 9.

D.8 OPEN SOUND CONTROL (OSC) FUNCTIONS

The `OscIn` and `OscOut` classes provide functionality for open sound control (OSC) communication between programs running on your computer and other OSC devices. As opposed to MIDI input and output, where devices need to be physically connected (via wire) to your computer, OSC devices (by the nature of the OSC protocol) may be located anywhere on the Internet (in the same room, next room, or anywhere on the planet).

This Internet connectivity opens endless possibilities for developing interactive applications and installations.

D.8.1 The `OscIn` Class

`OscIn` objects (OSC servers) may be used in your programs to receive incoming OSC messages from OSC devices (clients) that send out OSC messages (e.g., an OSC-enabled smartphone or control surface).

The `OscIn` class is included in the OSC library, so you need the following in your program:

```
from osc import *
```

Use the following function to create an `OscIn` (server) object:

Function	Description
OscIn(port)	Creates a new OscIn (server) object to receive incoming messages from an OSC device (such as a smartphone, or tablet) on the given port. The port number is an integer from 1024 to 65535 that is not being used by another program.

Once an `OscIn` object, `oscServer1`, has been created, the following functions are available:

Function	Description
oscServer1.onInput(address, function)	When an OSC message with the given address arrives, call function. OSC addresses look like a URL, e.g., "/first/second/third". The function should expect one parameter, the incoming OSC message.

D.8.2 The OscOut Class

`OscOut` objects (OSC clients) may be used in your programs to send OSC messages to OSC devices (servers) that listen for incoming OSC events. For instance, you could use a collection of objects to synchronize (or share data between) two different programs on the same computer or several computers used in an installation. These computers do not have to be physically connected (only to be on the Internet). Again, the possibilities are endless.

The `OscOut` class is included in the OSC library, so you need the following in your program:

```
from osc import *
```

Use the following function to create an `OscOut` (client) object:

Function	Description
OscOut(IPaddress, port)	Creates a new OscOut (client) object to send messages to another OSC (server) device (such as a smartphone, or tablet) at the given IPaddress (a string, e.g., "192.168.1.223") and port (an integer in the range 1024 to 65535).

Once an `OscOut` object, `oscClient1`, has been created, the following functions are available:

Function	Description

oscClient1.
  sendMessage(address,
  arg1, arg2, ...)
	Sends an OSC message with `address` and 0 or more arguments to the OSC device associated with `oscClient1` (when it was created).

Note: For more information, see Chapter 9.

D.9 ZIPF LIBRARY

Zipf's law models the scaling (fractal) properties of many phenomena in human ecology, including natural language and music (Zipf 1949). Zipf's law is one of many related laws that describe scaling properties of phenomena studied in the physical, biological, and behavioral sciences. These include Pareto's law, Lotka's law, power laws, Benford's law, Bradford's law, Heaps' law, etc.

The zipf library provides functions that may be used to calculate the fractal dimension of any natural or human phenomenon, including musical pieces. To use these functions, you need the following in your program:

```
from zipf import *
```

The following functions allow you to calculate Zipf slopes and R squared (R^2) values from event counts (and, potentially, event sizes) of natural and human phenomena (e.g., see Zipf 1949, p. 337).[*]

Function	Description
byRank(counts)	Returns the Zipf slope and R^2 values generated from plotting the counts (y-axis) against the ranks of the values from largest to smallest (x-axis) in log-log scale. The ranks are generated automatically.
bySize(sizes, counts)	Returns the Zipf slope and R^2 values generated from plotting the counts (y-axis) against the corresponding sizes (x-axis) in log-log scale. The two lists, sizes and counts, are parallel. That is, each value of counts indicates how many fractal subdivisions of the corresponding sizes value are presented in the phenomenon (e.g., music piece) being studied.

For more information, see Chapter 11.

---

[*] Through this technique, we may discern the composer, genre, and even popularity of musical pieces (Manaris et al. 2005, 2007).

D.10 COLORGRADIENT FUNCTION

The `colorGradient` function returns a list of RGB colors creating a "smooth" gradient between two colors. To use this function, you need the following in your program:

```
from gui import *
```

This function may be used in conjunction with code that gives different GUI elements (such as points or circles) different colors to simulate a gradient. This may be useful in various interactive applications and installations.

Function	Description
`colorGradient(color1, color2, steps)`	Returns a list of RGB colors creating a "smooth" gradient between `color1` and `color2`.

For more information, see Chapter 11.

Index